3rd International Conference on Manufacturing, Material and Metallurgical Engineering (ICMMME 2018)

IOP Conference Series: Materials Science
and Engineering Volume 409

Kuala Lumpur, Malaysia
17 – 19 March 2018

ISBN: 978-1-5108-7443-5
ISSN: 1757-8981

Printed from e-media with permission by:

Curran Associates, Inc.
57 Morehouse Lane
Red Hook, NY 12571

Printed by Curran Associates, Inc. (2019)

For permission requests, please contact the Institute of Physics at the address below.

Institute of Physics
Dirac House, Temple Back
Bristol BS1 6BE UK

Phone: 44 1 17 929 7481
Fax: 44 1 17 920 0979

techtracking@iop.org

Additional copies of this publication are available from:

Curran Associates, Inc.
57 Morehouse Lane
Red Hook, NY 12571 USA
Phone: 845-758-0400
Fax: 845-758-2633
Email: curran@proceedings.com
Web: www.proceedings.com

TABLE OF CONTENTS

POLYMERS AND COMPOSITES

DEVELOPEMENT OF COMPOSITE ARMORS USING NATURAL RUBBER REINFORCED WITH STEEL WIRE MESH FOR BALLISTIC RESISTANCE 1
Seksak Asavavisithchai, Teeranan Marlaiwong, Nuthaporn Nuttayasakul

A WIRELESS POWERED ELECTROACTIVE POLYMER USING MAGNETIC RESONANT COUPLING 7
Hau Ran Cheong, Koon Chun Lai, Pei Song Chee

EFFECT OF FILLER CONTENT ON TRANSPORT PROPERTIES OF SULFONATED POLYETHER ETHER KETONE (SPEEK) COMPOSITE PROTON EXCHANGE MEMBRANES 13
Min Yan Chia, Hui San Thiam, Loong Kong Leong, Chai Hoon Koo

OPTIMIZATION OF THE ELECTRICAL CONDUCTIVITY AND TENSILE STRENGTH OF CONDUCTIVE POLYMER COMPOSITES USING THE TAGUCHI METHOD 20
Hendra Suherman, Erwin Julisman, Duskiardi, Mulyanef, Irmayani

EVOLUTION OF MICROSTRUCTURE AND MECHANICAL PROPERTIES OF AL-B_4C COMPOSITE AFTER RECYCLING 26
Zhenjiang Guo, Qiulin Li, Wei Liu, Guogang Shu

ADSORPTION EFFECT OF CETYLTRIMETHYL-AMMONIUM BROMIDE MODIFIED FLY ASH ON METHYL ORANGE WASTE WATER 34
Dongji Liu, Yuling Yang, Fengqing Zhao

NANO-MATERIALS AND BIOMEDICAL MATERIALS

GROWTH AND STRUCTURAL PROPERTIES OF GRAPHENE OXIDE THIN FILM WITH SPRAY PYROLYSIS TECHNIQUE 38
Anoud Alshammari, Mohd Mahadi Halim, Yam Fong Kwong, Noor Haida Mohd Kaus, Saleh I. Alhiassah, Naser Mahmoud Ahmed, Lim Hwee San

SUNLIGHT PHOTODEPOSITION OF GOLD NANOPARTICLES ONTO GRAPHITIC CARBON NITRIDE (G-C_3N_4) AND APPLICATION TOWARDS THE DEGRADATION OF BISPHENOL A 44
Chen Hong Hak, Lan Ching Sim, Kah Hon Leong, Yik Heng Chin, Pichiah Saravanan

PHYSICAL MIXING OF N-DOPED GRAPHENE QUANTUM DOTS FUNCTIONALIZED TIO_2 FOR SUSTAINABLE DEGRADATION OF METHYLENE BLUE 50
Kah Hon Leong, Sean Yang Fong, Ping Feng Lim, Lan Ching Sim, Pichiah Saravanan

ELECTROSPUN PVDF NANOFIBROUS MEMBRANES EFFECTIVELY CAPTURING PM2.5 AND RELEASING NEGATIVE IONS 56
Z L Liu, W Zhang

MATERIALS PHYSICS AND CONDENSED MATTER PHYSICS

RESONANT OPTICAL TRANSMISSION THROUGH THE SUB-WAVELENGTH AIR-HOLE ARRAYS IN A GOLD THIN FILM FOR SENSING APPLICATIONS 61
Yuan-Fong Chou Chau, Chee Ming Lim, Nyuk Yoong Voo, Noormariah Muslim, Muhammad Nur Syafi'Ie Md Idris

A COMPARATIVE STUDY OF VISIBLE RAMAN SCATTERING OF CERIA PREPARED BY SOL-GEL AND HYDROTHERMAL TECHNIQUES WITH GOLD NANOPARTICLES 67
Muhammad Nur Syafi'Ie Md Idris, Hai-Pang Chiang, Noormariah Muslim, Yuan-Fong Chou Chau, Abdul Hanif Mahadi, Nyuk Yoong Voo, Chee Ming Lim

SHOCK WAVE EFFECT ON MICROMECHANICS PROPERTIES OF SN-AG-CU SOLDER (SAC 387) VIA NANOINDENTATION 76
Nur Shafiqa Safee, Wan Yusmawati Wan Yusoff, Ariffin Ismail, Norliza Ismail, Emee Marina Salleh, Maria Abu Bakar, Azman Jalar

MATERIAL PROCESSING

MONITORING OF VARIOUS MINERALS OF FLASH FURNACE CHARGE AND THEIR IMPACTS ON FURNACE OPERATION 82
Ali Manoochehri, Mahmoud Khodadadi, Mehdi Balochi, Roohallah Abbasi, Hossin Mahmoudi, Ali Karimi Far, Peyman Jamali, Ahmad Karami

PRODUCTIVITY IMPROVEMENT THROUGH LINE BALANCING AT ELECTRONIC COMPANY – CASE STUDY 88
A. F. H Fansuri, A. N. M Rose, M. F. F Ab Rashid, N. M. Z Mohamed Nik, H Ahmad

DEVELOPMENTS OF THERMOCHROMIC VO_2–BASED THIN FILMS FOR ARCHITECTURAL GLAZING 94
Noormariah Muslim, Ying Woan Soon, Chee Ming Lim, Nyuk Yoong Voo

ANALYSIS OF DEFORMATION AND TEXTURE GRADIENT IN SHOT PEENED PURE TITANIUM 102
H P Yang, Y M Wang

MECHANICAL ENGINEERING AND MANUFACTURING

APPLICATION OF TAGUCHI METHOD IN INVESTIGATING THE EFFECT OF LAYER THICKNESS AND FILL ANGLE ON FDM PARTS 108
Siti Maryam Md Nor, Mohd Nizam Sudin, Muhamad Farid Bin Razali

SUPPRESSION OF STRUCTURE-BORNE NOISE IN A RECTANGULAR ENCLOSURE 114
N. Fathiah Waziralilah, Aminudin Abu, Lee Kee Quen, M Hishamuddin Unuh, Lim Meng Hee

ENERGY HARVESTING, AN INCREDIBLE SOLUTION FOR STRUCTURAL HEALTH MONITORING SYSTEMS OF BRIDGE 121
Waheed Gul

PROPERTIES AND COMMON INDUSTRIAL APPLICATIONS OF POLYVINYL FLUORIDE (PVF) AND POLYVINYLIDENE FLUORIDE (PVDF) 127
M. H. Alaaeddin, S. M. Sapuan, M. Y. M Zuhri, E. S Zainudin, Faris M. Al-Oqla

A STUDY ON THE RELATIVE SLIP CHRACTERISTICS IN CONTINUOUS ROLLING PROCESS OF RAIL WITH FORWARD PULL 134
Jian-Feng Song, Hui Yu, Heng Zhu, Yong-Gang Dong

THE PERFORMANCE OF A PIEZOELECTRIC MICRO-PUMP FOR 3D ADDITIVE MANUFACTURING 141
Kai Li, Jun-Kao Liu, Wei-Shan Chen, Lu Zhang

METALLURGY AND ALLOY PROPERTIES

ELECTRIC DISCHARGE MACHINING OF TUNGSTEN CARBIDE USING UNCOATED AND COATED WIRE ELECTRODES: ANALYSIS OF CUTTING SPEED AND WORKPIECE GEOMETRICAL ACCURACY 147
Rakhshanda Naveed, Nadeem Ahmad Mufti

CRYSTAL STRUCTURE, SURFACE TOPOGRAPHY, SURFACE MORPHOLOGY AND OPTICAL PROPERTIES OF DC MAGNETRON SPUTTERED VO_2 THIN FILMS USING VO_2 TARGET 153
Noormariah Muslim, Muhammad Nur Syafi'Ie Md Idris, Ying Woan Soon, Yuan-Fong Chou Chau, Chee Ming Lim, Nyuk Yoong Voo

MODELING AND EXPERIMENTAL VALIDATION OF DEFORMATION OF AL-33CU DROPLET DURING IMPINGEMENT ON A SUBSTRATE 159
Amitesh Kumar

THE ROLE OF $CACO_3$ IN THE EXTRACTION OF VALUABLE METALS FROM LOW-NICKEL MATTE BY CALCIFIED ROASTING—ACID LEACHING PROCESS 166
Yong Zhao, Shuhua Geng, Hongwei Cheng, Guangshi Li, Xionggang Lu, Qian Xu

RESEARCH ON MICROSTRUCTURE AND WEAR RESISTANCE OF FE-BASED HARDFACING LAYER USING PLASMA SURFACING TECHNOLOGY 173
Lingling Song, Jian Guo, Ma Xiaolei, Jingnan Zhao

BUILDING MATERIALS AND STRUCTURES

EFFECT OF AIR ENTRAINING AGENT ON UNIAXIAL TENSILE PROPERTIES OF PVA-ECC 179
Jincai Zhu, Zigeng Wang, Yue Li, Zhanguo Li

STUDY ON AXIAL COMPRESSION TEST OF CORRODED REINFORCED CONCRETE COLUMNS REINFORCED BY MPC BONDING CFRP 183
Chang Zhao, Yue Li, Zigeng Wang, Zhanguo Li

STUDY ON THE IMPROVEMENT OF WATER RESISTANCE OF MAGNESIUM PHOSPHATE CEMENT BY STEEL SLAG 188
Mengyang Xie, Zhanguo Li, Yue Li, Zigeng Wang

ENERGY AND COLLIERY ENGINEERING

EFFECTS OF SUBSTRATE ON THERMAL AND OPTICAL CHARACTERISTICS OF HIGH-POWER THINGAN WHITE LED 194
Muna E. Raypah, Mutharasu Devarajan, Fauziah Sulaiman

INVESTIGATION ON ENERGY ABSORPTION PROPERTIES OF AL-FOAM, FOAM FILLED AND EMPTY MS TUBE UNDER 3-POINT LOADING CONDITION AT ROOM TEMPERATURE 201
Sanchit Yogesh Joshi, V. A. Kolhe

Author Index

IOP Conf. Series: Materials Science and Engineering **409** (2018) 011001 doi:10.1088/1757-899X/409/1/011001

PREFACE

It was a great pleasure to welcome all the participants in the 2018 3rd International Conference on Mining, Materials and Metallurgical Engineering (ICMMME 2018) held in Kuala Lumpur, Malaysia, 17 - 19 March 2018. The ICMMME is dedicated to new research developments and advances in the fields of Mining, Materials and Metallurgical Engineering.

The key goal of the conference is aimed at bringing together academicians, scientists, engineers, and researchers working in various disciplines of these fields to exchange views as well as to share their expertise, experience and research results, and discuss the challenges and future directions in their specialized areas of research in these fields.

Prof. Ramesh K. Agarwal from Washington University in St. Louis, USA, Prof. Changduk Kong from Chosun University, South Korea and Prof. Seong Soo Choi from Sun Moon University, South Korea delivered the invited keynote lectures about state of the art research in their areas of expertise. The conference provided a good opportunity for the students, researchers and engineers to interact with the experts and specialists to discuss with them on technical matters and future directions of research.

The proceeding of the conference consists of papers from the universities, research institutes and industry, and it was composed of 7 chapters such as Chapter 1of Polymers and Composites, Chapter 2 of Nano-materials and Biomedical Materials, Chapter 3 of Material Physics and Condensed Matter Physics, Chapter 4 of Material Processing, Chapter 5 of Mechanical Engineering and Manufacturing, Chapter 6 of Metallurgy and Alloy Properties, Chapter 7 of Energy and Colliery Engineering.

The papers were selected after a rigorous peer-review by conference committee members and international reviewers. The papers should provide the readers an overview of many recent advances in the fields related to materials especially the composite materials and metallurgical engineering including the modeling, analysis, computation and optimization as well as the experimental methods.

I would like to thank all the authors who have contributed to this volume and also to the members of the organizing committee, reviewers, speakers, chairpersons, sponsors and the conference participants for their support to ICMMME 2018.

Advisory Chair of ICMMME 2018, ***Prof. Changduk Kong****, Chosun University, South Korea*

Advisory Chair
Prof. Changduk Kong, Chsoun University, South Korea

Conference Chairs
Prof. Ramesh Singh, University of Malaya, Malaysia
Prof. Ramesh K. Agarwal, Washington University in St. Louis, USA

Program Chairs
Prof. Norhamidi Muhamad, Universiti Kebangsaan Malaysia, Malaysia
Prof. Md Mustafizur Rahman, Universiti Malaysia Pahang, Malaysia
Prof. Seong Soo Choi, Research Center for Nanobio Scince, SunMoon University, South Korea

Technical Committee
Assoc. Prof. Rusdi Nur, Politeknik Negeri Ujung Pandang,Indonesia
Assoc. Prof. Rosli Ahmad, Universiti Tun Hussein Onn Malaysia
Assoc. Prof. Leong Loong Kong, Universiti Tunku Abdul Rahman, Malaysia
Dr. Yu Lung Chiu, University of Birmingham, UK
Prof. Poi Sim Khiew, The University of Nottingham, Malaysia
Dr. Faiz Redza Ramli, Universiti Teknikal Malaysia Melaka, Malaysia
Dr. Hoong Thiam Toh, Universiti Teknologi Malaysia
Prof. Gelu-OvidiuTirian, Politehnica University of Timisoara, Romania
Prof. Nicoleta GILLICH, "Eftimie Murgu" University of Resita, Romania
Prof. M. MANZOORHUSSAIN, Jntuh College of Engineering Sultanpur, India
Dr. Cristian Paul Chioncel, "Eftimie Murgu" University of Resita, Romania
Dr. MELLAL Mohamed Arezki, M‟Hamed Bougara University(UMBB), Algeria
Dr. Amar Rezoug, Advanced Technology Development Center, Algeria
Dr. Yung-Yao Chen, National Taipei University of Technology, Taiwan
Assoc. Prof. Chou Chau Yuan Fong, Universiti Brunei Darussalam, Brunei
Dr. Elammaran Jayamani, Swinburne University of Technology Sarawak Campus, Malaysia
Prof. S. G. Ponnambalam, Monash University Malaysia Campus, Malaysia
Prof. Dr. Megat Mohamad Hamdan bin Megat Ahmad, Universiti Pertahanan Nasional Malaysia
Dr. Hong Geok Soon, National University of Singapore
Assoc. Prof. Thanakom Soontornchainacksaeng, King Mongkut's University of Technology North Bangkok, Thailand
Assist. Prof. AN Liang, Hong Kong Polytechnic University

IOP Conf. Series: Materials Science and Engineering **409** (2018) 011003 doi:10.1088/1757-899X/409/1/011003

Peer review statement

All papers published in this volume of *IOP Conference Series: Materials Science and Engineering* have been peer reviewed through processes administered by the proceedings Editors. Reviews were conducted by expert referees to the professional and scientific standards expected of a proceedings journal published by IOP Publishing.

IOP Conf. Series: Materials Science and Engineering **409** (2018) 012001 doi:10.1088/1757-899X/409/1/012001

Developement of Composite Armors Using Natural Rubber Reinforced with Steel Wire Mesh for Ballistic Resistance

Seksak Asavavisithchai*, Teeranan Marlaiwong, Nuthaporn Nuttayasakul

Innovative Metals Research Unit, Department of Metallurgical Engineering, Faculty of Engineering, Chulalongkorn University, Bangkok 10330, Thailand
Department of Civil Engineering, Chulachomklao Royal Military Acadamy, Nakornnayok, 26001, Thailand

E-mails:seksak.a@chula.ac.th;teeranan.mar@gmail.com; nuthaporn.crma@gmail.com

Abstract. A protective ballistic armor used for military buildings, that will meet the anti-terrorism and force protection demands, are increasingly needed, particularly in Thailand's southernmost provinces. A new type of ballistic armor system has been developed, composing of composite layers, using natural rubber reinforced with steel wire mesh, and Al plate. The composite layers were used as front material while the Al plate was used as backing plate. The armors were tested against NIJ-Type III threats using NIJ 0108.01 ballistic test standard. The armor, with 12 mm Al backing plate, can successfully resist the NIJ-Type III treat with only combination of two composite layers. Different than hard and brittle plate, such as ceramic, the use of composite layer at the front did not blunt and erode the projectile, due to its ductile property. However, the composite layer effectively reduces the velocity of projectile, due to high fracture toughness, as a result of reinforcing steel wire mesh layers in the composites. The projectile was finally stopped by the Al backing plate.

1. Introduction

In past decades, there have been continuously shooting and bombing attacks in Thailand's southernmost provinces from a long-dormant separatist insurgency [1]. The terrorist attacks resulted in loss of lives and properties in both government officers and civilians. To minimise the effect of attack, high efficiency protection systems have been developed and installed on building walls and vehicles. Currently, most of ballistic protection walls are the reinforce concrete system, due to relatively high compressive strength and low cost. However, drawbacks of this system is low tensile strength, ductility and toughness.

High-performance hard armor systems for ballistic protection typically consists of multiple layers [2]. It usually uses a hard material, such as ceramic or hardened steel plate, at the strike face, coupled with a ductile material, such as steel, aluminium or fiber reinforced composite [3, 4]. Upon the projectile impact, the hard material can blunt and erode the projectile, due to its high hardness, resulting in fracture of bullet. The ductile material, as backing plate, at the rear face then absorb the residual kinetic energy of the fractured or deformed projectile through plastic deformation [5]. High-performance composites, such as Kevlar™ and Twaron™, or Spectra™ or Dyneema™, are frequently used as backing plate to absorb the kinetic energy of the projectile [6-8]. Nevertheless, the major drawbacks of hard armor protection system are weight and cost. The use of ceramic or hard steel plate result in increasing weight to the armor. Development of light-weight protection system is then necessary, especially for transportation vehicles, in order to maintain high mobility. In addition, the

IOP Conf. Series: Materials Science and Engineering **409** (2018) 012001 doi:10.1088/1757-899X/409/1/012001

use of high-performance composites also results in high armor costs, contributing to limit in widespread use of the armor.

Alternatively, rubber composite materials can be used as backing plate in ballistic protection system applied in tanks, infantry fighting vehicles, and shipbuilding [9, 10]. The rubber composite provides high mechanical properties, such as low density, high specific strength and high specific modulus. It is reported that the rubber composite materials have high anti-eroding performance on bullets, grenade fragments, and kinetic energy projectiles. Compared with other high-performance composites, the rubber one is superior in terms of safety and environmental effects [11].

The present research aims to develop a protective ballistic armor used for military buildings that will meet the anti-terrorism and force protection demands. The armor is made of composite materials, composing of A5083 Al plate and natural rubber reinforced with steel wire mesh. Ballistic testing was performed using U.S. National Institute of Justice (NIJ) standard 0108.01 for 5.56 mm NATO (Type III High Power Rifle) [12].

2. Experimental Procedure

The ballistic composite armor (BCA) panel, with a size of 30.5 × 30.5 cm, is manufactured, using either A5083 aluminium (Al) plate, with a thickness of 6, 8, 10, 12 mm, or 400 stainless steel (SS) plate, with a thickness of 8, 10, 12 mm, and natural rubber reinforced with steel wire mesh (SWM). The Al plate is obtained from META Metal Enterprise (Thailand). The steel wire sheets, with a mesh size of 12 × 12 mm, is supplied by Global House. The mesh was sectioned to 30.5 × 30.5 cm and embedded to pre-vulcanised natural rubber (NR) latex. The NR latex, provided by Rubber Authority of Thailand, has chemical composition as shown in TABLE I. The mesh was orderly layered up to 5 layers. The NR, reinforced with steel wire meshes, and either A5083 Al plate or 400 SS was used as backing and strike plates, respectively. The mechanical bonding, using bolt and nut fasteners, were performed around the corner of BCA panels. The total thickness of BCA panels was varied between 30 and 48 mm.

Tensile testing of NR embedded with different steel wire meshes was conducted, following ASTM D3039 standard, using a universal testing machine (UTM), in order to examine mechanical behavior of samples. The testing was performed, using speed of 2 mm/s. Ballistic testing of BCA panels was conducted, using the guidelines included in the National Institute of Justice (NIJ) standard 0108.01 for Type III [12]. The test was performed at Military Explosive Factory. The projectile speeds were measured, using velocity chronographs BETA Model Cl35-0005, placed before and after the BCA panels. The shooting distance was maintained at 15 m between the gun and the panels with a zero angle of obliquity the gun. The tests were performed with combined two or three composite layers at the front and with Al plate at the back for different thicknesses, as presented in Fig. 1.

Table 1. Natural Rubber (Nr) Latex Composition

Chemical agents	**Content**	**Phr: (Part per hundred of rubber)**
STR 5L	1,000	100
Zno	50	5
$C_{18}H_{36}O_2$ (Stearic acid)	20	2
$CaCO_3$	3,000	300
Color agent	50	5
WSL (Wing Stay L)	10	1
N-Cyclohexylbenzothiazole-2-sulphenamide	15	1.5
Tetra methyl thiuramdisulphide	2	0.2
S (Sulfur)	25	2.5

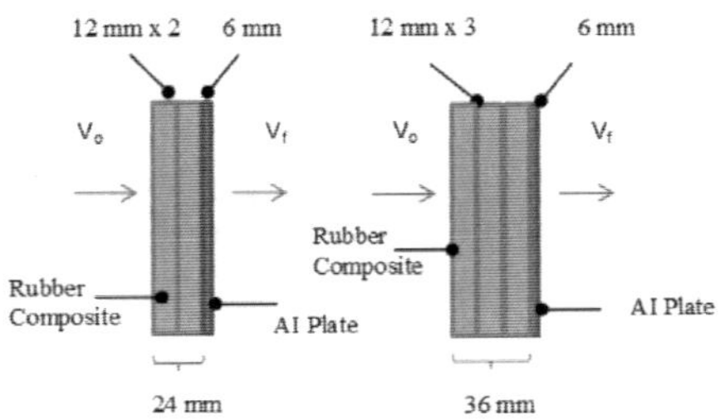

Figure 1. Arrangement of BCA plates with different composite layers at the front and the Al plate at the back.

3. Results And Discussion

Fig. 2 presents densities and mechanical stress-strain curves of NR reinforeced with different SWM layers. It is clear that the composite density increases when the layers of SWM in the composites increase. The highest density of 1,976 kg/m^3 is then obtained from the composite with five SWM layers. The addition of SWM layers in the composites results in increasing stresses. Similar to the effect of SWM addition to composite density, the more the addition of SWM, the higher the resistant stress. The maximum stress of 0.11 MPa is therefore found in the composite with the addition of five SWM layers.

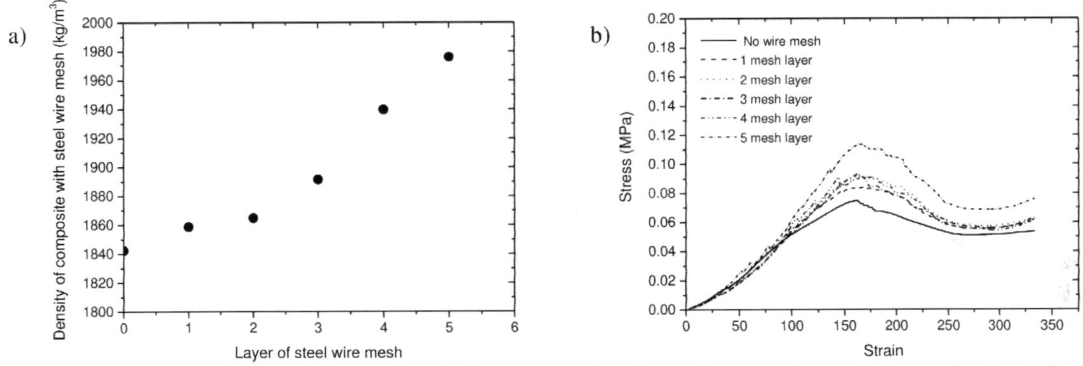

Figure 2. (a) Densities and (b) mechanical stress-strain plots of composites reinforeced with different steel wire mesh layers

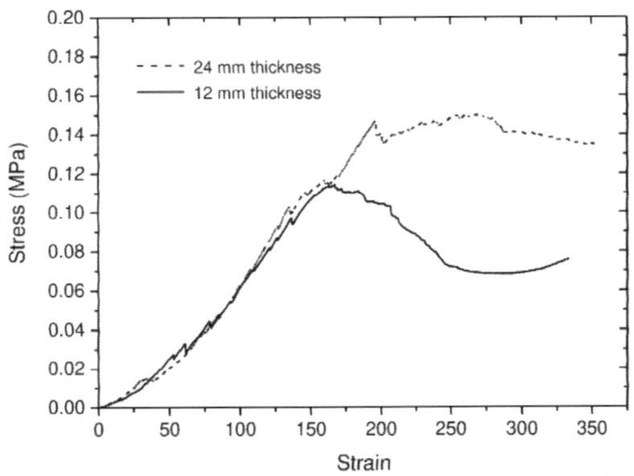

Figure 3. Mechanical stress-strain plots of composites at different thicknesses.

Fig. 3 shows the stress-strain plots of composites with different thicknesses. The thickness of 24 mm is estimated from placing two bolted composite samples under penetration testing. Due to the limit of compression machine capability, the maximum thickness of NR reinforced with SWM can be produced up to five layers, which is equivalent to approximately 12 mm thickness. It can be seen that the stress increases when thickness of composite sample increases. The larger the thickness, the higher the stress, as well as energy absorption. The energy absorption is calculated from the area under the stress-strain curves. The composites with thickness of 12 and 24 mm show the energy absorption of 9.7 and 11.4 J, respectively. An increase in composite thickness results in increasing stress.

Armor panel	1st composite layer	2nd composite layer	3rd composite layer	A5083 Al backing plate thickness
	Front face	Front face	Front face	
1	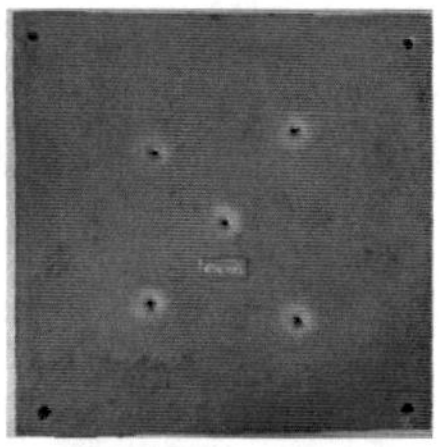	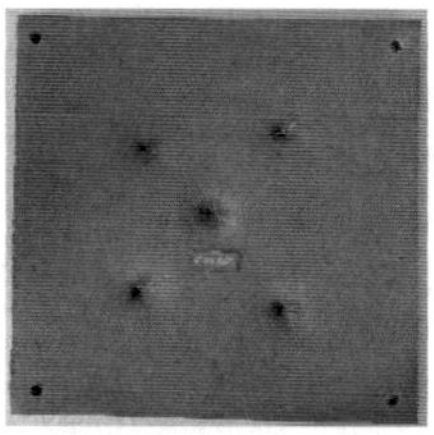	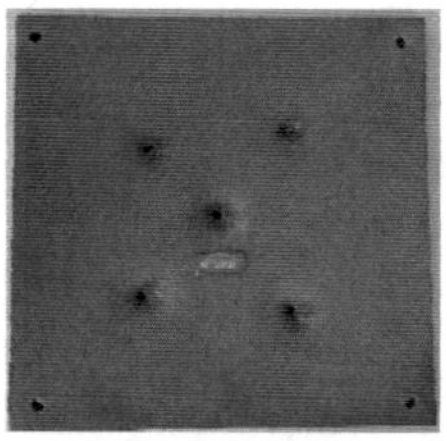	6 mm
2	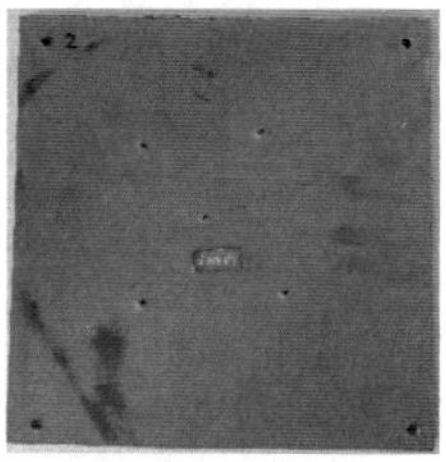	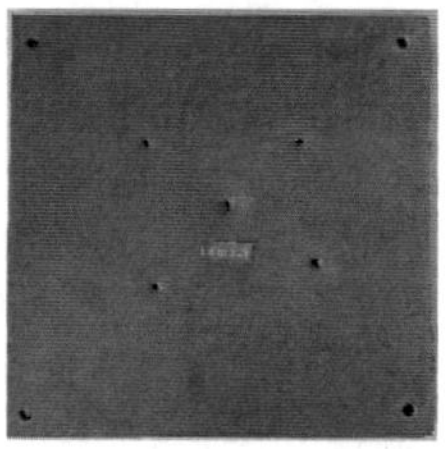	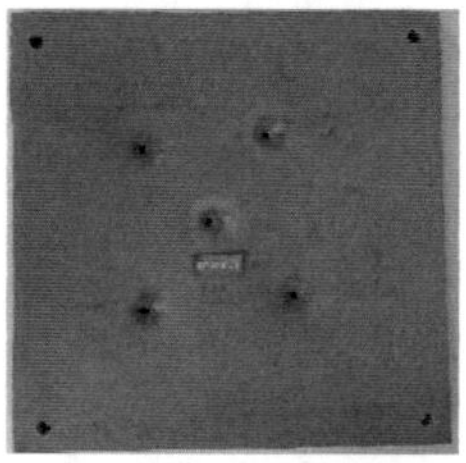	8 mm
3	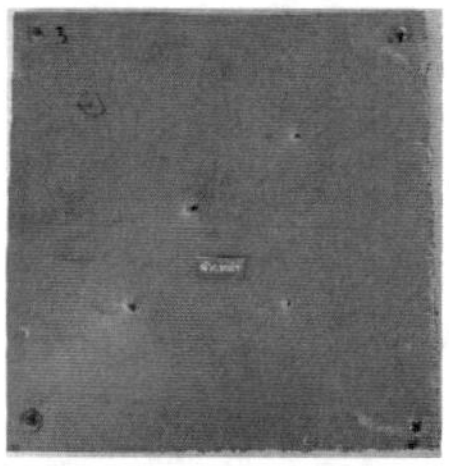			10 mm
4		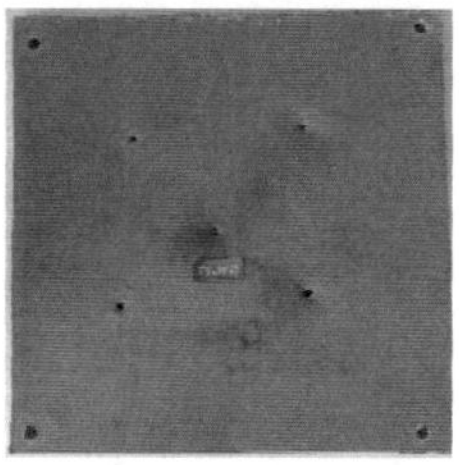		12 mm

Figure 4. Damages in front face strike and rear face of impact area of NIJ-Type III tests

The images of the front faces of individual composite layer assembled to BCA panel after NIJ-Type III ballistic tests are shown in Fig. 4. In all cases, the BCA panels can stop the projectiles upon impact, meeting NIJ-Type III standard requirements. In all cases, the projectiles were found to reside in Al backing plates, suggesting that the bullets were slowed and stopped by the plates. The BCA panels, which composed of three composite layers, with at least 6 mm Al backing plate, are required to resist the NIJ-Type III treat. However, the BCA panels, with 12 mm Al backing plate, can successfully resist the NIJ-Type III treat with only combination of two composite layers. In all cases, no fragment of projectiles was found outside of the BCA panels.

Different than hard and brittle plate, such as ceramic, the use of composite layer at the front show no radial crack at the impact area, due to its ductile property. Only holes in small craters were formed at the impact areas, indicating full penetration of projectiles through the composites. In all cases, no fragment of bullets is observed in all composites layers, suggesting that the composites did not cause any physical damage to the bullets. It is expected that the composite layers can effectively reduce the velocity of bullet and absorb its most kinetic energy through plastic deformation. The inspection of Al backing plates, after the NIJ-Type III ballistic tests, showed that they had deformed bullets, with small fragments, embedded at the front surface. This indicates that the Al backing plate had successfully absorbed the residual kinetic energy of the bullet through compression.

Generally, in ballistic impacts, the kinetic energy of the projectile is transformed into plastic deformation and fracture of the projectile and armor plate, and heat. In this system, upon the impact, the ductile composite layers effectively slow the projectile, due to high fracture toughness, as a result of reinforcing SWM layers in the composites. Fig. 5 presents the fracture surface of composite, reinforced with two SWM layers, after tensile test. It can be seen that the NR latex is still attached to the surface of steel wire, even after the severe damage. The steel wire has much higher strength than the NR latex, resulting in increasing resistance to deformation developed in the composite during the penetration of projectile.

Figure 5. Fracture surface of composite sample reinforced with two steel wire mesh layers.

4. Conclusion

Composite using natural rubber reinforced with steel wire mesh have been successfully produced. Two or three composite layers, used as front material, were attached to A5083 aluminium plate, used as backing plate. A combination of the composite layers and Al plate results in a new ballistic composite armor system. The armors were tested against NIJ-Type III threats using NIJ 0108.01 ballistic test standard. The armor, with 12 mm Al backing plate, can successfully resist the NIJ-Type III treat with only combination of two composite layers. Upon the impact, the ductile composite layers effectively slow the projectile, due to high fracture toughness, as a result of reinforcing steel wire mesh layers in the composites. The projectile was finally stopped by the Al backing plate. The study of these composite armors can lead to improvement in more effective and sustainable ballistic armors with lower material cost.

Acknowledgment
The authors would like to acknowledge Chulalongkorn University for its financial support.

References

[1] C. Nurakkate, The Conflict in Southern Thailand, Shedden Papers, Centre for Defence and Strategic Studies, 2012.
[2] M. Sadighi, R. C. Alderliesten and R. Benedictus, "Impact resistance of fiber metal laminates: A review", Int. J. Impact Eng., vol. 49, pp. 77-90, November 2012.
[3] H.L. Gower, D.S. Cronin and A. Plumtree, "Ballistic impact response of laminated composite panels", Int. J. Impact Eng., vol. 25, pp. 1000-1008, September 2008.
[4] M.J.N. Jacobs and J.L.J. Van Dingenen, "Ballistic protection mechanisms in personal armour", J. Mater. Sci., vol. 36, pp.3137-3142, July 2001.
[5] Y. Wang, X. Chen, R. Young, I. Kinloch and G. Wells, "A numerical study of ply orientation on ballistic impact resistance of multi-ply fabric panels", Compos. Part B-Eng, vol. 68, pp. 259-265, January 2015.
[6] Y. B. S. Sastry, P. R. Budarapu, Y. Krishna and S. Devaraj, "Studies on ballistic impact of the composite panels", Theor. Appl. Fract. Mec., vol. 72, pp. 2-12, August 2014.
[7] M.R. Abdullah and W.J. Cantwell, "The impact resistance of fiber-metal laminates based on glass fiber reinforced polypropylene", Polym. Composite, vol. 27, pp. 700-708, December 2006.
[8] M. Colakoglu, O. Soykasap and T. Ozek, "Experimental and numerical investigations on the ballistic performance of polymer matrix composites used in armor design", Appl. Compos. Mater., vol. 14, pp. 47-58, January 2007.
[9] N. Hassim, M.R. Ahmad, W.Y.W. Ahmad, A. Samsuri and M.H.M. Yahya, "Puncture resistance of natural rubber latex unidirectional coated fabrics", J. Ind. Text., vol. 42, pp. 118-131, October 2012.
[10] X. Zu, Z. Huang and X. Jia, "Study on rubber composite armor anti-shaped charge jet penetration", J. Energ. Mater., vol. 38, pp. 695-702, October 2013.
[11] M. R. Ahmad, W. Y. W. Ahmad, J. Salleh and A. Samsuri, "Effect of fabric stitching on ballistic impact resistance of natural rubber coated fabric systems", Mater. Design, vol. 29, pp. 1353-1358, 2008.
[12] Ballistic Resistant Protective Materials: NIJ Standard 0108.01. U.S. National Institute of Justice, US, 1985.

IOP Conf. Series: Materials Science and Engineering **409** (2018) 012002 doi:10.1088/1757-899X/409/1/012002

A wireless powered electroactive polymer using magnetic resonant coupling

Hau Ran Cheong, Koon Chun Lai, Pei Song Chee*

Department of Electrical and Electronic Engineering, Lee Kong Chian Faculty of Engineering and Science, Universiti Tunku Abdul Rahman, Sungai Long Campus, Malaysia.
Department of PetroChemical Engineering, Faculty of Engineering and Green Technology, Universiti Tunku Abdul Rahman, Kampar Campus, Malaysia

E-mails: hauran92@1utar.my, laikc@utar.edu.my, *cheeps@utar.edu.my

Abstract. This paper reports an ionic polymer-metal composite (IPMC) soft actuator that is controlled and powered in a wireless manner. IPMC is an ionic electroactive polymer (EAP) that can be actuated at a low voltage, fast response and miniaturized, therefore, attracts significant interest that have led to extensive investigations, especially for biomedical application. The device design utilizes magnetic resonant coupling as the wireless powering scheme. A planar based LC receiver circuit and DC rectifier circuit are fabricated using double sided and single sided copper-clad flexible polyimide sheets, respectively. The fabricated IPMC cantilever is bonded to the DC rectifier circuit and performs unimorph motion when the exposed magnetic field (generated by the transmitter circuit) matches the resonant frequency of the LC receiver circuit, ~25 MHz. Experiment wireless operation of the fabricated LC receiver circuit shows a maximum induced DC voltage of ~2.3 V_{dc}. The cantilever deflection is measured as 66 μm at ON-OFF activation cycle with 0.6 W RF output power.

1. Introduction

Electroactive polymers (EAPs) [1] based muscle-alike actuator offers a broad opportunity for flexible and more degree of freedom (DOF) in biomedical robotic design. Compared to other types of smart polymer actuators such as shape memory polymer, EAPs pose larger strain, faster speed and good efficiency. Ionic polymer metal composite is an EAP, consists of a thin ion exchange polymer membrane sandwiched between two metal electrodes. IPMC can be actuated at very low voltage, fast response and ease of miniaturization [2]. Moreover, the biocompatible and soft muscle-alike actuation properties allow it to be incorporated into biomedical device. Earlier researches involved a battery powered and wired interface circuitry to control the IPMC movement [3-4], which could limit its mobility. In addition, it is not practical to be utilized in implantable biomedical device where the battery leakage is hazardous to human health. Wireless powered IPMC actuator eliminates the need for battery might be a solution which potentially advantageous in term of size, flexibility and robustness. Lee et al. [5] reported high frequency (9 GHz) microwave powered IPMC actuator using microstrip patch antenna patterned on the EAP material. This approach, despite eliminating battery involvement, high frequency microwave transfer scheme is sensitive to the environments change, such as temperature and humidity. High frequency microwave also poses practical issue for implantable purpose. Wireless radio-frequency (RF) magnetic field power transfer scheme might be a solution which potentially advantageous in term of power consumption and stability. K. Abdelnour et al. [6] reported a resonant coupling as the wireless powering scheme incorporated with a bridge rectifier to convert AC signal into DC to control the IPMC. This RLC receiver coil and bridge rectifying circuit, nonetheless, tends to have relatively large size and dimension. Planar based receiver and rectifier circuit could be a promising approach to address the mentioned drawback. This paper reports a wireless radio frequency (RF) magnetic field controlled IPMC cantilever actuator using planar LC receiver circuit. Resonant coupling is selected as the wireless powering scheme due to its endearing performance, such as long wireless transmission distance, as discussed in [7]. The planar based LC receiver circuit and DC rectifier circuit are fabricated using double sided polyimide copper sheets. The wireless responses of the developed IPMC actuator are experimentally characterized.

2. Design and Working Principle

Table 1. Parameter Of The System

Parameter	Representation	Value	Parameter	Representation	Value
V_s	AC Voltage input	15 V_{ac}	C_1	Source capacitor	23.053 pF
L_1	Source loop inductance	2503.176 nH	C_2	Load capacitor	27.85 pF
L_2	Source resonator inductance	1791.462 nH	C_3	Rectifier capacitor	220 µF
L_3	Load resonator inductance	1482.834 nH	D_1	Rectifier diode	n/a
L_4	Load loop inductance	151.962 nH	D_2	Rectifier diode	n/a

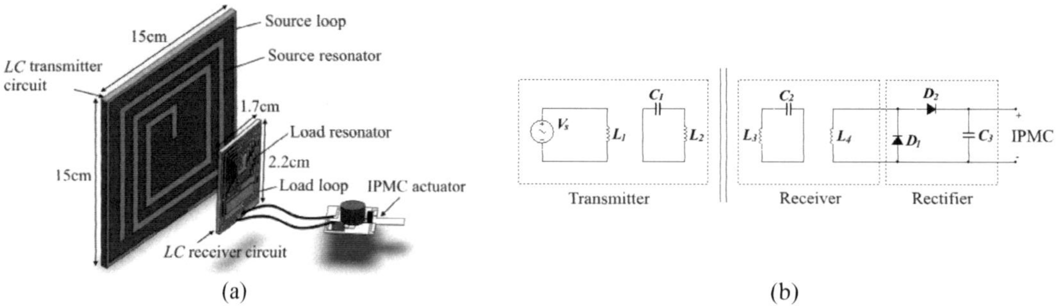

Figure 1. (a) Design and working principle of the wireless powered IPMC actuator and (b) Lumped mass model of the resonant coupling system.

The wireless power transfer system utilizes a planar inductor-capacitor (LC) circuit serves as the transmitter and receiver to power the IPMC actuator, as shown in Fig. 1 (a). The resonant coupling system consists of a LC transmitter and receiver circuits that are designed to have two rectangular-spiral coils as the resonator and source or load loops, respectively with integrated capacitor to form LC tank. The load loop induces AC voltage when it is exposed to the external magnetic field generated by the LC transmitter circuit when the field frequency matches its resonant frequency. To demonstrate IPMC cantilever in unimorph motion, the LC tank is coupled with a class- D half wave rectifier circuit [8]. The lumped mass model representing the resonant coupling system is shown in Fig. 1 (b). In this figure, L_1 and L_2 form the transmitter circuit and L_3 and L_4 form the receiver circuit. The AC voltage from the LC receiver is converted into DC voltage via rectifier circuit to activate the IPMC cantilever. Table 1 shows the parameter representation and the calculated components values using equation 1.

The self-inductance of each loop in the LC circuits were calculated using modified Wheeler Formula [9] defined as

$$L_{mw} = K_1\mu_o \left(n^2 d_{avg} / 1+K_{2p} \right) \qquad (1)$$

where is the fill ratio. The coefficients K_1 and K_2 are layout dependent, n is number of turn, d_{avg} is average diameter within the planar coil and $_o$ is the permeability in air. Resonator in the transmitter and receiver are resonantly coupled and the resonant frequency is achieved by tuning the shunting capacitor C_1 *and* C_2,

$$1/ L_2C_1 \approx 1/ L_3C_2 \qquad (2)$$

The diameter of resonator coil in transmitter circuit (100 mm) is designed larger than the receiver circuit (13 mm) in order to enhance the transmission distance as discussed in [10]. The distance of transmission, D is estimated through

$$D = \sqrt{D_{o\_Tx} \times D_{o\_Rx}} \qquad (3)$$

where D_{o_Tx} is diameter of the transmitting resonator and D_{o_Rx} is diameter of the receiving resonator. The resonant frequency of the system is designed at 25 MHz with the maximum transfer distance of ~5 cm. The IPMC cantilever bends at resonant frequency (maximum voltage) due to the migration of mobile cations to the cathode in the ion exchange membrane. This operation creates a pressure across the cantilever and causes the IPMC to bend in the anode direction [11].

3. Fabrication

The planar LC circuit was formed by using double sided copper-clad polyimide (PI) film (AP7156E, DuPont, USA) via dry film photolithography process. First, both top and bottom sides of the LC receiver circuit (consists of two rectangular-spiral coils as the resonator and load loop and a capacitor plate) were fabricated by wet etching the copper-clad layers. Next, to connect the top and bottom Cu side of the PI film, vial contacts were etched at the PI film and covered with silver conducting epoxy. The fabricated LC receiver circuit is illustrated in Fig. 2 (a). A 0.11 mm thick DC rectifier circuit was fabricated using the similar method with bonded power electronic components to form class- D half wave rectifier circuit. To fabricate the IPMC actuator, a 183 μm thick Nafion 117 (Dupond, USA) and was cut into T-shape to form 8 mm long cantilever. Next, silver was chemically deposited at the photolithography-patterned Nafion 117 through silver mirror method [12] to form different polarity electrode pads. The chemical electroless plating might induce high internal stress to the IPMC and result to serious curliness. Therefore, low temperature annealing was performed to remove the internal stress. The IPMC cantilever was placed in a vacuum drying oven and heated at 140 °C for 3 hours. Fig. 2 (b) shows the IPMC cantilever after the annealing process. The IPMC cantilever is flatten due to the removal of internal stress. Finally, the developed IPMC cantilever was bonded to the DC rectifier circuit using silver conducting epoxy.

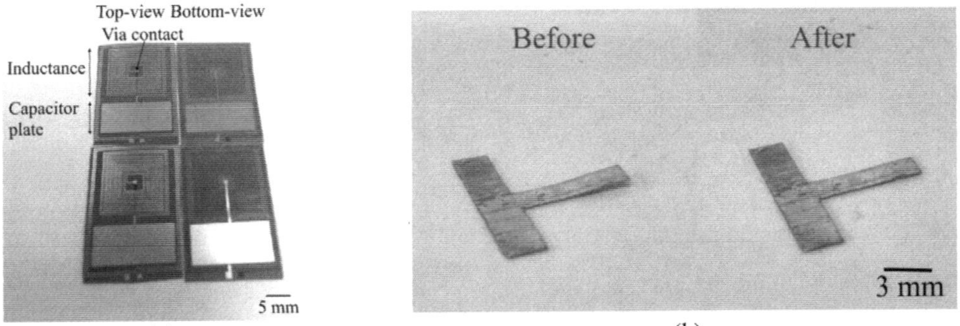

(a) (b)

Figure 2. (a) Top and bottom view of the fabricated LC receiver circuit. (b) IPMC actuator before and after the low temperature annealing process.

4. Experimental Result And Discussion

To demonstrate the proposed wireless power cantilever IPMC, the experimental setup in Fig. 3 (a) and (b) were constructed. A square wave function generator (SFG-830, GW Instek, Taiwan) and an op-amp amplifier (output power of 0.6 W) were implemented to produce magnetic field with the LC transmitter circuit connected. An oscilloscope (MDS1102CA, Matrix, China) was used to determine the electrical response of the LC receiver circuit and DC rectifier circuit. The setup also involved a laser displacement sensor (CD5-85, Optex, Japan) to characterize the cantilever tip displacement. The constructed wireless power transfer system was designed to be operated at ~25 MHz, based on the tunable value shown in Table 1. To features biocompatibility of the developed LC receiver circuit, the LC receiver circuit was encapsulated with a layer of 1 mm thick polydimethylsiloxane (PDMS) material. Fig. 4 (a) illustrates the induced DC output voltage at the DC rectifier circuit with the LC receiver circuit without PDMS encapsulation (direct exposed to air) and with PDMS encapsulation. A maximum DC output voltage of ~2.3 V_{dc} was noted in both cases at the same resonant frequency, ~25 MHz. This shows that the PDMS medium do not cause significant change in the device's resonant frequency. In addition, it also shows a good agreement with the theoretical calculated value, where a maximum output voltage is attained at resonant frequency of 25 MHz. To study the IPMC cantilever deflection profile, the laser displacement sensor was located above the tip of IPMC cantilever, as shown in Fig. 3 (b), to measure the tip displacement of the IPMC cantilever. The result presented in Fig. 4 (b) shows the temporal response of the IPMC cantilever activated in wireless ON-OFF cycle: 10 seconds ON, 10 seconds OFF with ~ 3 V_{dc} output voltage. As can be seen in Fig 4 (b), 66 m of cantilever displacement is measured at the 10th second of the ON activation. The cantilever deflection, however, experiences slight increase to 70 m at the end of OFF cycle. To verify this deflection behavior, the voltage profile of the LC receiver circuit was examined. The voltage slightly decreases after the 3th second at the ON cycle due to the capacitive effect in the DC rectifier circuit. On the other hand, the voltage decreases abruptly at the OFF cycle to ~2 V_{dc}. As discussed in [13], IPMC working voltage ranges from 1 V_{dc} to 3 V_{dc}. Therefore, the IPMC cantilever will have slight increase at the OFF cycle. IPMC cantilever will remain at its end position if no voltage is induced.

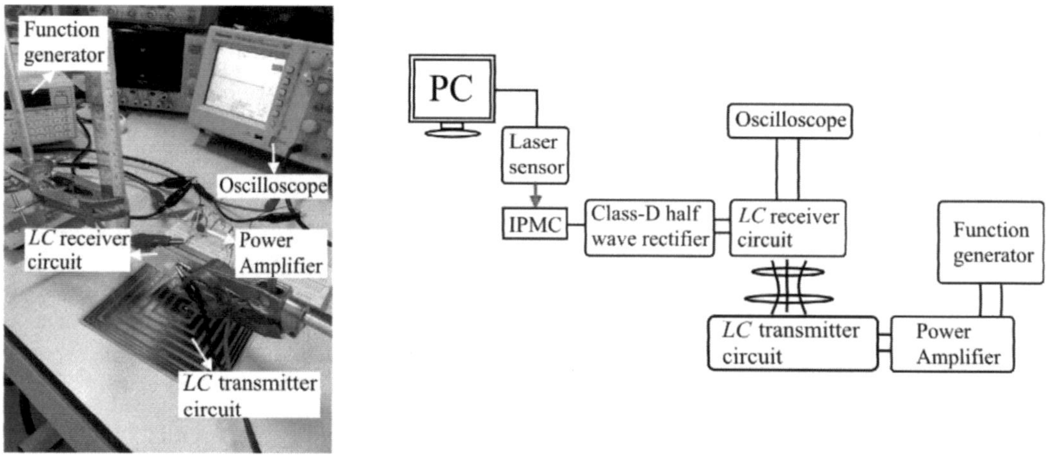

(a) (b)

Figure 3. Experimental setup: (a) transmitter and receiver setup and (b) block diagram of whole setup.

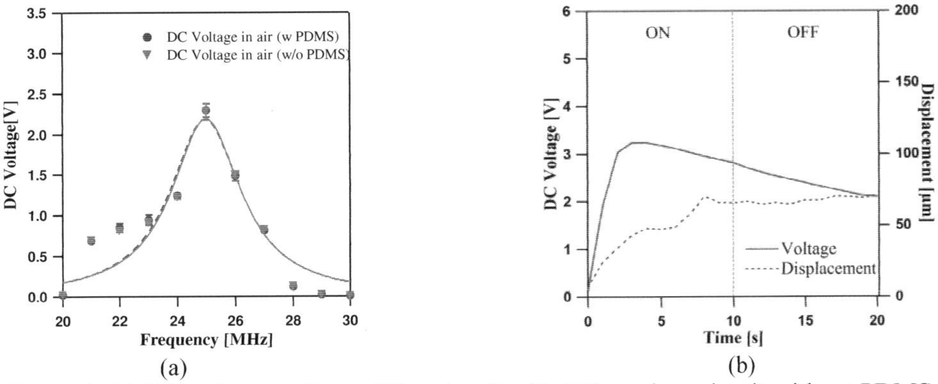

Figure 4. (a) DC voltage at the rectifier circuit with LC receiver circuit without PDMS encapsulation and with PDMS encapsulation. (b) IPMC actuator tip displacement and DC voltage versus time with responded to 10 s ON/ 10 s OFF activation.

5. Conclusion

In this research, a wireless RF magnetic field powered IPMC cantilever has been designed, fabricated and demonstrated. Resonant coupling has been selected as the wireless powering scheme. The planar based LC receiver circuit and DC rectifier circuit have been constructed to miniaturize the size of the whole antenna. The wireless coupling effect of the developed LC receiver circuit with PDMS encapsulated was examined and analyzed. The developed wireless RF magnetic field successfully activated the fabricated IPMC cantilever with a RF power 0.6 W at DC output voltage of ~3 V_{dc}. Future works will encompass the optimization of the LC receiver circuit design and fabrication procedure for miniaturization, and to use lower loss wireless transfer system to increase the power transfer efficiency.

Acknowledgment

The authors acknowledge the financial support from university fund, UTARRF Vote No: 6200/CC8 (Project No. IPSR/RMC/UTARRF/2017-C1/C01).

References

[1] F. A. M. Ghazali, C. K. Mah, A. AbuZaiter, P.S. Chee, and M. S. M. Ali, “Soft dielectric elastomer actuator micropump,” Sensors and Actuators A: Physical, vol. 263, pp. 276-284, 2017.

[2] B. -K. Fang, M. -S. Ju, and C. -C. K. LIN, “A new approach to develop ionic polymer–metal composites (IPMC) actuator: Fabrication and control for active catheter systems,” Sensors and Actuators A: Physical, vol. 137, pp. 321-329, 2007. (references)

[3] W. Jiaqi, A. Mcdaid, C. Lu, and K. Aw, “A Compact Ionic Polymer-Metal Composite (IPMC) Actuated Valveless Pump for Drug Delivery,” IEEE/ASME Transactions on Mechatronics, 2016.

[4] S. Ford, G. Macias, and R. Lumia, “Single active finger IPMC microgripper, ” Smart Materials and Structures, vol. 24, p. 025015, 2014.

[5] J. S. Lee, S. Gutta, W. Yim, and K. J. Kim, “Preliminary study of wireless actuation and control of IPMC actuator,” Advanced Intelligent Mechatronics (AIM), 2010 IEEE/ASME International Conference on, 2010, pp. 157-162.

[6] K. Abdelnour, A. Stinchcombe, M. Porfiri, J. Zhang, and S. Childress, “Wireless powering of ionic polymer metal composites toward hovering microswimmers,” IEEE/ASME Transactions on Mechatronics, vol. 17, pp. 924-935, 2012.

[7] S. K. Rahman, O. Ahmed, M. S. Islam, A. R. Awal, and M. S. Islam, “Design and construction of wireless power transfer system using magnetic resonant coupling,” American Journal of

IOP Conf. Series: Materials Science and Engineering **409** (2018) 012002 doi:10.1088/1757-899X/409/1/012002

Electromagnetics and Applications, vol. 2, pp. 11-15, 2014.
[8] G. Kkelis, D. C. Yates, and P. D. Mitcheson, “Comparison of current driven Class-D and Class-E half-wave rectifiers for 6.78 MHz high power IPT applications,” Wireless Power Transfer Conference (WPTC), 2015 IEEE, 2015, pp. 1-4.
[9] S. S. Mohan, M. Del Mar Hershenson, S. P. Boyd, and T. H. Lee, “Simple accurate expressions for planar spiral inductances,” IEEE Journal of solid-state circuits, vol. 34, pp. 1419-1424, 1999.
[10] Y. Z. Cheng, J. Jin, W. L. Li, J. F. Chen, B. Wang, and R. Z. Gong, “Indefinite-permeability metamaterial lens with finite size for miniaturized wireless power transfer system,” AEU-International Journal of Electronics and Communications, vol. 70, pp. 1282-1287, 2016.
[11] M. Shahinpoor, and K. J. Kim, “Ionic polymer-metal composites: I. Fundamentals,” Smart materials and structures, vol. 10, p. 819, 2001.
[12] Q. Chen, K. Xiong, K. Bian, N. Jin, and B. Wang, “Preparation and performance of soft actuator based on IPMC with silver electrodes, ” Frontiers of Mechanical Engineering in China, vol. 4, pp. 436-440, 2009.
[13] M. F. Shaari, S. Saw, and Z. Samad, “Fabrication and characterization of IPMC actuator for underwater micro robot propulsor,” Applied Mechanics and Materials, vol. 575, pp. 716-720, 2014.

IOP Conf. Series: Materials Science and Engineering **409** (2018) 012003 doi:10.1088/1757-899X/409/1/012003

Effect of Filler Content on Transport Properties of Sulfonated Polyether Ether Ketone (SPEEK) Composite Proton Exchange Membranes

Min Yan Chia, Hui San Thiam*, Loong Kong Leong, Chai Hoon Koo

Department of Chemical Engineering, Lee Kong Chian Faculty of Engineering and Science, Universiti Tunku Abdul Rahman, Bandar Sungai Long, Kajang, Selangor, Malaysia
Department of Civil Engineering, Lee Kong Chian Faculty of Engineering and Science, Universiti Tunku Abdul Rahman, Bandar Sungai Long, Kajang, Selangor, Malaysia

Email: chiamy94@gmail.com, thiamhs@utar.edu.my, leonglk@utar.edu.my, kooch@utar.edu.my

Abstract. Proton exchange membrane for direct methanol fuel cell was developed by the incorporation of silicotungstic acid supported on silica (SiO_2-SiWA) into SPEEK polymer. The properties of composite membranes with various silica/SiWA ratios were investigated and compared in this study. It was found that high silica to SiWA weight ratio (2:1) resulted in porous membrane with high proton conductivity and moderate methanol permeability. Reduction in the aforementioned ratio to 1:1 resulted in a more compact structure with better thermal stability, however, leaching of SiWA from the membrane occurred due to insufficient support. SPEEK membrane incorporated with 10 wt% SiO_2 and 5 wt% SiWA showed the highest selectivity of 6.44×10^4 S.s/cm^3 and performed better than that of the pure SPEEK membrane. These results suggested that these membranes have potential to be considered for DMFCs applications.

1. Introduction

The high energy conversion efficiency of fuel cell results in its emergence as one of the potent power sources. According to Kolli and co-workers [1], proton exchange membrane fuel cell (PEMFC) is the most widely used fuel cell technology due to its high power density at low volume and weight. Conventional PEMFC which uses hydrogen as fuel presented some problems related to the portability and physical properties of hydrogen fuel. This has sparked the researches on liquid organic hydrogen carrier such as direct methanol fuel cell (DMFC) [2].

At present, Nafion membrane is considered as the standard in DMFC industry due to its good thermal, chemical and mechanical properties. However, high methanol permeability of Nafion membrane and its low resistance towards high temperature operation have restricted its commercialization [3]. Modifications were done to improve the performance of Nafion membrane in DMFC by incorporating foreign materials [4]. Nevertheless, high cost of Nafion germinates the initiative to discover new DMFC membrane materials. In terms of proton conductivity to methanol permeability selectivity, sulfonated poly ether ether ketone (SPEEK) membrane is regarded as the best candidate to replace Nafion membrane due to its low methanol permeability and comparable proton conductivity with Nafion [5].

In this study, silicotungstic acid (SiWA), a crystalline material with high proton conductivity and thermal stability is incorporated into SPEEK polymer to enhance the membrane's performance. In fact, SiWA is reported to have the best proton conductivity enhancement among other heteropoly acids (HPA) [6]. However, SiWA is very soluble in water and prone to leaching. Therefore, silica is used as an inorganic support to immobilise SiWA while improving the thermal property and methanol rejection ability of SPEEK composite membrane [7]. Solution casting method is used to prepare the SPEEK composite membrane. In this method, mixing of additives into polymer solution can be achieved by ultrasonic treatment [8]. It is believed that under ultrasonic treatment, multiple effects such as dispersion and crushing can be obtained from cavitation which involves rapid formation, growth and collapse of microbubbles. This can help to ensure better dispersion and stabilisation of inorganic nanoparticles in the polymer matrix [9].

2. Experimental

2.1. Materials

Poly ether ether ketone (PEEK), SiWA and N-methyl-2-pyrolidone (NMP) were received from Sigma Aldrich. 95 – 98 % sulphuric acid (H_2SO_4), tetraethyl orthosilicate (TEOS, > 90 %), methanol (> 99.9 %) and ethanol (95 %) were obtained from Merck. All chemicals were used without further purification.

2.2. Synthesis of SPEEK and preparation SPEEK/SiO_2-SiWA Composite Membrane

1 g of PEEK was dissolved in 35 mL concentrated H_2SO_4. The reaction lasted for 48 hours at room temperature. Then the polymer-acid solution was quenched into ice-cold water to obtain solid SPEEK polymer and the precipitated polymer was left overnight. SPEEK polymer was washed with excess deionised water until the washing water becomes neutral (above pH 6.0) and dried in oven before use. DS of the synthesised SPEEK polymer was determined by titration method using 2 M sodium chloride solution. The DS of SPEEK polymer used in this study ranged from 55 % to 60 %.

SPEEK polymer was dissolved completely in NMP to form a 5 wt% polymer solution. SiWA was dissolved in deionised water and ethanol to form a homogeneous solution. Then, TEOS was added where TEOS:water:ethanol molar ratio was kept at 1:4:4. The solution was ultrasonicated for 30 minutes at room temperature. Polymer solution was added to the inorganic additive solution and ultrasonicated for 30 minutes at room temperature. The final solution was casted to a petri dish and dried in oven for 24 hours at 80 °C. The membrane was then peeled off and activated by immersing it in 1 M H_2SO_4 solution at room temperature for 24 hours. Finally, the membrane was washed several times and stored in deionised water for at least 24 hours before use. The membranes were labelled in SPEEK-X-Y where X and Y indicating the weight percentage of silica and SiWA, respectively.

2.3. Characterization

The composition of synthesised composite membranes was confirmed using energy dispersive X-ray spectroscopy (EDX, Amatek). Surface morphologies of the membranes were studied using scanning electron microscopy (SEM, Hitachi S-3400N). Water uptake and methanol uptake were measured by first immersing the membrane in water or 1 M methanol solution for 24 hours and then drying it to a constant weight. The percentage of the weight difference between wet and dry membranes to the dry weight was used as an indicator of water uptake (or methanol uptake).

Methanol permeability of the membranes was determined using diffusion method. A diffusion cell with two compartments was used where one compartment contained 1 M methanol solution and the other compartment contained deionised water. The membrane was fixed in between two compartments and the solution in each compartment was stirred constantly to ensure homogeneity. At suitable intervals, samples were collected from water compartment and methanol concentration of the sample was determined using gas chromatography. Equation (2) was used to calculate the methanol permeability (P_M):

$$P_M\,(\mathrm{cm^2/s}) = (S \times V_B \times t)/(A \times C_{M0}) \qquad (2)$$

where S (M/s) is the slope of methanol concentration as a function of time, V_B (cm^3) is the volume of deionised water, t (cm) is the membrane thickness, A (cm^2) the effective contact area and C_{M0} (M) is the initial methanol concentration

Proton conductivity was measured using a four-electrode conductivity cell connected to a potentiostat (Zive SP1). In-plane conductivity (σ) of the membrane was calculated using Equation (3):

$$\sigma\,(\mathrm{S/cm}) = L/Rwt \qquad (3)$$

where L (cm) is the distance between two inner electrodes, R (Ω) is the membrane resistivity while w and t are the width and thickness, in cm of the membrane, respectively.

3. Results and Discussion

3.1. Membrane Morphology

Fig 1 depicts the surface morphology of different SPEEK/SiO_2-SiWA composite membranes. SPEEK-10-5 had the largest pore size and the size reduced subsequently with the increment of SiWA loading. SiWA was acidic in nature and different amount of SiWA added in fact provided different hydrolysis/condensation reaction conditions for TEOS precursor. Studies had showed that the acid catalysed hydrolysis/condensation reaction of TEOS precursor occurred much faster than neutral and basic reaction conditions [10]. However, this rapid reaction also led to the formation of agglomerated silica particles within the polymer matrix.

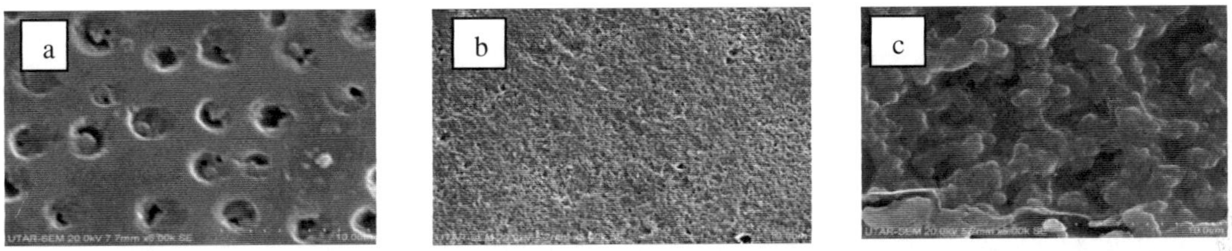

Figure 1: Surface Morphology of (a) SPEEK-10-5 (b) SPEEK-10-10 (c) SPEEK-10-20

In this study, agglomeration of silica particles was less of a problem due to the application of ultrasound during the reaction. Therefore, the amount of reacted TEOS was the key factor which would affect the surface morphology. From the presented SEM micrographs, the mentioned reducing pore size could be due to the increasing amount of silica produced from TEOS precursor. As shown Fig. 1a, there was some particles filling or clogging in the pores. More silica produced would enhance the blocking effect and hence resulted in a more compact morphology as shown in Fig. 1b for SPEEK-10-10 and Fig. 1(c) for SPEEK-10-20. Note that in SPEEK-10-20, there was no visible pores from the captured image due to the complete coverage of the silica on the surface of the membrane sample which led to the flaky morphology as shown.

EDX analysis results of the SPEEK/SiO_2-SiWA composite membranes were tabulated in Table I. From the results, it could be observed that as the ratio of silica to SiWA became smaller (from 10:5 to 10:20), the ability of SiWA to be incorporated into the polymer matrix (represented by weight composition of tungsten) became weaker. In SPEEK-10-5, the elemental weight ratio of silicon to tungsten was about 2, which was quite close to the theoretical value. However, in SPEEK-10-10, although there was an increase in tungsten loading, the ratio between silicon and tungsten could not match the theoretical 1:1. This effect became worse when the SiWA weight composition was raised to 20 %, where the amount of tungsten present in the membrane sample was too low to be detected. This was also accompanied by the increase in silicon elemental composition resulted from the decrease of another component in the sample. The reason for such phenomena could be due to the insufficient amount of silica available to secure all the added SiWA. Silica loading should be in excess of SiWA in order to immobilise high amount of the leach-prone additive successfully.

Table 1. Elemental Composition Of SPEEK/Sio2-Siwa Composite Membranes

Membrane	Elemental Compositions (wt%)	
	Si	***W***
SPEEK-10-5	9.72	4.81
SPEEK-10-10	21.77	5.74
SPEEK-10-20	24.21	0.00

3.2. Water and Methanol Solution Uptake

Water uptake and 1 M methanol solution uptake of SPEEK/SiO_2-SiWA composite membranes with different compositions were presented in Table II. Both results showed similar trend, where the water/methanol solution uptake experienced a drop from SPEEK-10-5 to SPEEK-10-10 before gaining some increment in SPEEK-10-20. Such trend could be explained by the microstructure of the membrane resulted from different membrane composition. In SPEEK-10-5, low level of silica and SiWA resulted in relatively porous and loose-packed structure. This would ease the swelling and thus water/methanol solution uptake ability of the membrane itself. As SiWA has terminal oxygen atoms which could form bonds with polar molecules such as methanol, successful incorporation of SiWA into SPEEK polymer matrix would also increase the methanol uptake of the composite membrane.

Table 2. Comparison Of Water Uptake, Methanol Uptake, Methanol Permeability, Proton Conductivity And Selectivity Of Different SPEEK/SiO_2-SiWA Composite Membranes

Membrane	Water Uptake (%)	Methanol Uptake (%)	Methanol Permeability ($\times 10^{-7}$ cm^2/s)	Proton Conductivity (S/cm)	Selectivity ($\times 10^4$ $S.s/cm^3$)
SPEEK-10-5	50.98	55.74	7.45	0.0480	6.44
SPEEK-10-10	33.33	35.29	5.21	0.0293	5.63
SPEEK-10-20	50.00	46.81	8.72	0.0291	3.34

As for SPEEK-10-10, there were significant reductions in both water and methanol solution uptake. Such phenomena could be attributed to the compact structure of the membrane as shown in Fig. 1b. The compact structure allowed reasonable water and methanol solution uptake by the hydrophilic domain of the polymer matrix. At the same time, the expansion of hydrophilic region was greatly restricted by the presence of silica which provided better mechanical integrity to the membrane. However, the trend did not persist when higher SiWA loading was added to the membrane. In SPEEK-10-20, the water and methanol solution uptake increased once again. As stated earlier, inability of silica to entrap all added SiWA had caused the latter additive to leach from the polymer matrix. Although the SEM image showed that the surface of the membrane did not have visible pores, however, it was believed that the internal structure of the membrane was relatively porous after the leaching of SiWA. The location where SiWA was originally being situated now became a void area which could be utilised by the hydrophilic domain for its expansion. It was also at this loading where methanol solution uptake was less than water uptake. The very low loading of SiWA within the SPEEK polymer matrix reduced the ability of the membrane to capture polar methanol and hence causing some depression in the methanol solution uptake.

3.3. Methanol Permeability

Methanol permeability of different SPEEK/SiO_2-SiWA membranes were also reported in Table II. The trend of methanol permeability was similar to that of water and methanol solution uptake. For SPEEK-10-5, the methanol permeability was 7.45 × 10^{-7} cm^2/s, which was considerably lower than that of conventional Nafion membrane, which was 1.23 × 10^{-6} cm^2/s [11]. This successfully showed that SPEEK based composite membrane had better methanol rejection capability as compared to Nafion. Incorporated silica particles blocked the pores available for methanol transport, as shown by the SEM image in Fig. 1. When the amount of SiWA was increased to 10 wt%, incorporation of silica in the SPEEK polymer matrix at a more acidic condition had caused more silica occupying hydrophilic domain of the composite membrane, specifically the ionic clusters and resulted a remarkable drop in methanol permeability.

Moving on to SPEEK-10-20, there was an increase in methanol permeability when compared to other SPEEK composite membranes. Such increment was due to the leaching of SiWA which created voids within the membrane's internal structure. This allowed methanol molecules to pass through the membrane more easily, thereby deteriorating its methanol rejection ability. Furthermore, silica content in SPEEK-10-20 was higher when compared with SPEEK-10-5 and SPEEK-10-10 as shown by the results of EDX analysis (Table I). In this case, there was a possibility that high silica content would not only present in the hydrophilic region, instead, it might appear in the hydrophobic backbone as well [12]. As silica was hygroscopic in nature, this would provide extra hydrophilic pathway which would be used by methanol for diffusion. By combining the two effects above, the SPEEK-10-20 experienced a significant increase in methanol permeability which might affect its overall performance.

3.4. Proton Conductivity

Proton conductivity values of different SPEEK/SiO_2-SiWA composite membranes were tabulated in Table II. SPEEK-10-5 had the highest proton conductivity of 0.048 S/cm, while SPEEK-10-10 and SPEEK-10-20 had similar proton conductivity which were 0.0293 S/cm and 0.0291 S/cm, respectively. Incorporation of SiWA in SPEEK especially in SPEEK-10-5 managed to improve the proton conductivity of the SPEEK membrane, which is 0.011 S/cm according to previous study [13]. This was because of the extra proton exchange sites available after the modification. At the same time, lower silica content and larger pore size allowed easy transportation of proton by vehicle diffusion or Grotthüss mechanism. The high proton conductivity was also aided by sufficient water uptake of the membrane. In SPEEK-10-10, higher amount of silica incorporated produced a relatively compact structure. This was proved by the low water and methanol solution uptake. The compact structure reduced the passing of proton through the membrane by the formation of smaller ionic clusters. However, SPEEK-10-20 which had a more porous membrane structure attained similar proton conductivity as SPEEK-10-10. It was because of the loss of SiWA through leaching reduced the membrane's capability to conduct proton. This result also justified the importance of SiWA in promoting and enhancing the transport of proton through SPEEK composite membrane.

By considering both proton conductivity and methanol permeability, selectivity of the membrane could be deduced and shown in Table II. SPEEK-10-5 (with reasonable amount of incorporated silica and SiWA) had the highest selectivity, with the reported value of 6.44 × 10^4 $S.s/cm^3$. It was followed by SPEEK-10-10 and SPEEK-10-20 with the values of 5.63 × 10^4 $S.s/cm^3$ and 3.34 × 10^4 $S.s/cm^3$ respectively. For summary, the outcome of this study was satisfactory with the realization of incorporation of silica and SiWA into SPEEK polymer to enhance its properties.

4. Conclusion

SiWA was successfully incorporated into SPEEK polymer by using silica as support to fabricate SPEEK/SiO_2-SiWA composite membrane. Effect of inorganic additive loadings and their interaction were deduced from SEM images and EDX analysis. High TEOS to SiWA ratio (SPEEK-10-5) would result in porous structure, while low TEOS to SiWA ratio (SPEEK-10-20) would cause leaching of SiWA alongside with the increase in compactness of membrane by formation of more silica. SPEEK-

10-5 had the highest selectivity because of its high proton conductivity coupled with moderate methanol permeability. The results showed that the proton conductivity of the composite membrane dropped with the increase in SiWA loading due to the increase in membrane's compactness (SPEEK-10-10) and leaching of SiWA (SPEEK-10-20). Similar reason was also used to explain the decrease in methanol permeability in SPEEK-10-10 and the increment of the same metric in SPEEK-10-20. Further works are in progress to use the SPEEK-10-5 composite membrane in membrane electrode assembly (MEA) for DMFC performance evaluation.

Acknowledgment
This work was supported by the Universiti Tunku Abdul Rahman (UTAR) through UTAR Research Fund (UTARRF) under project number IPSR/RMC/UTARRF/2016-C1/T8.

References

[1] A. Kolli, A. Gaillard, A. De Bernardinis, O. Bethoux, D. Hissel, and Z. Khatir, "A review on DC/DC converter architectures for power fuel cell applications," Energy Conversion and Management, vol. 105, Nov. 2015, pp. 716-730, doi: 10.1016/j.enconman.2015.07.060.

[2] G. L. Soloveichik, "Liquid fuel cells," Beilstein Journal of Nanotechnology, vol. 5, Aug. 2014, pp. 1399-1418, doi: 10.3762/bjnano.5.153.

[3] H. S. Thiam, M. Y. Chia, Q. R. Cheah, C. C. H. Koo, S. O. Lai, and K. C. Chong, "Proton conductivity and methanol permeability of Nafion-SiO_2/SiWA composite membranes," Proc. AIP International Symposium on Green and Sustainable Technology (ISGST 17), AIP Conference Proceedings, Apr. 2017, pp. 020007, doi: 10.1063/1.4979378.

[4] H. S. Thiam, W. R. W. Daud, S. K. Kamarudin, A. B. Mohammad, A. A. H. Kadhum, K. S. Loh, et al., "Overview on nanostructured membrane in fuel cell applications," International Journal of Hydrogen Energy, vol. 36, Feb. 2011, pp. 3187-3205, doi: 10.1016/j.ijhydene.2010.11.062.

[5] N. Intaraprasit and P. Kongkachuichay, "Preparation and properties of sulfonated poly(ether ether ketone)/Analcime composite membrane for a proton exchange membrane fuel cell (PEMFC)," Journal of the Taiwan Institute of Chemical Engineers, vol. 42, Jan. 2011, pp. 190-195, doi: 10.1016/j.jtice.2010.05.002.

[6] K.-M. Lee, J.-Y. Woo, B.-C. Jee, Y.-K. Hwang, C.-h. Yun, S.-B. Moon, et al., "Effect of cross-linking agent and heteropolyacid (HPA) contents on physicochemical characteristics of covalently cross-linked sulfonated poly(ether ether ketone)/HPAs composite membranes for water electrolysis," Journal of Industrial and Engineering Chemistry, vol. 17, Jul. 2011, pp. 657-666, doi: 10.1016/j.jiec.2011.02.017.

[7] Z. Li, W. Dai, L. Yu, J. Xi, X. Qiu, and L. Chen, "Sulfonated poly(ether ether ketone)/mesoporous silica hybrid membrane for high performance vanadium redox flow battery," Journal of Power Sources, vol. 257, Jul. 2014, pp. 221-229, doi: 10.1016/j.jpowsour.2014.01.127.

[8] J. Wang, H. Bai, H. Zhang, L. Zhao, H. Chen, and Y. Li, "Anhydrous proton exchange membrane of sulfonated poly(ether ether ketone) enabled by polydopamine-modified silica nanoparticles," Electrochimica Acta, vol. 152, Jan. 2015, pp. 443-455, doi: 10.1016/j.electacta.2014.11.165.

[9] Q. Wang, H. Xia, and C. Zhang, "Preparation of polymer/inorganic nanoparticles composites through ultrasonic irradiation," Journal of Applied Polymer Science, vol. 80, Mar. 2001, pp. 1478-1488, doi: 10.1002/app.1239.

[10] A. H. E. Yousif, O. Y. O. Alhussein, and M. S. A. Eltoum, "Characterization of hydrolyzed products of tetra ethoxy silane prepared by sol-gel method," International Journal of Multidisciplinary Science and Engineering, vol. 6, Jan. 2015, pp. 19-22.

[11] H. S. Thiam, W. R. W. Daud, S. K. Kamarudin, A. B. Mohamad, A. A. H. Kadhum, K. S. Loh, et al., "Nafion/Pd–SiO2 nanofiber composite membranes for direct methanol fuel cell

applications," International Journal of Hydrogen Energy, vol. 38, Jul. 2013, pp. 9474-9483, doi: 10.1016/j.ijhydene.2012.11.141.

[12] R. Jiang, H. R. Kunz, and J. M. Fenton, "Composite silica/Nafion® membranes prepared by tetraethylorthosilicate sol–gel reaction and solution casting for direct methanol fuel cells," Journal of Membrane Science, vol. 272, Mar. 2016, pp. 116-124, doi: 10.1016/j.memsci.2005.07.026.

[13] H. Zhang, T. Zhang, J. Wang, F. Pei, F. He, and J. Liu, "Enhanced proton conductivity of sulfonated poly(ether ether ketone) membrane embedded by dopamine-modified nanotubes for proton exchange membrane fuel cell," Fuel Cells, vol. 13, Oct. 2013, pp. 1155-1165, doi: 10.1002/fuce.201300130.

IOP Conf. Series: Materials Science and Engineering **409** (2018) 012004 doi:10.1088/1757-899X/409/1/012004

Optimization of the electrical conductivity and tensile strength of conductive polymer composites using the Taguchi method

Hendra Suherman[1*], Erwin Julisman[1], Duskiardi[1], Mulyanef[1], Irmayani[2]

[1]Department of Mechanical Engineering, Universitas Bung Hatta, Padang, Indonesia
[2]Department of Industrial Engineering, Universitas Ekasakti, Padang, Indonesia

E-mails: hendras@bunghatta.ac.id;irmahendras@gmail.com

Abstract. This research focused on optimization of the an electrical conductivity and tensile strength of G5/G74/epoxy composites using the Taguchi method. Graphite with particle size of 5 µm (G5) and 74 µm (G74) were used as a second and as a main conductive filler. The mixing parameter used as control factor in Taguchi method are the composition of the second filler, mixing rotation, and mixing time. The Taguchi array with three mixing parameter were used in this research. The signal to noise (S/N) ratio was used to obtaine the results of the research. The results of the research showed that the Taguchi method has succeeded in increasing the electrical conductivity and tensile strength of G5/G74/epoxy composites produced up to 18.28% and 18.42%, respectively.

1. Introduction

The conductive polymer composite (CPCs) material is potentially used as a material having high electrical coductivity and good mechanical properties. Compared to metals, this material is lighter, cheaper and has a better corrosion resistance, but it has a low electrical conductivity. To increase the electrical conductivity, a high content of conductive filler is required in the polymer matrix. On the other hand, the high content of the conductive filler material in matrix will also cause the poor mechanical properties of conductive polymer composites (CPCs) obtained [1-3]. Graphite is potentially used as a conductive material in a CPCs material, due to the high electrical conductivity, and it has various shapes and sizes. Beside that, this material is easily made by conventional processes such as casting or injection molding [4].

The Taguchi method is an effective method for obtaining optimum mixing parameters to produce a good properties of CPCs. This method can analyze each of the factors used [5]. The combination of the optimum factors was chosen based on the largest S/N ratio on all factors. Furthermore, the validity test on the best combination obtained was performed, to get the optimum results.

The optimum conditions for the electrical conductivity and tensile strength of the resulting material were obtained in separate studies [5, 6]. Therefore, it is necessary to optimize the electrical conductivity and tensile strength in one study by taking the highest S/N ratio value of each parameter value to obtain both desirable properties. Taguchi array L9 (3^3) with nine samples used to obtain the optimum mixing parameters. Although nine experiments were the minimum number in the Taguchi array for three factors and three levels, but this minimum number was able to produce the optimum combination parameters [7-12]. Therefore, this study used the Taguchi array L9 (3^3) for optimization of the electrical conductivity and tensile strength of conductive polymer composites.

IOP Conf. Series: Materials Science and Engineering **409** (2018) 012004 doi:10.1088/1757-899X/409/1/012004

2. Experimental

2.1. Materials
This research uses two conductive fillers of different size, ie graphite size 74 μm (G74) and graphite size 5 μm (G5). G74 is used as the main conductive filler while G5 is used as a second conductive filler. Both graphite materials are used in the form of flake. The gap formed from a larger conductive filler will be filled by a smaller conductive filler, so it is expected to increase the conductive network formed thus increasing the electrical conductivity produced. G74 was obtained from Asbury carbon, New Jersey while G5 was purchased from FRIway Industry, China. 635 thin epoxy resin obtained from the US Composites, with a viscosity of 6 Poise. The particle size and viscosity of the epoxy resin were obtained from the manufacturer.

2.2. Fabrication of G5/G74/epoxy composites
This research use G5 as a second filler (5, 7.5, 10 wt.%), G74 as a main filler (45, 42.5, 40 wt.%), and epoxy (50 wt.%) as the matrix. The fabrication process done by mixing epoxy resin and hardener using mechanical stirring machine (IKA RW 20 Digital) with mixing rotation of 25 rpm at 10 minutes. Next, the graphite (G5 and G74) at predefined composition is incorporated into the epoxy resin mixture based on control factor and level of mixing parameters used (see Table 1). Then the formed G5/G74/epoxy composites mixture is poured into the aluminum mold. Molds are inserted into the oven using a fixed temperature of 150 °C at 90 min.

2.3. Characterization
The electrical conductivity of G5/G74/epoxy composites with different composition of conductive fillers were measured by ASTM C 61 method [13]. Tensile strength of G5/G74/epoxy composites were measured using ASTM D3039. The scanning electron microscopy (SEM) of Hitachi S-3400 N used to observe the dispersion and voids of the conductive fillers within the polymer matrix through fractured surfaces of G5/G74/epoxy composites.

2.4. Selection of Control Factors
The Taguchi Method used on this study, to optimize control parameters to get the best results through the Design of Experiments of orthogonal array. The orthogonal arrays (OA) profide a set of well balanced (minimum) experiments and signal to noise ratios (S/N), help in data analysis and prediction of optimum results. The control factors of mixing parameter are composition of the second filler, mixing rotation, and mixing time. Table. 1 show the three control factors with three levels of mixing parameters.

Table 1. Control Factor And Level Of Mixing Parameters

Symbol	Control Factors	Unit	Level		
			1	2	3
A_{cf}	Composition of second filler	wt.%	5	7.5	10
B_{mr}	Mixing rotation	rpm	30	50	70
C_{mt}	Mixing time	Minutes	15	25	35

3. Researchs and Discussion

3.1. Characterization of conductive fillers
Figure 1. shows the conductive filler type used to produce a G5/G74/epoxy composites material. Scanning Electron Microscopic (SEM) images shows that G74 as the main conductive filler and G5 as the second filler filler have the different size and same shape (flake shape).

G74 (74μm) G5 (5μm)

Figure 1. SEM images: (a) G74 and (b) G5 conductive fillers

3.2. The electrical conductivity of G5/G74/epoxy composites of the mixing parameters

The electrical conductivity of G5/G74/epoxy composites on each experiment based on the Taguchi array L9 (3^3) that has been performed is shown in Table 2. The lowest electrical conductivity of G5/G74/epoxy composites was obtained on the experiment 2 (0.3 S/m), while the highest value was on the experiment 7 (3.80 S/cm). From this condition, it can be seen that the weight percentage (wt. %) of the second filler, mixing rotation and mixing time are effective influence the electrical conductivity of G5/G74/epoxy composites.

Table 2. The Electrical Conductivity Of G5/G74/Epoxy Composites Based On Taguchi Array L9 (33)

Experiment	Control Factors and Levels			Electrical conductivity
	A_{cf} (wt.%)	B_{mr} (rpm)	C_{mt} (minutes)	(S/m)
1	10	30	15	2,90
2	10	50	25	0,30
3	10	70	35	3,05
4	7.5	50	15	2,60
5	7.5	70	25	0,70
6	7.5	30	35	2,70
7	5	70	15	3,80
8	5	30	25	2,80
9	5	50	35	3,35

3.3. Taguchi response to the electrical conductivity in mixing parameters

Figure 2. Shows S/N ratio of mixing parameters comprising of second filler composition, mixing rotation; and mixing time. Based on the desired S/N ratio of the larger the better, then the optimum mixing parameter of G5/G74/epoxy composites obtained was at A_3 (5 wt.% second fillers), B_1 (30 rpm mixing rotation), and C_3 (35 minutes mixing time). The results show that high content of second conductive filler and high mixing rotation is not required to get the high electrical conductivity of G5/G74/Epoxy composites. The high content of the second conductive filler in the matrix will cause the number of voids to occur, so that the conductive filler is not well dispersed, thus causing in low electrical conductivity of G5/G74/Epoxy composites. Similarly, high rotational mixing will damage the conductive network formed, thereby lowering the electrical conductivity [14, 1]

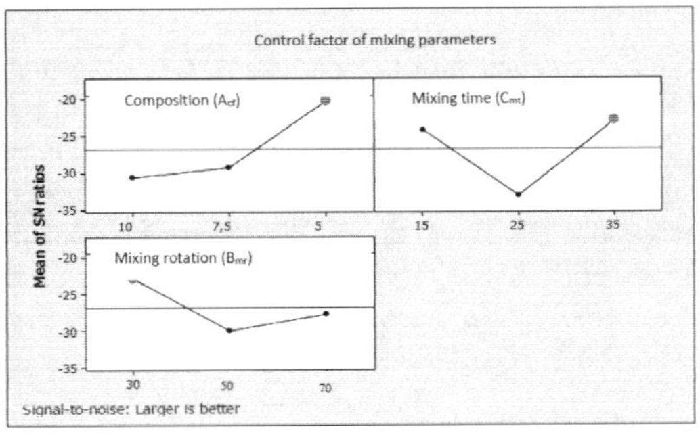

Figure 2. S/N ratio of G5/G74/epoxy composites materials to the mixing parameters

3.4. The electrical conductivity and tensile strength based on recommendation of Taguchi method
The recomended optimum parameters of the electrical conductivity based on S/N ratio of G5/G74/epoxy composites to mixing parameters were used to produce the G5/G74/epoxy composites new specimen. At this stage, the electrical conductivity and tensile strength were measure to see how effective the Taguchi array L9 (3^3) to increase the electrical conductivity and tensile strength of G5/G74/epoxy composites. The results show that the specimen produced based on Taguchi recommendation has a better properties of the electrical conductivity and tensile strength compared to the value obtained in the experiment 1 of mixing parameters (see Table.3).

Table 3. Optimization Of The Electrical Conductivity And Tensile Strength Using Mixing Parameters Based On The Taguchi Array L9 (33) Recommendation

No	CPCs material	Electrical Conductivity (S/m)	Tensile Strength (N/mm^2)
1	Experiment 1	2,90	123,8
2	Taguchi recomendation	3,43	146,6

The dispersion of the conductive fillers within polymer matrix were greatly determines the electrical conductivity of G5/G74/epoxy composites [14, 2]. Figure 3. shows scanning electron microscopic (SEM) images of fractured surface of G5/G74/ epoxy composites. The conductive fillers in the polymer matrix (epoxy resin) of Taguchi recomendation (Figure 3.b) were dispersed better, compared than the G5/G74/epoxy composites material produced on the experiment 1 (Figure.3a). As well as voids, which were successfully reduced by using the recomended parameters of the Taguchi method. This condition generate a better electrical conductivity and tensile strength of G5/G74/epoxy composites (see figure 3) [15]

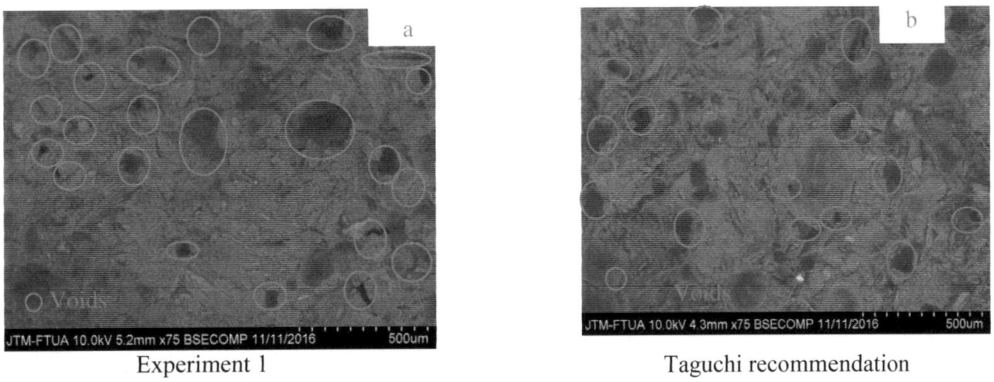

Experiment 1 Taguchi recommendation

Figure 3. SEM images of fracture surface of G5/G74/epoxy composites (a) Experiment 1, (b) Taguchi recomendation

4. Conclussion

Taguchi Method has been used to optimize mixing parameters to obtain better electrical conductivity and tensile strength for G5/G74/epoxy composites. The conclusions can be summarized as follows:

1. The Taguchi array of L9 (3^3) is able to optimize the optimum parameters to increases the electrical conductivity and tensile strength of resulting G5/G74/epoxy composites material effectively.
2. The optimum parameter for the electrical conductivity and tensile strength is on combination of A_3 (5 wt.% second fillers), B_1 (30 rpm mixing rotation), and C_3 (35 minutes mixing time).
3. The electrical conductivity and tensile strength of G5/G74/epoxy composites material produced are sucsessfully increase from 2,9 S/cm to 3,43 S/cm (18.28 %), and from 123,8 N/mm^2 to 146,6 N/mm^2 (18.42 %) respectively.

Acknowledgment

The authors gratefully acknowledge the financial support given for this work by the Directorate General of Higher Education through the University Exelent Applied Research on 2018.

References

[1] P.C. Ma, M.Y. Liu, H. Zhang, S.Q. Wang, R.Wang, K. Wang, Y.K.Wong, B.Z. Tang, S.H. Hong, K.W. Paik, and J.K. Kim, "Enhanced electrical conductivity of nanocomposites containing hybrid fillers of carbon nanotubes and carbon black," ACS Applied Materials & Interface, vol. 1, 2009, pp. 1090–1096, doi: 10.1021/am9000503

[2] H. Suherman, A.B, Sulong, and J. Sahari," Electrical properties of carbon nantubes-based epoxy nanocomposites for high electrical conductive plate," Advanced Materials Research, vol. 264-265, 2011, pp. 559-564, doi: 10.4028/www.scientific.net/AMR.264-265.559

[3] N,R, Rajendran Royan, A.B. Sulong, J. Sahari, H. Suherman," Effect of acid- and ultraviolet/ozonolysis-treated mwcnts on the electrical and mechanical properties of epoxy nanocomposites as bipolar plate applications," Journal of Nanomaterials, vol. 2013, 2013, Article ID 717459, pp. 8 pages, doi: 10.1155/2013/717459

[4] R.B. Mathur, S.R. Dhakate, D.K. Gupta, T.L. Dhami, and R.K. Aggarwal," Effect of different carbon fillers on the properties of graphite composite bipolar plate," Journal of Materials Processing Technology, vol. 203, 2008, pp. 184-192, doi: 10.1016/j.jmatprotec.2007.10.044

[5] S.H. Park, Robust design and analysis for quality engineering. London: Chapman & Hall, 1996.

[6] R.K. Roy, Design of experiment using taguchi approach. New York: John Wiley & Sons, 2001.

[7] M. Nalbant, H. Go¨kkaya, and G. Sur," Application of taguchi method in the optimization of cutting parameters for surface roughness in turning," Materials and Design vol. 28, 2007, pp. 1379–1385, doi: 10.1016/j.matdes.2006.01.008

[8] A. Hasçalık, and U. Çaydaş," Optimization of turning parameters for surface roughness and tool life based on the taguchi method," International Journal Advances Manufacturing Technology, vol. 38, 2008, pp. 896–903, doi: 10.1007/s00170-007-1147-0

[9] Y.T. Liu, W.C. Chang, Y. Yamagata," 2010. A study on optimal compensation cutting for an aspheric surface using the taguchi method," CIRP Journal of Manufacturing Science and Technology, vol. 3, 2010, pp. 40–48, doi: 10.1016/j.cirpj.2010.03.001

[10] A.K. Lakshminarayanan, and V. Balasubramanian," Process parameters optimization for friction stir welding of rde-40 aluminium alloy using taguchi technique," Trans. Nonferrous Metal Society, vol. 18, 2008, pp. 548-554, doi: 10.1016/S1003-6326(08)60096-5

[11] R. Surace, L.A.C. Filippis, A.D. Ludovico, and G. Boghetich," Application of taguchi method for the multi-objective optimization of aluminium foam manufacturing parameters," International Journal of Materials Forming vol. 3, 2010, pp. 1–5, doi: 10.1007/s12289-009-0409-9

[12] A.B. Sulong, J. Park, C.H. Azhari, and K. Jusoff," Process optimization of melt spinning and mechanical strength enhancement of functionalized multi-walled carbon nanotubes reinforcing polyethylene fibers," Composites: Part B, vol. 42, 2011, pp. 11–17, doi:

10.1016/j.compositesb.2010.09.014
[13] B.K. Kakati, D. Sathiyamoothy, A.Verma," Electrochemical and Mechanical Behaviour of Carbon Composite Bipolar Plate for Fuel Cell," International Journal of Hydrogen Energy, vol. 35, 2010, pp. 4185-4194, doi: 10.1016/j.ijhydene.2010.02.033
[14] N. Hu, Z. Masuda, G. Yamamoto, H. Fukunaga, T. Hashida, and J. Qiu," Efect of fabrication process on electrical properties of polymer/mwnt nanocomposite," Composites Part A vol. 39, 2008, pp. 893-903, doi: 10.1016/j.compositesa.2008.01.002
[15] C. Hui, L.H. Bo, Y. Li, L.J. Xin, and Y. Li," Study on the preparation and properties of novolac epoxy/graphite composite bipolar plate for PEMFC. International Journal of Hydrogen Energy, vol. 35, 2010, pp. 3105–3109, doi:10.1016/j.ijhydene.2009.08.030

IOP Conf. Series: Materials Science and Engineering **409** (2018) 012005 doi:10.1088/1757-899X/409/1/012005

Evolution of microstructure and mechanical properties of Al-B_4C composite after recycling

Zhenjiang Guo[1,2,a], Qiulin Li[1,b], Wei Liu[2], Guogang Shu[3]

[1]Graduate School at Shenzhen, Tsinghua University, Shenzhen, 518055, China
[2]School of Materials Science and Engineering, Tsinghua University, Beijing, 100084, China
[3]State Key Laboratory of Nuclear Power Safety Monitoring Technology and Equipment, China Nuclear Power Engineering Co. Ltd., Shenzhen of Guangdong Prov., 518172, China

E-mails: [a]gzj15@mails.tsinghua.edu.cn,[b]liql@sz.tsinghua.edu.cn

Abstract. A remelting method was used to recycle Al-15wt.% B_4C composite and the microstructure and mechanical properties of the composite were investigated before and after recycling. Scanning electronic microscope (SEM) showed that some of the particle clusters present in the original composite were broken down and the distribution of B_4C in the matrix was more uniform after recycling. The interface reaction was further intensified and the protection layer coated on the surface of B_4C was thicker after recycling. In addition, there were more TiB_2 particles which peeled off from the surface of B_4C and distributed around B_4C. Al_3Ti phase, which was present in the original composite, disappeared after recycling, while Al_3BC phase increased markedly. Tensile test and hardness test revealed that the mechanical properties were preserved well after recycling. Therefore, it is proposed that remelting method is a feasible method for Al-B_4C composite recycling.

1. Introduction

Aluminum matrix composites (AMCs), due to their advantageous properties such as high specific strength and modulus, unique wear resistance as well as corrosion resistance, are attractive structural and functional materials for automotive, aeronautical and many other applications[1-3]. As far as reinforcements are concerned, boron carbide (B_4C) are an attractive reinforcement because of their low density, high hardness, excellent chemical and thermal stability[4, 5]. More importantly, due to their specific capability of capturing neutrons, B_4C particle-reinforced AMCs have been mainly used in nuclear spent fuel storage and transportation applications[6, 7]. However, in the manufacturing processes of B_4C particle-reinforced AMCs, the process scraps from rolling can reach more than 20% of the total materials produced. The need of recycling Al-B_4C composites is urgent in terms of saving energy, reducing volumes of waste and decreasing pollutants emission.

There are several methods for aluminum matrix composites recycling, including electrolysis, salt fluxing, hot pressing, and remelting [8]. Electrolysis has been used to recycle SiC-reinforced AMCs[9]. The recovery of pure aluminium is achieved and the purity is over 98%, but this method is very time-and energy-consuming. Salt fluxing has low energy consumption and the key to this technique is finding suitable salts[10]. This method has been used to separate SiC particles from aluminum matrix[11]. But the yield is low and the recovered aluminum matrix contains a small amount of Si elements. Hot pressing, similar to powder metallurgy, has been used to recycle Al_2O_3-reinforced

IOP Conf. Series: Materials Science and Engineering **409** (2018) 012005 doi:10.1088/1757-899X/409/1/012005

AMCs through pressing and hot extrusion[12]. Due to the refinement of the microstructure and the dispersion of the aluminium oxide caused by the extrusion process, mechanical properties of recycled composite are improved compared to the original composite. But this method requires high performance equipment. Remelting technique is an attractive method because of its simple process and lower energy consumption. In one previous study, the changes in the properties of SiC_p-reinforced AMC after multiple remelting were studied, and the results showed that remelting times was crucial to the tensile properties of composite. As the remelting times increase, the mechanical properties decrease seriously[13]. It was also reported that with the increase of the stirring time, the fluidity of melt composite decreased and reaction products increased[14], which will have an impact on the casting and material properties. Though the remelting method has been employed to recycle aluminum matrix composites, its effect on microstructure and mechanical properties of B_4C particle-reinforced AMCs has not been fully studied.

In the present work, scrap materials produced during rolling Al-B_4C composite were recycled by a remelting method. Then the microstructure, including particles distribution, interfacial reaction and the reaction-induced particles, was characterized before and after recycling. In addition, mechanical properties such as tensile properties and hardness were also investigated.

2. Experimental Procedure

2.1. Recycling process

Processed scrap materials of the 1060Al-15 wt.% B_4C composite in the form of rolled sheets were used in this investigation, and the thickness of the sheets was 3 mm. The rolled sheets were cut into rectangles with a sectional dimension of 30×150 mm^2 by water jet cutter. After the sheets were cleaned, put them into a vacuum oven and dry at 120°C for 2 hours. A stirring-casting furnace under argon protection condition was used, the temperature was kept at 730°C for a long time holding. Put the composite into the furnace to heat for 45 minutes. Then the melted composite was agitated by a mechanical stirrer with the stirring speed constant at 320 r/min and the stirring time of 10 minutes. Finally, the fusing fluids was cast into a preheating steel mold in the form of slab.

2.2. Hot rolling

Hot rolling of the slab was carried out at 450 °C. The thickness reduction for each pass of rolling was kept at a lower value of 15%. For every rolling pass, the rolled sheet was reheated at 450 °C for 20 minutes. Finally, the slab was rolled into 3 mm sheets to evaluate their microstructure and mechanical properties.

2.3. Microstructure characterization

Samples with size $3\times10\times10$ mm^3 were polished by both mechanical polishing and ion beam (IB-09020CP, JEOL) polishing methods. Mechanical polishing was down to 1μm by diamond suspension. Optical microscope and a scanning electron microscope (SEM, SU8010, Hitachi, Japan) in back scattered electron (BSE) mode and equipped with Energy Dispersive Spectrometer (EDS) was applied to observe the particle distribution and interfacial microstructure along the rolling direction. X-ray diffraction (XRD, D/MAX-2500/PC, Rigaku, Japan) test was performed on samples to identify changes in reaction products.

2.4. Mechanical testing

The tensile samples and hardness samples were annealed at 400°C for 2 hours. The tensile tests were carried out according to the ASTM E8-04 standard (rectangular flat samples with a 12.5 mm width reduced section) at a test speed of $10^{-3}s^{-1}$ using a 50 mm gage extensometer, and each average value was obtained from three tensile tests. The Vickers hardness tests were carried out with a load of 300g and dwelling time of 15s on mechanical polishing samples based on the ASTM E92 standard, and each average value was obtained from 10 measurements for each sample.

IOP Conf. Series: Materials Science and Engineering **409** (2018) 012005 doi:10.1088/1757-899X/409/1/012005

3. Results And Discussion

3.1. Microstructure

Figure 1 presents the microstructure of original scrap material and recycled material observed using OM and SEM. In the original material, the distribution of B_4C in the matrix is inhomogeneous. Particle clusters, a dense solid particle complex consisting of tens to hundreds of particles, could be observed in the original materials although some particle clusters were broken down and banded distribution along the rolling direction due to the rolling deformation. After recycling, the distribution of B_4C in the matrix is more uniform and banded particle clusters is disappeared, and the amount and size of the clusters are considerably less and smaller compared with the original scrap materials. This may be attributed to the further agitation after remelting and further rolling deformation.

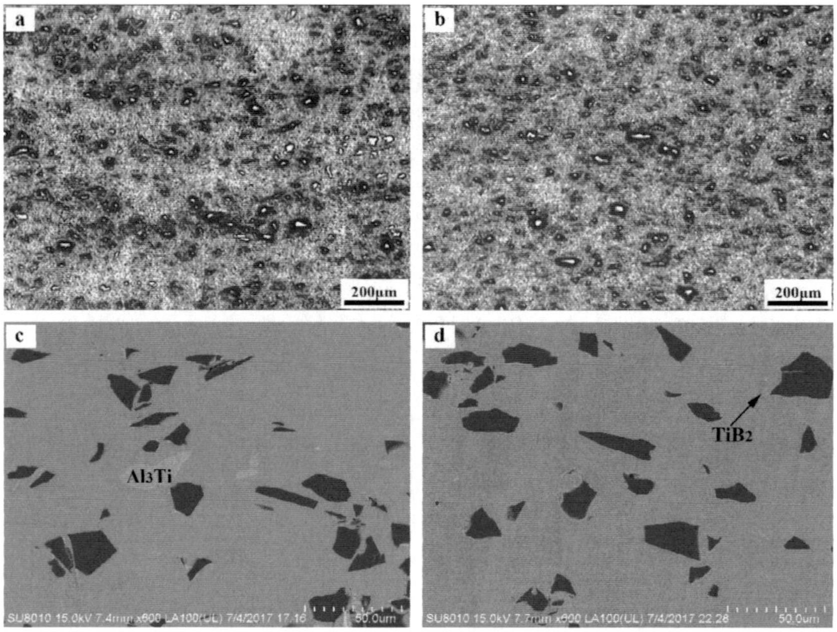

Figure 1. Microstructures of 1060Al-15%B_4C composite: (a, c) original material; (b, d) recycled material.

Figure 2 shows the EDS elemental mapping of B_4C particle and EDS line scan at the interface between Al matrix and B_4C particle at a high magnification. It can be seen that the Ti elements are enriched at the interface and form a protective layer around the B_4C. In the original composite, the distribution of the protection layer on the surface of B_4C is not uniform and the thickness is various at different positions, which is consistent with Li's studies[15]. This observation is proposed to be a result of the maldistribution of the nucleation sites for the protection layer. But after recycling, the protection layer is uniformly coated on the surface of B_4C and is thicker. The protection layer was identified as TiB_2 in previous studies[16]. This protective layer not only can protect B_4C from being eroded by liquid aluminum, but also it can improve the wettability of B_4C in liquid aluminum, and strengthen the bonding between Al and B_4C when Al-B_4C composites fabricated by stir casting technique[17-19].

EDS line scan of the samples was carried out under the same parameters. The intensity of different elements is presented. At the interface, it can be seen that the intensity of Ti element increases significantly after recycling, and B elements are also gathered at the interface and form a small peak on the curve. This further proves that the protective layer consists of B and Ti elements. Comparing the two interfaces, it reveals that after the transformation of the protective layer from a fine layer to

IOP Conf. Series: Materials Science and Engineering **409** (2018) 012005 doi:10.1088/1757-899X/409/1/012005

discretely coarse layer, the TiB_2 grew up at some discrete interface spots and formed needle-like crystals. This needle-like crystals peel off easily when subjected to external force. As shown in Figure 1, there are more white needle-like particles around B_4C after melting, which means thicker protective TiB_2 layers peeled off from the surface of B_4C attributed to the rolling deformation and liquid mechanical agitation.

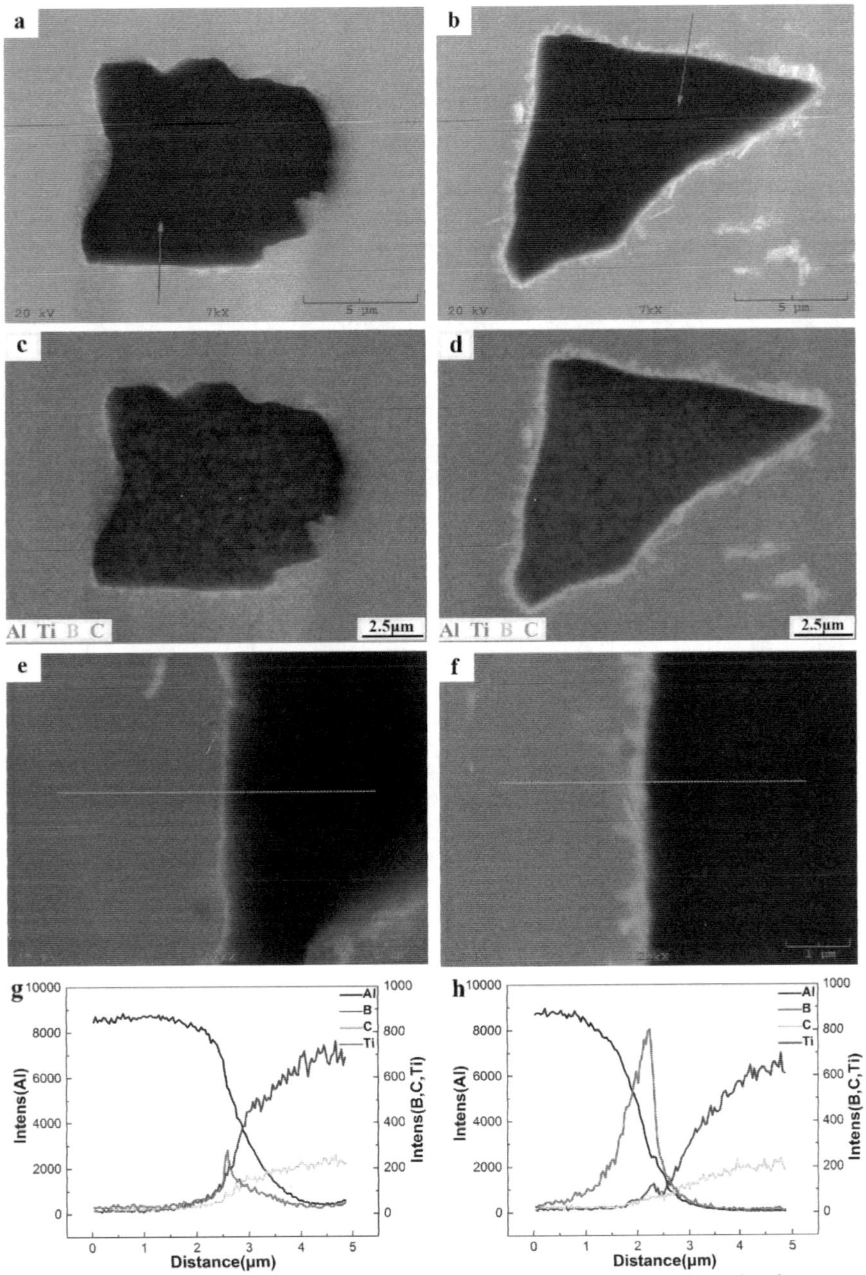

Figure 2. Microstructures of 1060Al-15%B_4C composite: (a, c, e, g) SEM backscattered electron image, EDS elemental mapping, EDS line scan of interface and the result of EDS line scan for the original material, respectively; (b, d, f, h) SEM backscattered electron image, EDS elemental mapping, EDS line scan of interface and the result of EDS line scan for the recycled material, respectively.

Figure 3 exhibits EDS pattern of large grey particle in Figure 1 and XRD patterns of Al-B_4C composites before and after recycling. The grey particle only contained Al and Ti elements, and atomic ratio of Al to Ti is approximately 3:1. The XRD patterns show that the Al_3Ti is present in the original composite. So it can be concluded that the grey particle in Figure 1 is Al_3Ti. This result shows that Ti elements is excessive in the primary production of composite materials. But it is disappeared after recycling. This can be attributed to the intensification of the interfacial reaction during the remelting process, which consumed Ti elements and lead to the decomposition of Al_3Ti[16, 20]. The interfacial reaction can be expressed as follows:

$$6Al+2B_4C+3Ti=2Al_3BC+3TiB_2$$

$$Al_3Ti=3Al+Ti$$

And the reaction products from the B_4C particle to Al matrix are Al_3BC, fine TiB_2 and coarse TiB_2. The XRD patterns also show that the peak of Al_3BC is obviously stronger compared with the composite, which indicates that Al_3BC increases after recycling. But in contrast to Chen's studies[14], we didn't find Al_3BC and AlB_2 particles around the B_4C, it may be due to the shorter stirring time and excessive Ti elements.

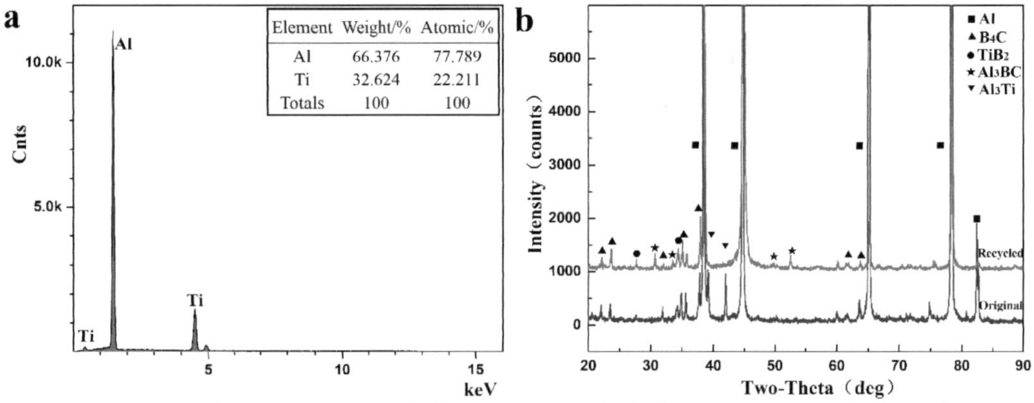

Figure 3. EDS and XRD pattern analysis: (a) EDS analysis for large grey particle in Fig.1(c); (b) XRD patterns for original and recycled materials.

3.2. *Mechanical properties*

Microstructure, such as particle distribution, is tightly associated with the mechanical properties of the metal matrix composites, such as fracture strength and plastic deformation[21, 22]. Cracks are easier to generate at clusters of particles due to the elastic misfit and the plastic constraint of particles. Damage accumulation also tended to occur in clustered regions ahead of a propagating crack. Therefore, we assesses the mechanical properties of the original materials and recycled materials.

Figure 4 presents the tensile properties of Al-B_4C composites before and after recycling. The average yield strength, ultimate tensile strength (UTS) and elongation of the original composite is 60.5MPa, 108.1MPa and 25.2%, respectively. After recycling, the average yield strength, ultimate tensile strength (UTS) and elongation is 62MPa, 109.5MPa and 26.6%, respectively. The Vickers hardness of Al-B_4C composites before and after recycling is 42.1 HV and 42.8 HV, respectively. The increase of tensile properties and hardness after recycling may be attributed to the further uniform distribution of B_4C and TiB_2 which peeled off from the surface of B_4C and serves as strengthening phase.

Figure 5 shows the SEM micrographs of tensile fracture surfaces of composites. In the original composite, there are many dimples on the fracture surface and the dimples are large and uniform. After recycling, there are many newly formed micro dimples in the big dimples. Further observation at a high magnification of SEM reveals that some regular shape particles were found in the micro dimples. Considering their morphology and size, these particles are assumed to be TiB_2 which peeled off from the surface of B_4C and dispersed in the matrix. It seems that these particles act as void initiation sites. The fractured B_4C particles can be seen clearly both in Fig. 5(a) and (b). Fig. 5(d) presents the fractured B_4C particle which is typical in other samples. The cross section is smooth and distributed with some steps, it is a typical brittle fracture. B_4C is surrounded by aluminum matrix and many dimples are around B_4C, it is a typical ductile fracture. It can be seen that the interface layer between the reinforcement and the matrix is stable enough, and the force from the matrix to the B_4C is well delivered.

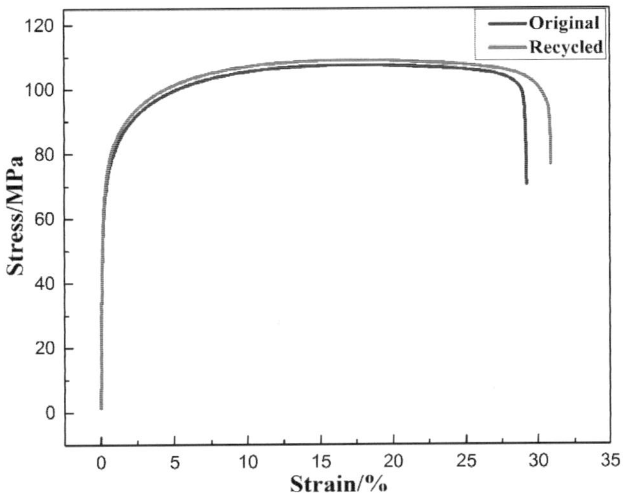

Figure 4. Stress-strain curves of original and recycled materials.

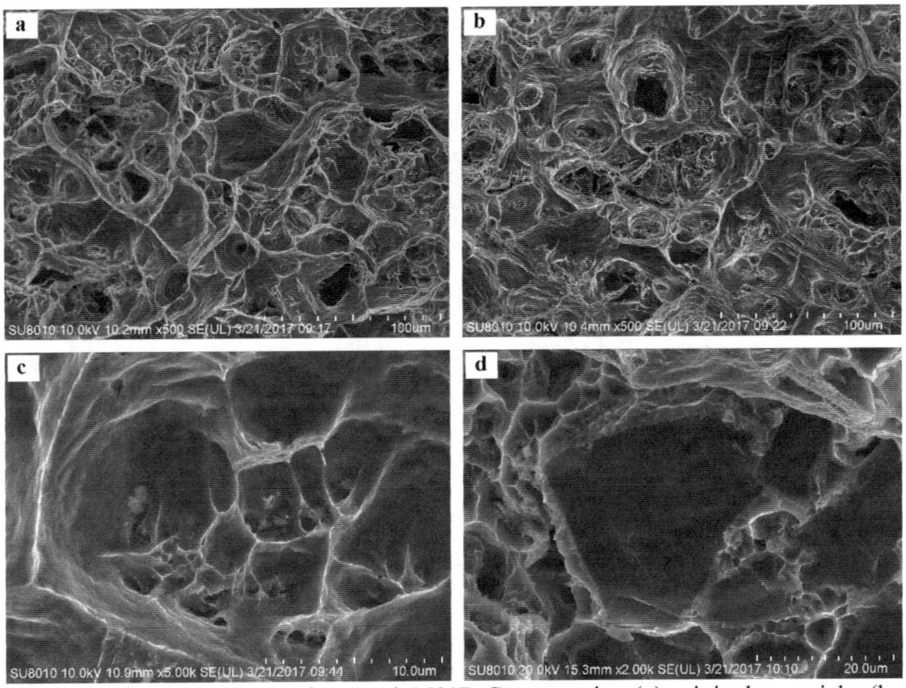

Figure 5. Tensile fracture surfaces of 1060Al-15%B_4C composite: (a) original materials; (b-d) recycled materials.

4. Conclusions

In the current study, a remelting method was successfully employed to recycle scrap materials of 1060Al-15wt.% B_4C composite. And the microstructure and mechanical properties were investigated.

It showed that after recycling, more particle clusters were broken down and B_4C distributed in the matrix more homogeneously. The interface reaction was further intensified, and the protection layer coated on the surface of B_4C was thicker. There were more TiB_2 particles peeled off from the surface of B_4C, which contributed to the good mechanical properties.

EDS and XRD showed that Al_3Ti phase was present in the original composite. But disappeared after recycling due to the consumption of Ti elements at the interface. The peak of Al_3BC phase was obviously stronger compared with original composite.

The average tensile properties and hardness of the recycled composite was comparable to those of the original composite suggesting that the mechanical properties of the recycled composite remained well. This work is of great use for the industrial recycling of Al-B_4C metal matrix composites.

Acknowledgments

This work was founded by Joint Laboratory of Nuclear Materials and Service Safety (2013966003), China and the authors thank Dr. Liang Wang for useful discussion. The authors also thank Mr. Shangyang Feng and Mr. Xiang Zeng at Tsinghua University for experimental assistance.

References

[1] B. V. Ramnath, C. Elanchezhian, M. Jaivignesh, S. Rajesh, *C. Parswajinan*, A. S. A. Ghias, “Evaluation of mechanical properties of aluminium alloy-alumina-boron carbide metal matrix composites,” *Materials & Design,* vol. 58, pp. 332-338, Jun, 2014.

[2] K. Shirvanimoghaddam, H. Khayyam, H. Abdizadeh, M. Karbalaei Akbar, A. H. Pakseresht, E. Ghasali *et al.*, “Boron carbide reinforced aluminium matrix composite: Physical, mechanical characterization and mathematical modelling,” *Materials Science and Engineering A-Structural Materials Properties Microstructure and Processing,* vol. 658, pp. 135-149, Mar 21, 2016.

[3] A. J. Knowles, X. Jiang, M. Galano, F. Audebert, “Microstructure and mechanical properties of 6061 Al alloy based composites with SiC nanoparticles,” *Journal of Alloys and Compounds,* vol. 615, pp. S401-S405, Dec 5, 2014.

[4] F. Thévenot, “Boron carbide—A comprehensive review,” *Journal of the European Ceramic Society,* vol. 6, no. 4, pp. 205-225, 1990.

[5] V. Domnich, S. Reynaud, R. A. Haber, M. Chhowalla, “Boron Carbide: Structure, Properties, and Stability under Stress,” *Journal of the American Ceramic Society,* vol. 94, no. 11, pp. 3605-3628, Nov, 2011.

[6] P. Zhang, Y. L. Li, W. X. Wang, Z. P. Gao, B. D. Wang, “The design, fabrication and properties of B_4C/Al neutron absorbers,” *Journal of Nuclear Materials,* vol. 437, no. 1-3, pp. 350-358, Jun, 2013.

[7] G. Bonnet, V. Rohr, X.-G. Chen, J.-L. Bernier, R. Chiocca, H. Issard, “Use of Alcan's Al-B_4C metal matrix composites as neutron absorber material in TN International's transportation and storage casks,” *Packaging, Transport, Storage & Security of Radioactive Material,* vol. 20, no. 3, pp. 98-102, 2009.

[8] Y. X. Yang, R. Boom, B. Irion, D. J. V. Heerden, P. Kuiper, H. D. Wit, “Recycling of composite materials,” *Chemical Engineering and Processing,* vol. 51, pp. 53-68, Jan, 2012.

[9] V. Kamavaram, D. Mantha, and R. G. Reddy, “Recycling of aluminum metal matrix composite using ionic liquids: Effect of process variables on current efficiency and deposit characteristics,” *Electrochimica Acta,* vol. 50, no. 16-17, pp. 3286-3295, May 30, 2005.

[10] Y. Nishida, “Recycling of metal matrix composites,” *Advanced Engineering Materials,* vol. 3, no. 5, pp. 315-317, May, 2001.

[11] K. R. Ravi, R. M. Pillai, B. C. Pai, M. Chakraborty, “Separation of matrix alloy and

reinforcement from aluminum metal matrix composites scrap by salt flux addition," Bulletin of Materials Science, vol. 30, no. 4, pp. 393-398, Aug, 2007.

[12] J. B. Fogagnolo, E. M. Ruiz-Navas, M. A. Simon, M. A. Martinez, "Recycling of aluminium alloy and aluminium matrix composite chips by pressing and hot extrusion," *Journal of Materials Processing Technology,* vol. 143, pp. 792-795, Dec 20, 2003.

[13] A. Klasik, J. Sobczak, and K. Pietrzak, "Changes in properties of aluminium matrix composite reinforced with SiC particles after multiple remelting," *Materials Research Innovations,* vol. 15, pp. S249-S252, Feb, 2011.

[14] C. J. Shi, Z. Zhang, and X. G. Chen, "Microstructure and fluidity evolution during recycling of Al-B_4C metal matrix composites," *Journal of Composite Materials,* vol. 46, no. 6, pp. 641-652, Mar, 2012.

[15] Y. Li, Q. L. Li, W. Liu, G. G. Shu., "Effect of Ti content and stirring time on microstructure and mechanical behavior of Al-B_4C composites," *Journal of Alloys and Compounds,* vol. 684, pp. 496-503, Nov 5, 2016.

[16] Z. Zhang, K. Fortin, A. Charette, X.-G. Chen, "Effect of titanium on microstructure and fluidity of Al-B_4C composites," *Journal of* Materials *Science,* vol. 46, no. 9, pp. 3176-3185, May, 2011.

[17] D. A. Weirauch, W. J. Krafick, G. Ackart, P. D. Ownby, "The wettability of titanium diboride by molten aluminum drops," *Journal of Materials Science,* vol. 40, no. 9-10, pp. 2301-2306, May, 2005.

[18] D. Kocaefe, A. Sarkar, and X. G. Chen, "Effect of Ti addition on the wettability of Al-B_4C metal matrix composites," *International Journal of Materials Research,* vol. 103, no. 6, pp. 729-736, Jun, 2012.

[19] Q. L. Lin, P. Shen, F. Qiu, D. Zhang, Q. C. Jiang, "Wetting of polycrystalline B_4C by molten Al at 1173-1473 K," *Scripta Materialia,* vol. 60, no. 11, pp. 960-963, Jun, 2009.

[20] J. Y. Zheng, Q. L. Li, W. Liu, G. G. Shu, "Microstructure evolution of 15wt% boron carbide/aluminum composites during liquid-stirring process," *Journal of Composite Materials,* vol. 50, no. 27, pp. 3843-3852, Nov, 2016.

[21] M. K. Akbari, H. R. Baharvandi, and K. Shirvanimoghaddam, "Tensile and fracture behavior of nano/micro TiB_2 particle reinforced casting A356 aluminum alloy composites," *Materials & Design,* vol. 66, pp. 150-161, Feb 5, 2015.

[22] A. Matin, F. F. Saniee, and H. R. Abedi, "Microstructure and mechanical properties of Mg/SiC and AZ80/SiC nano-composites fabricated through stir casting method," *Materials Science and Engineering A-Structural Materials Properties Microstructure and Processing,* vol. 625, pp. 81-88, Feb 11, 2015.

ICMMME 2018 IOP Publishing
IOP Conf. Series: Materials Science and Engineering **409** (2018) 012006 doi:10.1088/1757-899X/409/1/012006

Adsorption Effect of Cetyltrimethyl-ammonium Bromide Modified Fly Ash on Methyl Orange Waste Water

Dongji Liu[1, a], Yuling Yang[1, b], and Fengqing Zhao[2, c]

1 Department of Chemical Engineering, Hebei University of Science & Technology, Shijiazhuang, 050018, China
2 Hebei Engineering Research Center of Solid Waste Utilization, Shijiazhuang, 050000, China
E-mail: [a]504762121@qq.com, [b]2508876287@qq.com, [c]zhaofq3366@126.com

Abstract. Fly ash is a solid waste from power plant, which will pollute the environment if not handled properly. The modification of fly ash as an adsorbent is an ideal method of resource utilization. In this paper, fly ash was firstly treated with cetyltrimethy-lammonium bromide (CTAB), and then further stabilized to obtain the final adsorbent product. The proper treating conditions are obtained: mass ratio of CTAB to fly ash 0.017, modification temperature 32℃, and time 1 h. The removal rate for methyl orange with the modified fly ash stabilized by drying at 100 ℃ is up to 93.2 %, which is higher than that calcined at 500 ℃. The adsorption effect of the modified fly ash is much better than that of unmodified fly ash.

1.Introduction

Fly ash is a common solid industrial waste from coal-fired power plant. If not properly disposed, it will seriously pollute the environment and harm human and animals [1]. Fly ash contains more than 60 percent of SiO_2, Al_2O_3 and CaO, which is similar to natural zeolite. This makes it possible for fly ash as an adsorbent [2]. In 1975, Gagoli [3] et al. first used fly ash to remove heavy metals from industrial waste-water, and investigated the mechanisms. Generally, the adsorption effect cannot meet the actual needs. Several technologies have been developed these years to improve the adsorbing performance of fly ash by acid modification, alkali modification and metal modification.

Li Weiqing [4] and Peng Ronghua [5] used sulfuric acid to modify the fly ash, which is used for waste-water treatment in paper industry. It was found that the combined use of sulfuric acid-modified fly ash and carbonate has remarkable effect. Yang Jing [6] and Duan Xuemei [7] also modified fly ash with sulfuric acid. They found that the adsorption of fly ash is excellent and in accordance with the formula of Langmuir adsorption isotherm. Wang jingyun [8] used calcium hydroxide to modify fly ash, and used for the decoloration and adsorption treatment of active red dye waste-water. After adding calcium hydroxide, the dosage of fly ash is greatly reduced, and the adsorption chromaticity is obviously improved. Liu Jiali [9] et al. adopted $Ca(OH)_2$ to adsorb oily waste water, and got positive results. Tian Yongqi [10] used the mixture of NaOH and $NaHCO_3$ to modify fly ash to treat copper containing waste-water, and obtaining high removal rate. Bai Maojuan [11] used ferrous ions to modify fly ash and used for the treatment of phosphorus waste-water. The removal rate is over 90 %. Zhang Xin [12] et al. used two kinds of metal salt modifiers, $FeSO_4$ and $FeCl_3$, to modify fly ash, and obtained desired results. Xia Changbin [13] used metal cations Al^{3+} and Fe^{3+} to modify fly ash, which is used to treat the waste-water containing p-nitrophenol. The results showed that the adsorption property of the modified fly ash is obviously improved.

The aim of this work is to develop a suitable process for fly ash modification for the purpose of methyl orange waste-water treatment. We adopt two-step method to modify fly ash in this paper. Firstly, fly ash is modified with CTAB, and secondly stabilized by drying at 100 ℃ and calcining at 500 ℃ respectively, to investigate the adsorption effect for methyl orange.

2. Experimental

2.1. Materials

CTAB, sodium hydroxide, hydrochloric acid, hydrogen peroxide, copper chloride and methyl orange are bought on the market. Fly ash is provided by Shijiazhuang Jingxing Construction Material Corp. Ltd. Table 1 shows the chemical composition of fly ash.

Table 1. Chemical composition of fly ash (%)

SiO_2	MgO	CaO	Al_2O_3	Fe_2O_3	K_2O	SO_3	Others
49.82	0.66	5.96	33.90	4.40	1.24	2.05	1.97

2.2. Sample preparation

Fly ash is washed and dried. CTAB solution (2.5g/L) is prepared beforehand as a modifier. The optimal conditions for the modified fly ash are determined by changing the amount of CTAB, reaction time and reaction temperature. The modified fly ash samples are further treated by drying (at 100 ℃) and calcining (at 500 ℃) respectively, to obtain stabilized finish.

2.3. Adsorption test

Take a certain amount of distilled water (Keep the total volume of the solution at 25 mL), add 5 mL (100 mg/L) methyl orange solution. Then, add stabilized fly ash sample. After the reaction finished, centrifuges, the clarification liquid is used for the measurement of absorbance.

The removal rate of methyl orange is calculated as following.

$$R = (1 - \frac{A}{A_0}) \times 100\% \quad (1)$$

Where A_0 and A are the absorbance before and after the reaction. R is the removal rate of modified fly ash sample for methyl orange.

3. Results and discussion

3.1. Effect of mass ratio of CTAB to fly ash and stabilization process

Table 2 shows the effect of dosage of CTAB on the adsorption performance of modified fly ash stabilized by drying at 100 ℃ and calcining at 500 ℃. The modification temperature and time are fixed at 30 ℃and 1h.

Table 2. Effect of mass ratio of CTAB to fly ash(in different stabilization method)

Mass ratio	Temperature/℃	R/%
0.0083	100	66.2
	550	10.9
0.0125	100	76.8
	550	8.6
0.017	100	85.4
	550	11.4
0.025	100	89.2
	550	11.9
0.033	100	90.0
	550	12.3

IOP Conf. Series: Materials Science and Engineering **409** (2018) 012006 doi:10.1088/1757-899X/409/1/012006

The results show that with the increase of the mass ratio of CTAB to fly ash, the removal rate for methyl orange gradually increases. The adsorption effect of CTAB modified fly ash is much better than that of unmodified fly ash. The removal rate for methyl orange with the modified fly ash stabilized by drying at 100 ℃ is higher than that calcined at 550 ℃. Considering the cost-effectiveness, the optimal mass ratio of CTAB to fly ash is 0.017, and the removal rate for methyl orange reaches 85.4 %.

3.2. Effect of modification temperature

Table 3 shows the effect of modification temperature on the adsorption performance of modified fly ash stabilized by drying at 100 ℃ and calcining at 500 ℃. In the experiments, the mass ratio of CTAB to fly ash is 0.017, the modification time is 1h.

Table 3. Effect of modification temperature on the removal rate for methyl orange

Reaction temperature/℃	Temperature/℃	R/%
22	100	88.0
	550	11.8
32	100	92.1
	550	12.6
42	100	91.0
	550	10.2
52	100	83.5
	550	9.6

The results show that the removal rate for methyl orange with the modified fly ash dried at 100 ℃, is higher than that calcined at 550 ℃. At higher temperature, a large number of voids in the samples were sintered. The optimal modification temperature is 32 ℃, at which, the removal rate for methyl orange is up to 92.1%.

3.3. Effect of modification time

Figure 1 shows the effect of modification time on the adsorption performance of modified fly ash stabilized by drying at 100 ℃. In the experiment, the modification temperature is 32 ℃, the mass ratio of CTAB to fly ash is 0.017.

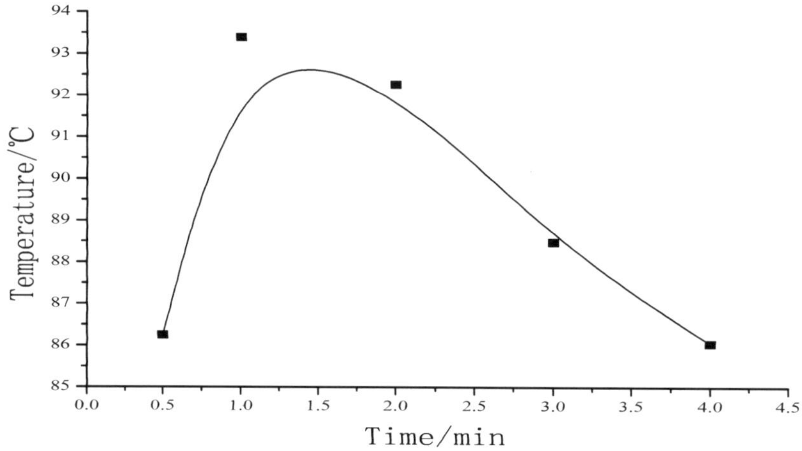

Figure 1. Effect of modification time on the adsorption for methyl orange

It is shown that the removal rate of methyl orange increases first and then decreases with the increase in modification time. The optimal modification time is 1 h and the maximum removal rate reaches 93.2%. The good adsorption effect of modified fly ash is mainly due to the application of modified particles on the surface of fly ash, which increases the specific surface area of fly ash and enhances the adsorption capacity of methyl orange. Besides, the modified fly ash changes from hydrophilic to oleophilic, and the hydration of CTAB is less than that of inorganic cation, which reduces the resistance of modified fly ash to methyl orange, thus improving the adsorption capacity for methyl orange.

4. Conclusions

- Two-step method to modify fly ash is developed, in which fly ash is firstly modified with CTAB, and secondly stabilized by drying at 100 ℃ and calcining at 500 ℃ respectively. The adsorption effect of CTAB modified fly ash is much better than that of unmodified fly ash.
- The optimized modification conditions are obtained: mass ratio of CTAB to fly ash 0.017, modification temperature 32 ℃, and time 1 h. Compared to calcining at 500 ℃, the modified fly ash stabilized by drying at 100 ℃ shows higher adsorption rate for methyl orange.
- The good adsorption effect of modified fly ash is mainly due to the application of modified particles on the surface of fly ash, which increases the specific surface area and reduces the resistance of modified fly ash to methyl orange, improving the adsorption capacity for methyl orange.

5. References

[1] F Q Zhao, W Ni and H J Wang 2008 *J. Resources Conservation & Recycling.* **52** 303-313.
[2] W. H. Chen 2008. *J. High and new technology enterprises of China.* **23** 65-65.
[3] N. Gangoli, D. C. Markey and C Thodos 1975 *J. Proceeding of the National Conference on Complete Waste reuse.* 270-275.
[4] W. Q. Li, G. Ma and L. J. Lu 2006 *J. Modern scientific instruments.* **16** 76-79.
[5] R. H. Peng, L. J. Chen and X. X. Li 2004 *J. Material protection.* **38** 48-50.
[6] J. Yang and X. Y. Liu 2002 *J. Environmental protection of coal mine.* **16** 18-22.
[7] X. M. Duan 2001 *D. Henan Agricultural University.*
[8] J. Y. Wang 2011 *J. Comprehensive utilization of fly ash.* **1** 33-35.
[9] J. L. Liu, Q. P. Wang, F. F. Min and Y. Q. Tian 2017 *J. New chemical materials of chemical industry.* **1** 243-245.
[10] Y. Q. Tian 2006 *D. Anhui University of Science And Technology.*
[11] M. J. Bai and Y. Y. Chu 2008 *J. Journal of Qingdao University of Science & Technology.* **3** 213-216.
[12] X. Zhang, Q. Y. Yue and J. Z. Zhang 2007 *J. Comprehensive utilization of fly ash.* **4** 44-46.
[13] C. B. Xia 2000 *J. Environmental Science and technology.* **3** 35-38.

IOP Conf. Series: Materials Science and Engineering **409** (2018) 012007 doi:10.1088/1757-899X/409/1/012007

Growth and Structural Properties of Graphene Oxide Thin Film with Spray Pyrolysis Technique

Anoud AlShammari[1,2*], Mohd Mahadi Halim[1], Yam Fong Kwong[1], Noor Haida Mohd Kaus[3], Saleh I. Alhiassah[1], Naser Mahmoud Ahmed[1], and Lim Hwee San[1]

[1]School of Physics, USM, 11800 Universiti Sains Malaysia, Pulau Pinang, Malaysia
[2]School of Physics, Northern Border University, Rafha, Saudi Arabia
[3]School of Chemical Sciences, USM, 11800 Universiti Sains Malaysia, Pulau Pinang, Malaysia

E-mail: [1,2]*Ano221@hotmail.com

Abstract. The spray pyrolysis technique (SPT) is one of the most useful and simple methods to grow graphene oxide (GO) thin film. Good-quality GO thin films were fabricated by spray coating with GO powder (prepared by the modified Hummers method) on glass substrates. Raman spectroscopy, Field Emission Scanning Electron Microscope (FESEM), X-ray diffraction (XRD) analysis, and UV–visible spectroscopy were done to observe the presence of functional groups in the growth thin films and to determine the structure of GO. From the result of XRD test, the GO demonstrates peak at 10.58°(2θ), while the Raman spectroscopy reveals the presence of two prominent peaks, known as D and G bands, as well as 2D band. The I_D/I_G ratio for spray-deposited GO film was calculated with the value of 0.84. FESEM images showed that the glass substrates were completely covered by GO at different magnifications with a sheet-like structure. The optical absorption of graphene oxide was also observed at a wavelength rang of 275-350 nm.

1. Introduction

Graphene-based materials exhibit remarkable electronic, optical, and mechanical properties, which has resulted in both high scientific interest and huge potential for a variety of applications. Graphene is recognized as an allotrope of carbon in the shape of 2D, atomic-scale, and honey-comb pattern according to which every vertex is formed by a single atom [1]. Graphene oxide (GO) is an intermediate in the chemical synthesis of graphene before the oxygen functional groups are removed thermally or chemically to obtain graphene. Hence, both graphene oxide (GO) and reduced graphene oxide (RGO) properties are interesting from a research perspective and have potential for the development of new applications [2]. The transition between the insulating GO and conducting graphene state is accompanied by a change in the optical properties of the material. The covalent oxygen functional groups in GO give rise to remarkable mechanical strength along with molecular-level chemical sensing capability. The presence of functional groups also implies modification of the graphene electronic structure. Therefore, the chemical composition of GO, which can be chemically, thermally, or electrochemically engineered, allows tunability of its optoelectronic properties [3, 4]. Recently, GO has been potentially used in many modern applications such as sensors and solar cell [5,6]. Spray pyrolysis technique (SPT) is well known techniques to grow nanoparticles [7]. With this technique, thin film material could be grown on glass substrate homogeneously. The advantage of this method is being simple, low operating cost as well as effectiveness [8]. In this study, SPT was used to grow graphene oxide (GO) thin films on glass substrates. The prepared thin films were characterized using different analytical techniques to study the properties of GO growth by SPT.

2. Experiment

2.1. Synthesis of GO

GO was synthesized according to the modified Hummers method. The steps of the experiment are as follows:

(1) A mixture of 120 ml sulphuric acid (H_2SO_4,96%) and 13.3 ml phosphoric acid (H_3PO_4,75%) was put in an ice bath.

(2) Next, 1 g of graphite was added with proper stirring, followed by the addition of 6 g of potassium permanganate ($KMnO_4$) in a very slow manner, keeping the temperature less than 20 °C. Then the solution was heated up to 50 °C and stirred for 8 h.

(3) The solution was cooled to room temperature and poured onto ice water (133.3 mL) with 1 mL of 30–35% hydrogen peroxide (H_2O_2).

(4) A filtration process was performed and followed by the centrifuge process (4,000 rpm for 1 h). Subsequently, the supernatant was decanted away.

(5) The remaining solid materials were washed twice in succession with 66.6 mL of water, 66.6 mL of 30% HCL, and 66.6 mL of ethanol. The obtained solid was dried under vacuum overnight at room temperature [7,8].

2.2. Cleaning of substrates

For the cleaning process, the glass substrates were put in a methanol bath, and the ultrasonication process was run for 15 to 20 min. The substrates were rinsed well with deionized water and finally dried using nitrogen gas [9].

2.3. Film preparation

For the spray deposition method, the instrument comprised an airbrush nozzle (0.3 mm aperture size), timer, beaker solution, thermocouple probe, hot plate heater, and air pump compressor (about 4 bars of pressure). Figure 1 shows a typical spraying system. The thin films were prepared by spray coating GO dispersion in deionized water (0.9 M) onto a glass substrate at 170 ^{0}C [10]. Figure 2 shows the prepared GO thin film sample on glass. The diameter of the nozzle was 0.3 mm, and the distance between the nozzle and substrate was 40 cm.

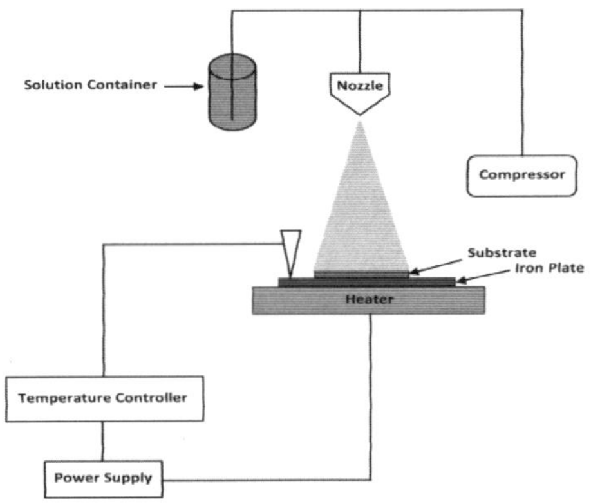

Figure 1. Schematic of spray pyrolysis system

Figure 2. The GO thin film on glass substrate prepared by SPT

IOP Conf. Series: Materials Science and Engineering **409** (2018) 012007 doi:10.1088/1757-899X/409/1/012007

2.4. Film stucture
X-ray diffraction (XRD) patterns were recorded using Philips X-ray diffractometer model X'Pert with CuKα (1.5406Å) radiation operated at 40 kV and 25mA. The patterns were recorded automatically with scanning speed of 2 deg/min.

Field emission Scanning Electron Microscope (FESEM) was recorded using SEM Model Quanta 250 FEG (Field Emission Gun) attached with EDX Unit (Energy Dispersive X-ray Analyses), with accelerating voltage up to 30 KV, FEI Company, Netherlands. The average film thickness of GO thin film (113 nm) was determined accurately after deposition by using FSEM as shown in Figure 3.

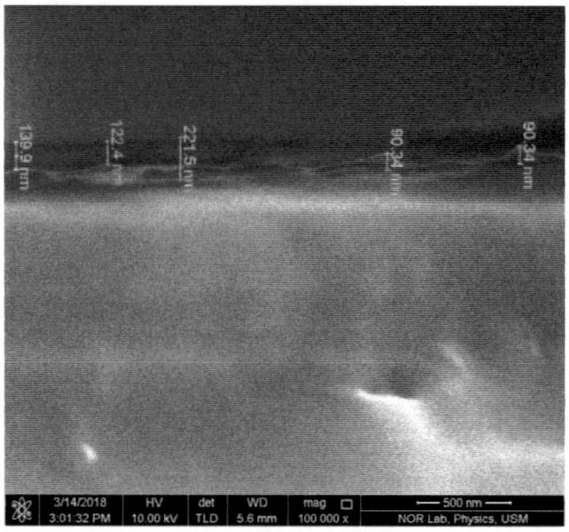

Figuer 3. A FESEM image shows thickness of GO thin film

Raman Spectroscopy analysis was done using Nicolet 6700 Spectrophotometer (Thermo Scientific) at room temperature in the spectral range of 1000-3500 cm^{-1}.

UV-visible absorbance (λ), was measured at normal incidence in the wavelength 200-800 nm by means of a double beam spectrophotometer (JASCO model V-670 UV-Vis-NIR).

3. Results And Discussion
After preparing the thin films of GO by using SPT, several thin films samples have been made. The best results were obtained for the following growth parameter; the temperature was 170 °C, diameter of the nozzle was 0.3 mm with distance of 40 cm from the surface of substrate. The samples then were analyzed further to examine the quality of the prepared films.

3.1. Raman Spectroscopy
Raman spectroscopy analysis determined the structural and chemical bonding of molecules. As shown in Figure 4, there are two prominent peaks, D and G bands [11]. The D band is located at 1,343 cm^{-1}, which is attributed to the graphitic domain (carbon, sp2). The G band is at 1,583 cm^{-1}, which is associated with disordered edges (sp^3). It is noted that the additional graphene peak on the 2D band is located at around 2,718cm^{-1}. The 2D band is highly sensitive to the number of graphene layers when a few layers of graphene are compared to GO. In this sample, the 2D band's intensity is attenuated, and its bandwidth is broader. The intensity ratio of the D band over the G band (ID/IG) was used to investigate the amount of structural defects and disorders [12,13]. For this sample, the ID/IG ratio has a value of 0.84, a result that confirms the fabrication of GO.

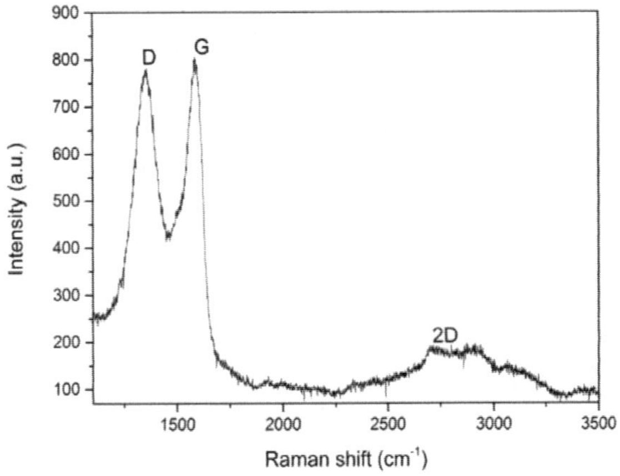

Figure4. Raman spectrum of graphene oxide thin film

3.2. X-ray Diffraction (XRD)

XRD technique is used to determine the presence of functional groups on the sample surface. After the chemical modification of graphite by the improved method to obtain the GO, it was shown that the peak (002) is at around 10.58° (2θ). As shown in Figure (5). Further analysis reveals that $D_{(002)}$=0.4 nm. It is clear that the measured values are due to the presence of graphene oxide on the surface of the substrate [14,15], and the intensity values indicate that few layers of graphene oxide are formed on the substrate. This result corresponds to the result obtained from Raman spectrum measurements.

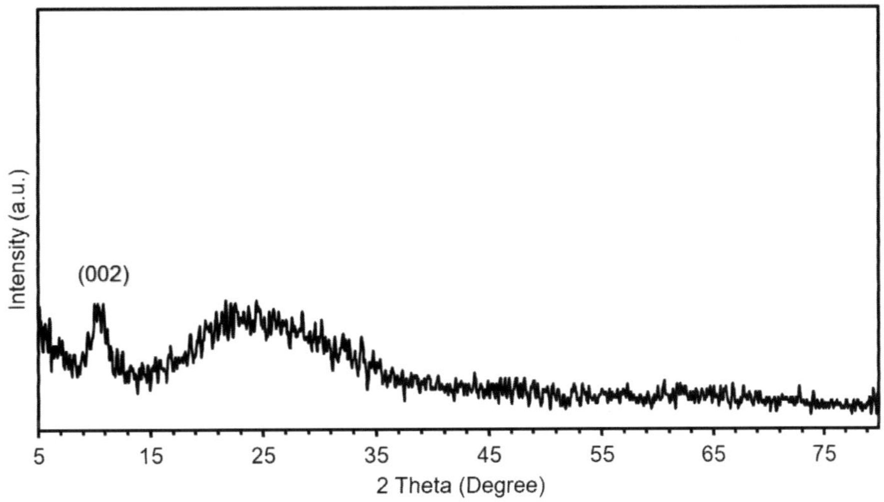

Figure 5. XRD patterns of graphene oxide

3.3. Scanning Electron Microscopy (SEM)

The SEM images of the prepared GO thin film with different magnifications are shown in Figure 5 (a) and (b). It is clear that we have a uniform GO coating with sheet-like structure. This sheet-like structure indicates that graphite has been exfoliated during the oxidation process. This homogeneous coating confirms the effectiveness of this technique to be used for large-scale applications.

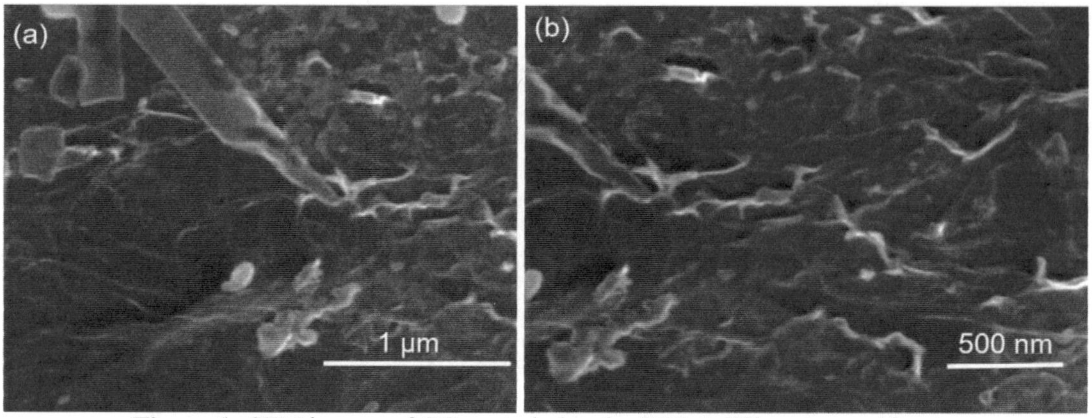

Figure 6. SEM images of GO-coated glass by SPT at different magnifications.

3.4. D. UV–*visible* spectroscopy

The optical absorbance spectrum of GO thin film with a thickness of 113 nm in UV-V is the spectral range shown in Figure 7. Graphene oxide exhibits a strong absorption edge in a wavelength range of 275-350 nm. Illakkiya et al. [16] reported a band at 230 nm, which is attributed to π-π* the transition of the aromatic C=C bond. A shoulder peak at 270 nm corresponds to n-π* the transition of the carbonyl group.

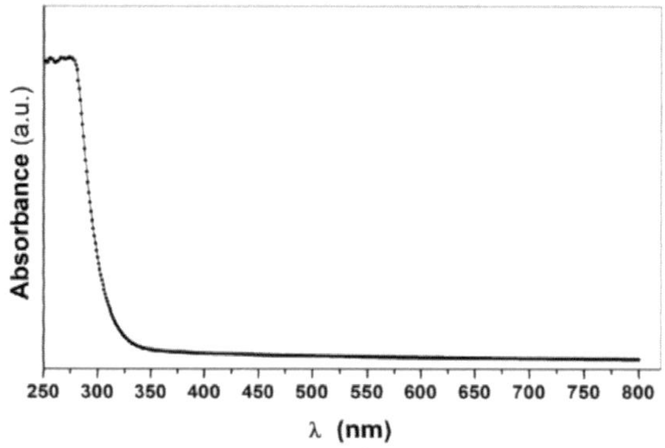

Figure 7. Optical absorbance spectra of GO thin film

4. Conclusion

The spray pyrolysis technique (SPT) is well known for growing nanomaterials. The prepared GO thin films can be used for varying applications such as gas sensors, solar cells, polymer-based composites, and photodetectors .The highlights of this method are that it is simple and effective, with a low operating cost .However, a thin film of GO was prepared on glass substrates by SPT in a controlled factor, including temperature at 170 °C, diameter of the nozzle at 0.3 mm, and substrate distance of 40 cm. The results of the Raman spectroscopy and XRD analysis reveal the structural properties and quality. The absorption spectrum of graphene exhibits a peak around 275-350 nm. This analysis suggests that the SPT can give good results and promising. These results may later be used in the wide range of advanced electronic, optic and material for the future.

References

[1] A.M. Affoune, B.L. V. Prasad, H. Sato, T. Enoki, Y. Kaburagi, Y. Hishiyama, Experimental evidence of a single nano-graphene, Chem. Phys. Lett. 348 (2001) 17–20.
[2] T. Tani, L. Mädler, S.E. Pratsinis, Homogeneous ZnO nanoparticles by flame spray pyrolysis, J.

Nanoparticle Res. 4 (2002) 337–343.
[3] S. Pei, H.M. Cheng, The reduction of graphene oxide, Carbon N. Y. 50 (2012) 3210–3228.
[4] D.A.C. Brownson, D.K. Kampouris, C.E. Banks, An overview of graphene in energy production and storage applications, J. Power Sources. 196 (2011) 4873–4885.
[5] J. Chen, C. Li, G. Eda, Y. Zhang, W. Lei, M. Chhowalla, W.I. Milne, W.-Q. Deng, Incorporation of graphene in quantum dot sensitized solar cells based on ZnO nanorods, Chem. Commun. 47 (2011) 6084–6086.
[6] L.D. Jadhav, A.P. Jamale, S.R. Bharadwaj, S. Varma, C.H. Bhosale, Synthesis and characterization of YSZ by spray pyrolysis technique, Appl. Surf. Sci. 258 (2012) 9501–9504.
[7] D.C. Marcano, D.V. Kosynkin, J.M. Berlin, A. Sinitskii, Z. Sun, A. Slesarev, L.B. Alemany, W. Lu, J.M. Tour, Improved synthesis of graphene oxide, ACS Nano. 4 (2010) 4806–4814.
[8] W. Cai, R.D. Piner, F.J. Stadermann, S. Park, M.A. Shaibat, Y. Ishii, D. Yang, A. Velamakanni, S.J. An, M. Stoller, J. An, D. Chen, R.S. Ruoff, Synthesis and solid-state NMR structural characterization of 13C-labeled graphite oxide, Science (80-.). 321 (2008) 1815–1817.
[9] G. Shugar, J.T. Ballinger, L.M. Dawkins, Chemical Technicians' Ready Reference Handbook, Illustrated, McGraw Hill Professional, New York, 1996.
[10] S.T. Jadhav, S.J. Rajoba, S.A. Patil, S.H. Han, L.D. Jadhav, Temperature-dependent photoluminescence of graphene oxide, J. Electron. Mater. 45 (2015) 11664.
[11] S.S. Maktedar, S.S. Mehetre, M. Singh, R.K. Kale, Ultrasound irradiation: A robust approach for direct functionalization of graphene oxide with thermal and antimicrobial aspects, Ultrason. Sonochem. 21 (2014) 1407–1416.
[12] L.M. Malard, M.A. Pimenta, G. Dresselhaus, M.S. Dresselhaus, Raman spectroscopy in graphene, Phys. Rep. 473 (2009) 51–87.
[13] P. Kaur, M.S. Shin, N. Sharma, N. Kaur, A. Joshi, S.R. Chae, J.S. Park, M.S. Kang, S.S. Sekhon, Non-covalent functionalization of graphene with poly (diallyl dimethylammonium) chloride: Effect of a non-ionic surfactant, Int. J. Hydrogen Energy. 40 (2015) 1541–1547.
[14] L. Li, R. Qin, H. Li, L. Yu, Q. Liu, G. Luo, Z. Gao, J. Lu, Functionalized graphene for high-performance two-dimensional spintronics devices, ACS Nano. 5 (2011) 2601–2610.
[15] L. Ji, Y. Wu, L. Ma, X. Yang, Noncovalent functionalization of graphene with pyrene-terminated liquid crystalline polymer, Compos. Part A Appl. Sci. Manuf. 72 (2015) 32–39.
[16] J. Tamil Illakkiya, P. Usha Rajalakshmi, Rachel Oommen, Nebulized spray pyrolysis: a new method for synthesis of graphene film and their characteristics, Surface & Coatings Technology 307 (2016) 65–72

Sunlight Photodeposition of Gold nanoparticles onto Graphitic Carbon Nitride (g-C_3N_4) and Application Towards the Degradation of Bisphenol A

Chen Hong Hak[1], Lan Ching Sim[1], Kah Hon Leong[1], Yik Heng Chin[1] and Pichiah Saravanan[2]

[1]Department of Environmental Engineering, Faculty of Engineering and Green Technology, Universiti Tunku Abdul Rahman, Kampar, Perak, Malaysia
[2]Department of Environmental Science and Engineering, Indian Institute of Technology(ISM), Dhanbad, Dhanbad-826004, Jharkhand, India

E-mails: simcl@utar.edu.my

Abstract. Gold (Au) nanoparticles (NPs) were successfully deposited onto the surface of graphitic carbon nitride (g-C_3N_4) by using natural sunlight. Bisphenol A (BPA) were adopted as a pollutant with different pH condition to indicate the photocatalytic performance under natural sunlight. The presence of Au NPs (8.8 nm) was confirmed by TEM and XPS characterization. The duration for degraded BPA in pH 3 was 30 minutes faster than the normal degradation rate which is pH 7. This is due to the point of zero charge (PZC) properties of g-C_3N_4, contributing to better adsorption capacity efficiency and promoted the contact between photocatalysts and pollutants. The pH condition of BPA will affect the degradation efficiency, hence lower down pH condition of BPA can increase the photocatalytic performance.

1. Introduction

With the betterment of technology, rapid development of industry and rise in population, the environment pollution caused by organic pollutant is considering as a significant problem for the global. Endocrine disrupting chemicals (EDCs) are kind of organic pollutants that have the ability to interfere with natural hormonal mechanisms and alter the function of the endocrine system and thus lead to adverse health effects in an organism [1]. BPA is a well-known EDCs and it is widely utilized as an antioxidant or a stabilizer for various types of plastics [2]. Improper disposal of not readily biodegradable EDCs are highly toxic to environment and will produce some potential carcinogens. They are extremely resistant to degradation due to its high degree of chemical and biological stability properties. Consequently, direct discharge of EDCs into water sources will be harmful to the aquatic life, humans, animals and whole ecosystems. However, the traditional treatment methods are non-destructive and only turn the pollutant to another phase, producing secondary pollutant and requiring further treatment [3].

Since the first discovery of photocatalytic water splitting with titanium dioxide (TiO_2) electrodes by Honda and Fujishima (1972), extensive researches have undergone to comprehend the principles of photocatalysis and enhance the photocatalytic performance. The photocatalytic process harvests the light illumination as the energy source. It is known that the sunlight is abundant, sustainable and green resource. However, TiO_2 is active only under Ultraviolet (UV) light and can only absorb about 5% of sunlight in the UV region [4]. Graphitic carbon nitride (g-C_3N_4) has recently attracted tremendous attention due to its high thermal and chemical stability, visible-light driven band gap (2.7 eV) and facile synthesis method [5]. However, the main shortcoming of g-C_3N_4 is the poor electron-hole pairs separation, contribute to the slow photocatalytic reaction rate [6]. The electron-hole pairs recombination rate of g-C_3N_4 need to be suppress in other to make g-C_3N_4 become a promising and adaptable photocatalyst for sunlight harvesting. Recently, doping with a noble metal is recognized as one of the most promising alternatives to overcome the recombination issue. Xie and co-workers [7] reported that doped Au NPs onto g-C_3N_4 will widen visible light response range and increased the hydrogen evolution rate. According to Cheng and co-workers [8], the Au/g-C_3N_4 composite demonstrated superior photocatalytic activities towards the degradation of methyl orange under visible light irradiation. Furthermore, a pH experiment was used to find the optimum condition for photocatalytic degradation of BPA due to the pH of the solution could affect the surface charges on the photocatalyst [9].

2. Experiments And Characterizations

2.1. Synthesis of g-C_3N_4 and Au/g-C_3N_4

30g of urea (R&M Chemicals) was dried in an oven at 80°C for 24h and then pyrolysed in a Muffle Furnace at 550°C for 3h. The obtained yellow-colored product was washed with nitric acid (0.1 M) and deionized water for several times to remove any residual alkaline species. For Au/g-C_3N_4, the prepared g-C_3N_4 (0.5 g) was added to 20 mL of ethylene glycol containing a chosen amount of gold (III) chloride trihydrate ($HAuCl_4$, Sigma–Aldrich). Then the resulting suspension was exposed under natural sunlight for 60 min with continuous stirring to promote photodeposition. The schematic diagram and photograph of experimental setup is shown in Fig. 1(a). Then the obtained precipitate was washed repeatedly with ethanol and deionized water and followed by dried in an oven for overnight at 80 °C.

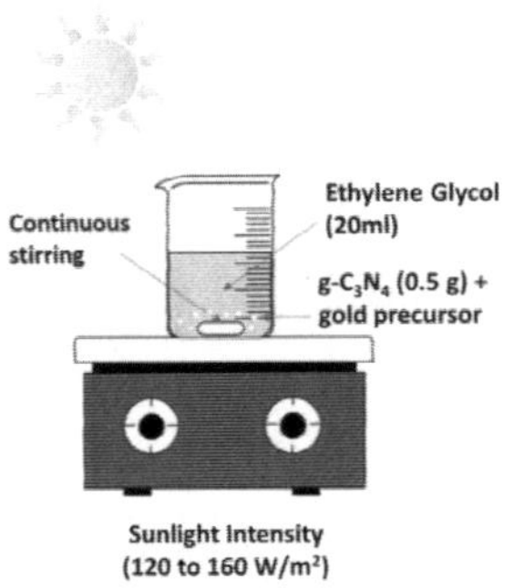

Figure 1a. Schematic diagram of photodeposition experiment setup

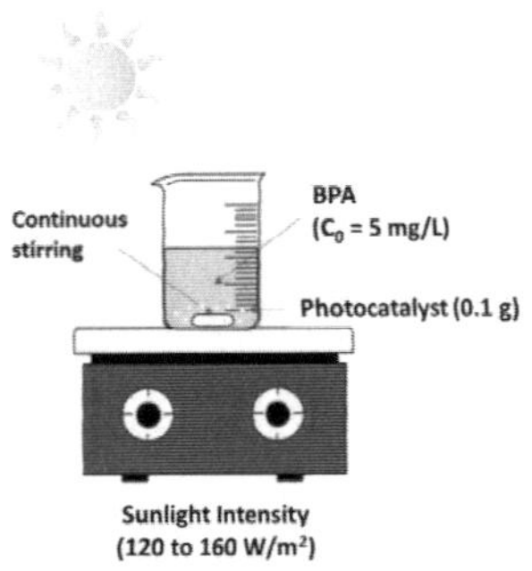

Figure 1b. photocatalysis experiment setup

2.2. Characterization

The surface chemical composition of samples was analyzed by X-ray photoelectron spectroscopy (XPS, PHI Quantera II, Ulvac-PHI, Inc.) with Al Kα radiation source. The morphology of the samples was observed using a field emission scanning electron microscope (FESEM, JSM-6701F, JEOL). High-resolution transmission electron microscope (HRTEM, FEI-TECNAI F20) images were obtained at 200 kV. Ultraviolet-visible diffuse reflectance spectra (UV-vis DRS) were performed using UV-vis NIR spectrophotometer (UV-260, SHIMADZU) with a wavelength range of 200– 800 nm.

2.3. Photocatalytic degradation of organic pollutants

0.1 g Au/g-C3N4 was adopted as photocatalyst in the degradation of BPA (5 ppm) under different pH condition as shown in Fig. 1(b). pH experiment was used to find the optimum condition for photocatalytic degradation of BPA. 0.1 M of sodium hydroxide (NaOH) was used for alkali condition and 0.1M of nitric acid (HNO_3) was used for acidic adjustment. The BPA solution was set at pH 3, 5, 7, 9 and 12 to identify the degradation efficiency at certain pH condition. The pH of BPA solution was adjusted and verified using HI2210-01 Benchtop pH Meter with 0.01 pH resolution. The experiment was carried out for 2 h under natural sunlight and sample was collected for high-performance liquid chromatography (HPLC, PerkinElmer Flexar) analyzer. The HPLC was equipped with fluorescence detector, a separation column used was Ascentis C18 column (Supelco Analytical, USA) (15 cm × 4.6 mm × 5 μm). The mobile phase was 40 % acetonitrile (ACN), 60 % water and the flow rate was 1.0 mL/min.

3. Results And Discussions

SEM and TEM were used to characterize the morphologies of synthesized samples and the obtained images are shown in Fig. 2. The obtained FESEM images for Au/g-C_3N_4 showed a characteristic lamellar structure, which is assembled of massed irregular nanosheets of g-C_3N_4 [10]. The dark contrast in TEM images indicated the presence of Au NPs (0.235 nm lattice fringes) and uniform distributes over the g-C_3N_4 nanosheets with average particles size 8.8nm.

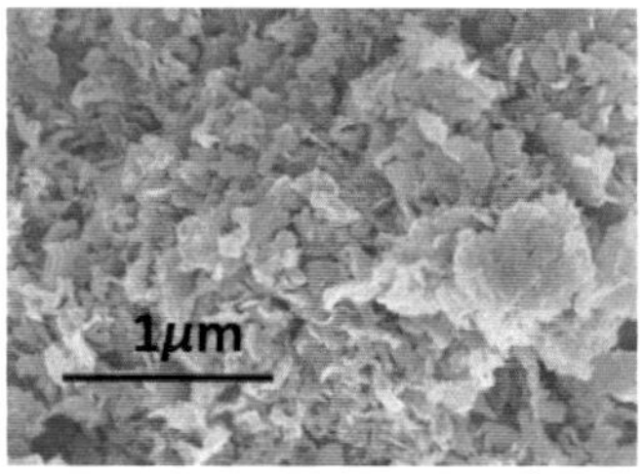

Figure 2a. SEM images of Au/g-C_3N_4

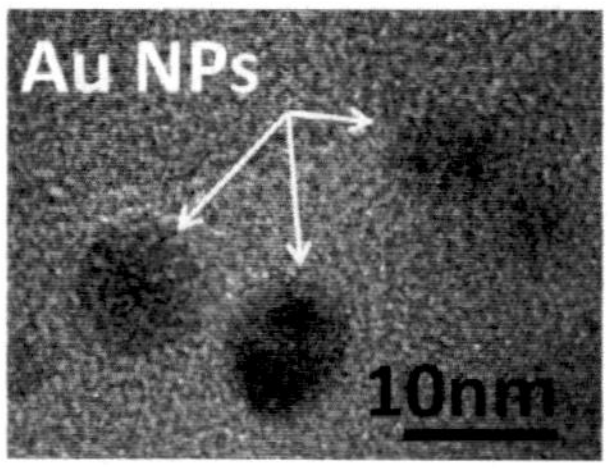

Figure 2b. TEM images of Au/g-C_3N_4

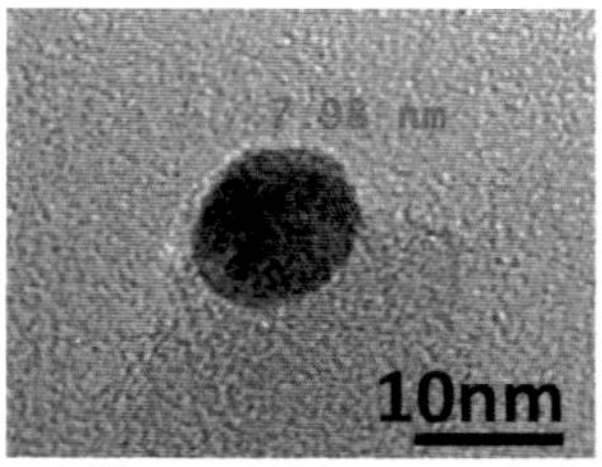

Figure 2c. TEM images of Au/g-C3N4

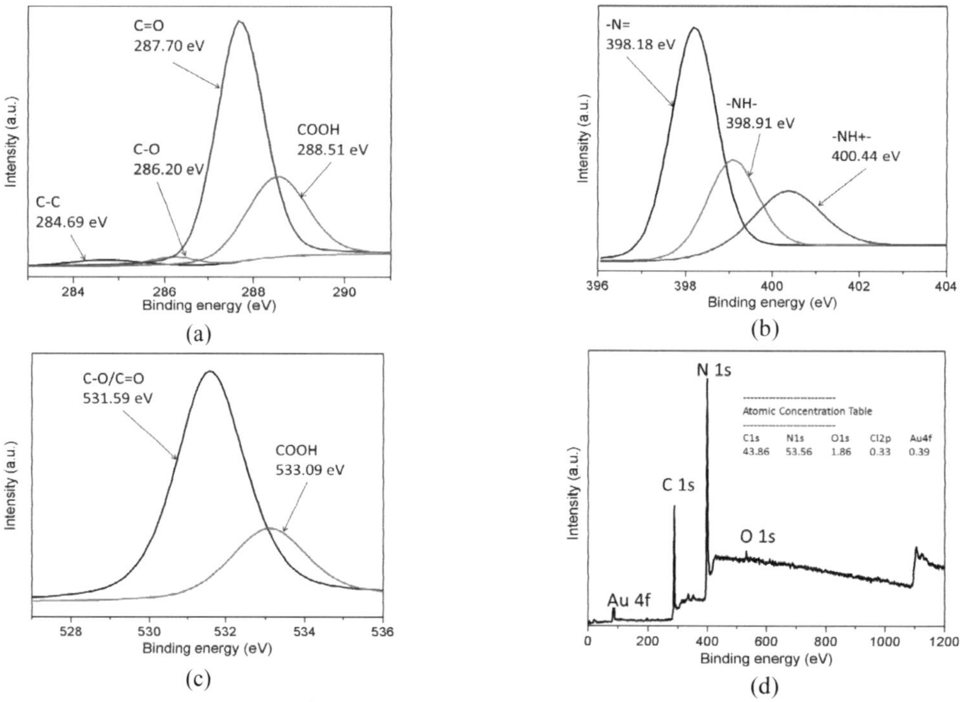

Figure 3. XPS Spectra of Au/g-C3N4 (a) C1s, (b) N1s, (c) O1s and (d) wide scan.

XPS was used to examine the chemical composition and chemical state of Au/g-C_3N_4. The XPS spectra in Fig. 3 reveal the existence of C, N, O and Au in the composite with the weight percentage of 37.74, 53.74, 2.13 and 5.56 wt%, respectively. The C1s peak was deconvoluted into four peaks which is C-C (284.69 eV), C-O (286.20 eV), C=O (287.70 eV) and COOH (288.51 eV). In the N1s spectrum, the main peak observed at 398.18 eV is ascribed to CN=C coordination which originates from the sp^2-bonded N in triazine rings, while the other two weak peaks observed at binding energies 398.91 and 400.44 eV can be attributed to N-$(C)^3$ and C-N-H species [11]. In Fig. 3(c), the peak observed at 531.59 eV was assigned to some hydroxyl groups (OH) attached on the surface of g-C_3N_4 [12]. A peaks located at around 85 eV observed in wide scan (Fig. 3(d)) was assigned to the Au 4f, indicated the presence of metallic Au NPs.

The visible light absorption capability of the Au/g-C_3N_4 was determined by UV–vis DRS and shown in Fig. 4(a). The absorption range from 200 to 450 nm UV was ascribed to the optical band gap of g-C_3N_4. Nevertheless, a hump was observed at absorption band ~550 nm was attributed to Au NPs on surface of g-C_3N_4, this is acknowledged to LSPR effect delivered by Au NPs, thus broaden the absorption band towards the visible region. The existence of Au NPs had generated a disturbance on dielectric constant of the surrounding matrix and contributed for visible light absorption [13]. Furthermore, the band gaps of Au/g-C_3N_4 were estimated to be 2.95 eV by using the Tauc plot as shown in Fig. 4(b).

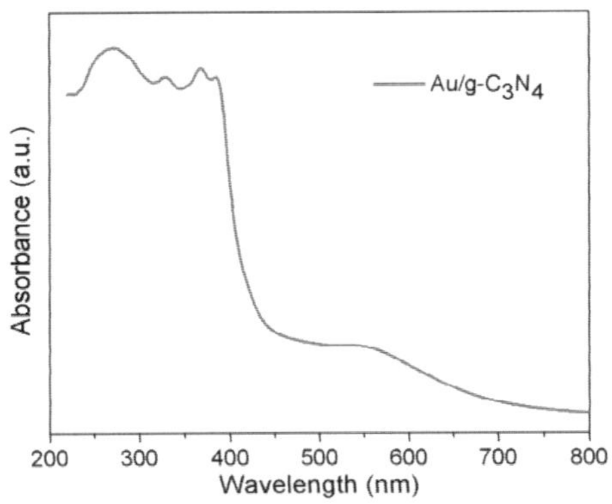

Figure 4a. UV-Visible Absorption Spectra of Au/g-C3N4.

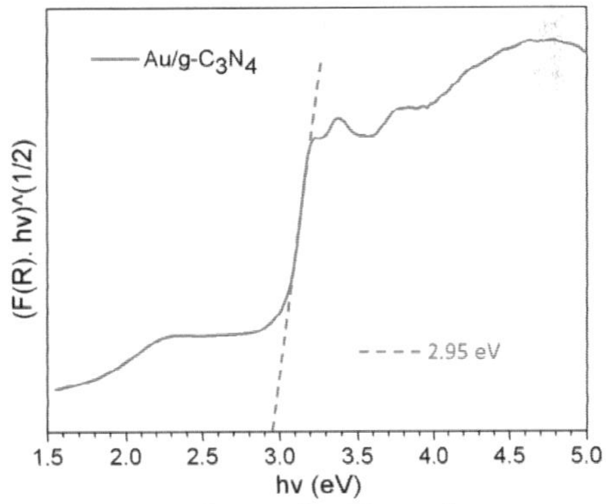

Figure 4b. Tauc plot

3.1. Photocatalytic Experiment

As the pH of the solution could affect the surface charges on the photocatalyst, the degradation of BPA was investigated under different condition. From Fig. 5, the highest degradation efficiency was pH 3 followed by pH 5, 7, 9 and 12. The BPA compound was negatively charged as it possesses two negative oxygen atoms at the hydroxyl groups and four negative carbon atoms ortho to the phenolic groups. The formation of hydroxyl radicals was due to the adsorption mechanism of BPA onto the surface of photocatalyst [14].

Furthermore, the dissociation constant of BPA pKa values within 9.6 and 10.2 where the dissociation constant representing the favorability of BPA separate into smaller molecules. When the pH > pKa, the ionic species was the predominant which means the surplus hydroxide ions in alkali inhibited the dissociation of BPA into simpler form. On the other hand, when pH < pKa the undissociated species of BPA became the predominant. Therefore, the degradation of BPA at pH 12 was much slower compared to pH 9 due to the hydroxyl radicals ($^{\bullet}OH$) scavenging by carbonate and bicarbonate ions [15].

The adsorption capacity eventually depends on the point of zero charge (PZC) of the photocatalyst. The PZC was an useful measurement in pH unit used to evaluate the net electric charges on the surface. From the report of Fronczak, et al. (2017) [16], the PZC of g-C_3N_4 tends to have better adsorption capacity in pH less than 7. High pH with more negative charge could neutralize the surface charge of photocatalyst in causing the electrostatic repulsion of itself and the organic pollutants [17]. Additionally, the electrostatic repulsion also retarded the adsorption ability of photocatalyst and thus decreasing lower degradation efficiency.

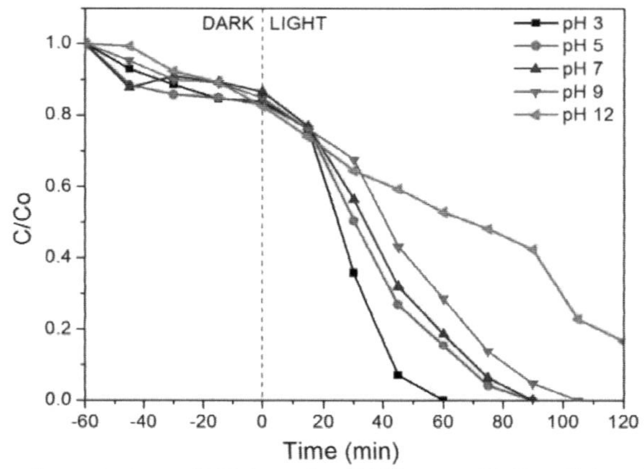

Figure 5. Photodegradation of BPA under Different pH Condition for Au/g-C3N4

4. Conclusion

In summary, the Au/g-C_3N_4 composite was successfully synthesized *via* a simple and sustainable photodeposition method. The composite performed better photodegradation of BPA in low pH condition. This is mainly due to the PZC of g-C_3N_4 incline to have better adsorption capacity in pH less than 7. pH 3 have the smallest electrostatic repulsion compare with others pH condition. Thus promoted the adsorption ability of photocatalyst and thus increased the degradation efficiency. Moreover, BPA is more favorable to separate into smaller molecules when the pH is lower than 9.6 contributed to the dissociation constant of BPA pKa values are within 9.6 and 10.2. Therefore, the PZC of photocatalysts and the dissociation constant of pollutant played an essential part in boosting the photocatalytic performance.

Acknowledgment

This work was supported by the Universiti Tunku Abdul Rahman Research Fund (IPSR/RMC/UTARRF/2017-C1/S04).

References

[1] J. C. Sin, S. M. Lam, A. R. Mohamed, and K. T. Lee, "Degrading Endocrine Disrupting Chemicals from Wastewater by TiO_2 Photocatalysis: A Review," Int. J. Photoenergy, vol. 2012, 2012, doi: 10.1155/2012/185159.

[2] S. Imai, A. Shiraishi, K. Gamo, I. Watanabe, H. Okuhata, H. Miyasaka, K. Ikeda, T. Bamba, and K. Hirata, "Removal of phenolic endocrine disruptors by Portulaca oleracea," J. Of Biosci. And Bioeng. vol. 103, Feb. 2007, pp. 420-426, doi: 10.1263/jbb.103.420.

[3] Z. Li, "Surface modified nanocarbons for photodegradation of organic pollutants," Master thesis, Curtin University, 2016.

[4] H. Dong, G. Zeng, L. Tang, C. Fan, C. Zhang, X. He, and Y. He, " An overview on limitations of TiO_2-based particles for photocatalytic degradation of organic pollutants and the corresponding countermeasures," Water Res., vol. 79, Aug. 2015, pp. 128-146, doi: 10.1016/j.watres.2015.04.038.

[5] H. Yan, Y. Chen, and S. Xu, "Synthesis of graphitic carbon nitride by directly heating sulfuric acid treated melamine for enhanced photocatalytic H_2 production from water under visible light," Int. J. hydrogen energy, vol. 37(1), Oct. 2011, pp. 125-133, doi:

10.1016/j.ijhydene.2011.09.072.
[6] J. Liu, H. Wang, and M. Antonietti, ". Graphitic carbon nitride "reloaded": emerging applications beyond (photo)catalysis," Chem. Soc. Rev., vol. 45(8), 2016, pp. 2308-2326, doi: 10.1039/c5cs00767d.
[7] L. Xie, Z. Ai, M. Zhang, R. Sun, and W. Zhao "Enhanced Hydrogen Evolution in the Presence of Plasmonic Au-Photo-Sensitized g-C_3N_4 with an Extended Absorption Spectrum from 460 to 640 nm,", PloS one, vol. 11(8), Aug. 2016, e0161397, doi:10.1371/journal.pone.0161397.
[8] N. Cheng, J. Tian, Q. Liu, C. Ge, X. Sun, A. Qusti, A. Asiri, and A. Al-Youbi, "Au-nanoparticle-loaded graphitic carbon nitride nanosheets: Green photocatalytic synthesis and application toward the degradation of organic pollutants," ACS Appl. Mater. Interfaces, vol. 5(15), July 2013, pp. 6815-6819, doi: 10.1021/am401802r.
[9] M. Yasmina, K. Mourad, S.H. Mohammed, and C. Khaoula, "Treatment heterogeneous photocatalysis; factors influencing the photocatalytic degradation by TiO_2," Enrgy. Proced. vol. 50, 2014, pp. 559-566, doi: 10.1016/j.egypro.2014.06.068.
[10] S. Liang, Y. Xia, S. Zhu, S. Zheng, Y. He, J. Bi, M. Liu, and L. Wu, "Au and Pt co-loaded g-C_3N_4 nanosheets for enhanced photocatalytic hydrogen production under visible light irradiation," Appl. Surf. Sci., vol. 358, Aug. 2015, pp. 304-312, doi: 10.1016/j.apsusc.2015.08.035.
[11] Y. Fu, T. Huang, L. Zhang, J. Zhu, and X. Wang, "Ag/g-C_3N_4 catalyst with superior catalytic performance for the degradation of dyes: a borohydride-generated superoxide radical approach," Nanoscale, vol. 7(32), 2015, pp. 13723-13733, doi: 10.1039/C5NR03260A.
[12] S. Samanta, S. Martha, and K. Parida, "Facile Synthesis of Au/g-C_3N_4 Nanocomposites: An Inorganic/Organic Hybrid Plasmonic Photocatalyst with Enhanced Hydrogen Gas Evolution Under Visible - Light Irradiation," Chem. Cat. Chem., vol. 6(5), 2014, pp.1453-1462, doi: 10.1002/cctc.201300949.
[13] K. H. Leong, B. L. Gan, S. Ibrahim, and P. Saravanan, "Synthesis of surface plasmon resonance (SPR) triggered Ag/TiO_2 photocatalyst for degradation of endocrine disturbing compounds," Appl. Surf. Sci., vol. 319, July 2014, pp.128-135, doi: 10.1016/j.apsusc.2014.06.153.
[14] E. B. Simsek, "Solvothermal synthesized boron doped catalysts : Photocatalytic degradation of endocrine disrupting compounds and pharmaceuticals under visible light irradiation," Appl. Catal. B, vol. 200, July 2016, pp. 309-322, doi: 10.1016/j.apcatb.2016.07.016.
[15] K. S. Tay, N. A. Rahman, M. Radzi, and B. Abas, "Degradation of bisphenol A by ozonation : rate constants , influence of inorganic anions , and by-products," J. Sci. Technol., vol. 6(1), Feb. 2012, pp. 77–94, doi: 10.14456/mijst.2012.7.
[16] M. Fronczak, M. Krajewska, K. Demby, and M. Bystrzejewski, "Extraordinary Adsorption of Methyl Blue onto Sodium-Doped Graphitic Carbon Nitride," J. Phys. Chem. C, vol. 121(29), July 2017, pp. 15756–15766, doi: 10.1021/acs.jpcc.7b03674.
[17] C. L. Wong, Y. N. Tan, and A. R. Mohamed, "Photocatalytic Degradation of Phenol Using Immobilized TiO_2 Nanotube Photocatalysts," J. Nanotechno., vol. 2011(2011), April 2011, pp. 1–9, doi: 10.1155/2011/904629.

IOP Conf. Series: Materials Science and Engineering **409** (2018) 012009 doi:10.1088/1757-899X/409/1/012009

Physical Mixing Of N-Doped Graphene Quantum Dots Functionalized TiO_2 For Sustainable Degradation Of Methylene Blue

Kah Hon Leong[1*], Sean Yang Fong[1], Ping Feng Lim[1], Lan Ching Sim[1], Pichiah Saravanan[2]

[1]Department of Environmental Engineering, Faculty of Engineering and Green Technology, Universiti Tunku Abdul Rahman, Perak, Malaysia
[2]Department of Environmental Science and Engineering, Indian Institute of Technology (ISM) Dhanbad, Jharkhand, India

E-mails: khleong@utar.edu.my; fseanyang.syf@gmail.com; raven_lim@hotmail.com; simcl@utar.edu.my; pichiahsaravanan@gmail.com

Abstract. An in-situ physical assisted mixing was applied for uniform incorporation of N-doped graphene quantum dots (N-GQDs) with TiO_2 nanoparticles to construct N-GDQs/TiO_2. The unique characteristics of N-doped graphene quantum dots enable it to act as a rich optical light harvesting. The strong absorption under the UV-Vis spectroscopy shows the efficacy in absorbing visible and Near Infrared (NIR) photon. The photocatalysis of the synthesized composite was evaluated through the efficiency of photodegradation of Methylene Blue in aqueous solution under direct sunlight irradiation. A complete degradation was achieved within 150 mins for 1.5% N-GQDs/TiO_2 composite as compare to pure. The vast improvement of this photocatalysis efficiency is much attributed to the strong light absorbance in the entire solar spectrum. Hence, the formation of N-GQDs/TiO_2 could be broadly used as future sustainable photocatalyst for treating various organic pollutants in the field of aquatic environmental remediation.

1. Introduction

With the rapid industrialization increases the water pollution where various toxic matters enter the water bodies. This directly influences the quality of water in our water bodies. One of the main pollutants that happen in Malaysia is dyes. These pollutants are manufactured from many sources such as textile, plastics, paper industries and many others. Methylene Blue (MB) is the most common industrial dye which causes severe environmental issues due to its high toxicity and accumulation in the environment. Therefore, it is important to address this issue effectively and photocatalysis nanotechnology could be a vital solution.

Semiconductor mediated photocatalyst has attracted enormous attention worldwide as it has proven its ability in eliminating environmental pollutants and generating energy. In the context of efficiently harness abundant sunlight for sustainable environmental remediation, the visible and near infrared light sensitive photocatalyst is necessary to be developed. This makes them an appropriate tool to address the alarming environmental issues such as dyes and organic pollutant in the near future by utilizing the entire solar spectrum. To date, titania (TiO_2) is considered a well-known and promising photocatalyst due to its several unique characteristics such as chemically stable, environmentally friendly product, cheap and etc [1,2]. Regrettably, it retarded its ability in extending the light absorption to longer than 390 nm due to its relatively wide band gap energy of 3.2 eV [3].

Massive modifications have been studied to extend to higher wavelength such as doping with metal and non-metal elements [4], depositing with noble metal [5,6], sensitizing with organic dyes [7,8] and so on. However, recently N-doped graphene quantum dots have turned out to be a new candidate in harvesting visible and near-infrared light [9]. N-doped graphene quantum dots, a zero dimensional allotrope of graphitic carbon have gradually become a promising photocatalysts due to its tunable band gap, stable photoluminescence and good biocompatibility [10]. It has been used in various applications such as biosensors, drug carriers, nuclear targeted drug delivery and other energy conversion [11-13]. Herein, we incorporate TiO_2 with N-doped graphene quantum dots through a facile physical mixing route to enhance its utilization of the entire solar spectrum for photodegradation of Methylene Blue.

2. Experimental

2.1. Materials

Titanium (IV) Oxide (Anatase TiO_2, >99%, Sigma Aldrich), urea powder and citric acid were purchased from Sigma Aldrich and ethanol was purchased from R&M Chemical. All chemicals were analytical grade and used as received without any further purification.

2.2. Synthesis of N-GQDs & N-GQDs/TiO_2

N-GQDs were synthesized by mixing 1 mmol of citric acid and 3 mmol of urea and dissolve into 5 ml of distilled water. Solution was then stirred gently until clear solution was formed. Then transfer it to an 80 ml Teflon-lined-stainless-steel autoclave and heated at 160°C for 8 hours in an oven. The obtained N-GQDs resulted in a green luminescence solution. Different percentage amount of obtained N-GQDs (0.5% and 1.5%) are physically mixed with 0.4 g TiO_2 under 2500 rpm for an hour. The resulted solution was then washed with ethanol and distilled water repeatedly. The samples were dried at 80°C overnight.

2.3. Characterization

The morphology of the composite was investigated by field emission scanning electron microscope (FESEM, Hitachi SU-8000) equipped with an energy dispersive X-ray spectroscopy (EDS, Zeiss Auriga). The images were recorded at an accelerating voltage of 20 kV. High resolution transmission electron microscope (HRTEM, JEM-2100F, Jeol) images were obtained at 200 kV. UV-Vis diffuse reflectance spectra (UV-DRS) were performed through Shimadzu UV-2600 spectrophotometer equipped with an integrating sphere attachment to characterize the optical properties. The absorption spectra were obtained with $BaSO_4$ as a reference.

2.4. Photocatalysis Experiment

The photocatalytic performance of the developed composites was studied by degrading Methylene Blue (MB). All the photocatalysis experiments were performed in a simple 500 ml borosilicate beaker with a working volume of 250 ml. The initial concentration of the dye was set to 20 mg/L with 1 g of photocatalyst under continuous stirring conditions. One hour dark reaction was performed to achieve the absorption and desorption equilibrium. Then, the experiment is continue under bright sunlight for a duration of 180 mins. Control experiment was carried out with the absent of photocatalysts. The residual concentrations of MB in the samples were estimated through UV-Vis spectrometer.

3. Results And Discussion

The morphology of the synthesized composites is illustrated in Fig. 1. As shown, it is clearly indicate the homogenous nature and well dispersed of TiO_2 with N-GQDs as captured by FESEM analysis. From the HRTEM image (Fig. 1b), the N-GQDs/TiO_2 photocatalysts are nearly spherical in shape with lattice spacing of 0.21 nm, which corresponds well to the graphitic carbon (100) plane [14]. This correlate well with the formation of sp^2 hybridized carbon nanostructures. Moreover, it proves that the simple physical mixing is capable in incorporating both TiO_2 and N-GQDs.

Figure 1. (a) FESEM and (b) HRTEM image of 1.5% N-GQDs/TiO_2.

Fig. 2 shows the obtained optical absorbance spectra of the synthesized composites. As shown in Fig. 2, pure TiO_2 exhibits low absorption in the region of 390 – 800 nm due to its wide band gap energy of 3.2 eV. This limiting the capability of TiO_2 in harvesting of UV light only. However, the composites show obvious enhancement in light absorption especially in the visible and NIR light region after incorporating with N-GQDs. This enhancement is contributed to the unique characteristics of N-GQDs that promoted red shift towards higher wavelength. This is attributed to the n-π^* transition of graphene quantum dots [15]. Moreover, the absorption intensity of N-GQDs/TiO_2 increases with increasing amount of N-GQDs.

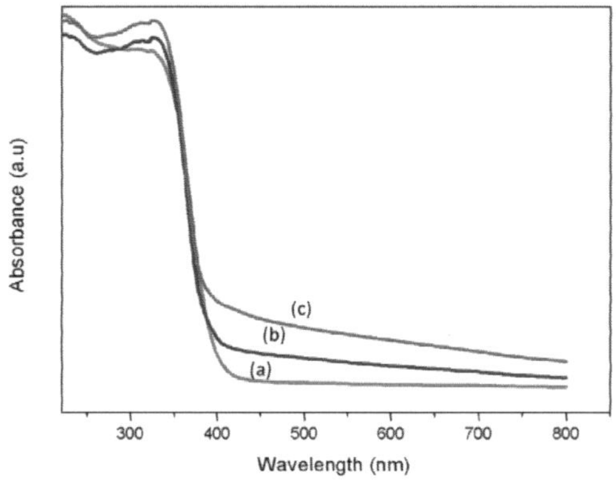

Figure 2. UV-Vis absorption spectra of (a) TiO_2, (b) 0.5% N-GQDs/TiO_2 and (c) 1.5% N-GQDs/TiO_2.

The photoluminescence of the synthesized composites was performed under the excitation wavelength of 315 nm and it is shown in Fig. 3. Pure TiO_2 exhibited a high and strong emission peaks at about 540-550 nm indicating its drawback of having high recombination rate of electron and hole pairs. However, this was eradicated with the incorporation of N-GQDs. The presence of N-GQDs has drastically separated the electron and hole pairs from recombine to generate heat. This enhanced separation had resulted to a low PL counts that was achieved by the composite (0.5% and 1.0% N-GQDs/TiO_2) and it is clearly shown in Fig. 3. The higher percentage of N-GQDs give rise to a better electron and hole pairs separation. Hence, its prolonged the lifetime of these charge carriers. The separated electrons and holes will then undergo redox reactions and formed axcess of active radicals. These active radicals are reponsible for the photodegradation of MB and will result to an enhanced photocatalytic activity.

IOP Conf. Series: Materials Science and Engineering **409** (2018) 012009 doi:10.1088/1757-899X/409/1/012009

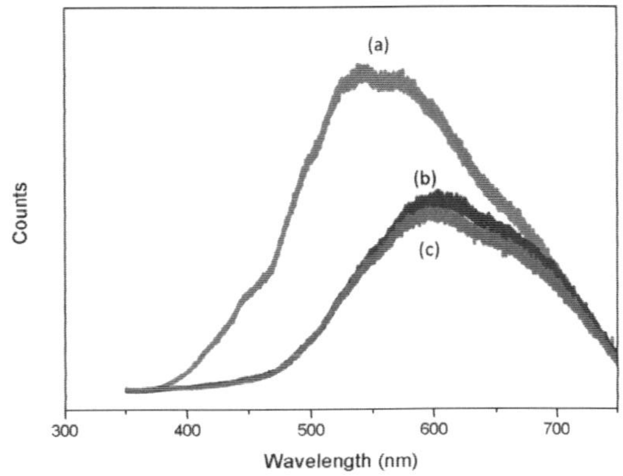

Figure 3. Photoluminescence spectra of (a) TiO_2, (b) 0.5% N-GQDs/TiO_2 and (c) 1.5% N-GQDs/TiO_2.

Fig. 4 portraits the photodegradation profile of MB with the presence of synthesized composite photocatalysts. The N-GQDs/TiO_2 shows great photocatalysis enhancement as compare to pure TiO_2. Result revealed that, 1.5% N-GQDs/TiO_2 only required 150 mins for a complete removal of MB under bright sunlight as compare to pure TiO_2 (180 mins). The superiority degradation efficacy of 1.5% N-GQDs/TiO_2 is due to the presence of N-GQDs that extended the light absorption range of pure TiO_2 to visible and near-infrared region as shown in the above UV-vis spectra analysis. With the strong light absorption capacity has increase the massive generation of electron and holes pairs in the synthesized photocatalysts. The free generated electrons will then formed the superoxide anion radicals ($\bullet O_2^-$) and followed by protonation that yields $\bullet HO_2$ radicals. These instable radicals will further form H_2O_2 and lead to the formation of hydroxyl radicals $\bullet OH^-$. Hence, this active radical are the key species in eliminating MB. However, 0.5% N-GQDs/TiO_2 could not achieve a complete degradation under the same duration is due to the weak absorption of visible and NIR light. This leads to a lower generation of free electrons in the photocatalysts. The low generation of free electron will resulted to minimum formation of active radicals to eliminate methylene blue dyes.

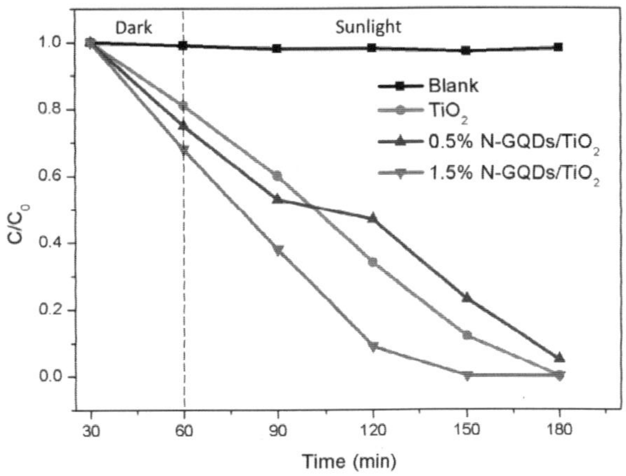

Figure 4. Photodegradation of Methylene Blue under dark and sunlight irradiation.

4. Conclusions

The present study shows the successful synthesized of N-GQDs functionalized TiO_2 through a simple physical mixing route to extend light absorption towards visible and near infrared light. Various characterizations proved the homogenous incorporation of N-GQDs with TiO_2. A complete removal of

Methylene Blue was achieved in 150 mins by 1.5% N-GQDs/TiO_2 under sunlight irradiation. The improved photocatalysis was contributed by the red shift toward higher wavelength that enhanced the light absorption capacity of TiO_2 after incorporation with N-GQDs. This resulted to the massive generation of active radicals that responsible for photodegradation of MB. The absorption over the full solar spectrum laid a sustainable pathway for future practical applications. Hence, the synthesized composite will play a dynamic role as a sunlight sensitive photocatalyst in promoting sustainable environmental remediation in the near future.

Acknowledgment
This research work was supported by Universiti Tunku Abdul Rahman Research Fund, UTARRF (IPSR/RMC/UTARRF/2016-C2/L05)

References

[1] M.N. Chong, B. Jin, C.W.K. Chow and C. Saint, "Recent Developments in Photocatalytic Water Treatment Technology: A review" Water Res 44, 2010, 2997-3027 doi: 10.1016/j.watres.2010.02.039.

[2] J. Zhang, X. Zhang, S. Dong, X. Zhou and S. Dong, "N-doped Carbon Quantum Dots/TiO_2 Hybrid Composites With Enhanced Visible Light Driven Photocatalytic Activity Toward Dye Wastewater Degradation And Mechanism Insight" J Photoch Photobio A 325, 2016, 104-110 doi: 10.1016/j.jphotochem.2016.04.012.

[3] H. Xie, C. Hou, H. Wang, Q. Zhang and Y. Li, "S, N Co-Doped Graphene Quantum Dot/TiO_2 Composites For Efficient Photocatalytic Hydrogen Generation" Nanoscale Res Lett 12, 2017, 400 doi: 10.1186/s11671-017-2101-1.

[4] F.T. Li, Y. Zhao, Y.J. Hao, X.J. Wang, R.H. Liu, D.S. Zhao and D.M. Chen, "N-Doped P25 TiO_2-amorphous Al_2O_3 composites: One-step solution Combustion Preparation And Enhanced Visible-Light Photocatalytic Activity" J Hazard Mater 239, 2012, 118-127 doi: 10.1016/j.jhazmat.2012.08.016.

[5] H. Wang, T.T. You, W.W. Shi, J.H. Li and L. Guo, "Au/TiO_2/Au As A Plasmonic Coupling Photocatalyst" J Phys Chem C 116, 2012, 6490-6494 doi: 10.1021/jp212303q.

[6] K.H. Leong, H.Y. Chu, S. Ibrahim and P. Saravanan, " Palladium Nanoparticles Anchored To Anatase TiO_2 For Enhanced Surface Plasmon Resonance-Stimulated, Visible-Light-Driven Photocatalytic" Beilstein J Nanotech 6, 2015, 428-437 doi: 10.3762/bjnano.6.43.

[7] W.J. Youngblood, S.H. Lee, Y. Kobayashi and E.A. Hernandez-Pagan, P.G. Hoertz, T.A. Moore, A.L. Moore, D. Gust, T.E. Mallouk, "Photoassisted Overall Water Splitting In A Visible Light-Absorbing Dye-Sensitized Photoelectrochemical Cell" J Am Chem Soc 131, 2009, 926-927 doi: 10.1021/ja809108y.

[8] J.X. Wang, J. Huang, H.L. Xie, A.L. Qu, "Synthesis of g-C_3N_4/TiO_2 With Enhanced Photocatalytic Activity For H_2 Evolution By A Simple Method" J Hydrogen Energy 39, 2014, 6354-6363 doi: 10.1016/j.ijhydene.2014.02.020.

[9] T. Majumder and S.P. Mondal, "Advantages Of Nitrogen-Doped Graphene Quantum Dots As A Green Sensitizer With ZnO Nanorod Based Photoanodes For Solar Energy Conversion" J Electroanal Chem 769, 2016, 48-52 doi: 10.1016/j.jelechem.2016.03.018.

[10] W. Liu, X. Yan, J. Chen, Y. Feng and Q. Xue, "Novel And High-Performance Asymmetric Micro-Supercapacitors Based On Graphene Quantum Dots And Polyaniline Nanofibers" Nanoscale 5, 2013, 6053-6062 doi: 10.1039/C3NR01139A.

[11] X. Yan, Y. Song, C. Zhu, J. Song, D. Du, X. Su and Y. Lin, "Graphene Quantum Dot-MnO_2 Nanosheet Based Optical Sensing Platform: A Sensitive Fluorescence "Turn Off-On" Nanosensor For Glutathione Detection And Intracellular Imagind" ACS Appl Mater Interfaces 8, 2016, 21990-21996 doi: 10.1021/acsami.6b05465.

[12] P. Zhang, X. Zhao, Y. Ji, Z. Ouyang, X. Wen, J. Li, Z. Su and G.Wei, "Electrospinning Graphene Quantum Dots Into A Nanofibrous Membrane For Dual-Purpose Fluorescent And

Electrochemical Biosensors" J Mater Chem B 3, 2015, 2487-2496 doi: 10.1039/C4TB02092H.

[13] D. Xiong, X. Li, H. Shan, B. Yan, D. Li, C. Langford and X. Sun, "Scalable Synthesis Of Functionalized Graphene As Cathodes In Li-ion Electrochemical Energy Storage Devices" Appl Energy 175, 2016, 512-521 doi: 10.1016/ j.apenergy.2016.03.105.

[14] A.B. Ganganboina, A.D. Chowdhury and R. Doong, "Nano Assembly Of N-Doped Graphene Quantum Dots Anchored Fe_3O_4/halloysite Nanotubes For High Performance Supercapacitor" Electrochim Acta 245, 2017, 912-923 doi: 10.1016/j.electacta.2017.06.002.

[15] D. Qu, Z.Sun, M. Zheng, J. Li, Y. Zhang, G. Zhang, H. Zhao, X. Liu, Z. Xie, "Three Colors Emission From S, N, Co-Doped Graphene Quantum Dots For Visible Light H2 Production And Bioimaging" Adv. Optic. Mater 3, 2015, 360-367 doi: 10.1016/j.jelechem.2016.03.018.

IOP Conf. Series: Materials Science and Engineering **409** (2018) 012010 doi:10.1088/1757-899X/409/1/012010

Electrospun PVDF nanofibrous membranes effectively capturing PM2.5 and releasing negative ions

Z L Liu, W Zhang

College of Textile and Garment, Hebei University of Science and Technology, 26 Yuxiang Road, Yuhua District, Shijiazhuang 050018, China
E-mail: liuzhaolin0508@163.com

Abstract. Polyvinylidene fluoride (PVDF) nanofibrous membranes doped with negative ion powder were fabricated by electrospinning, which were aimed to efficiently capture PM2.5 and robustly release negative ions. Different fiber morphologies were formed by regulating the solution concentration. The effect of solution concentration on air filtration property and negative ion release capacity of the PVDF membranes was also investigated. The results show that with the increase of PVDF concentration, bead-on-string fibers are formed firstly and then smooth fibers with larger diameters can be obtained. The filtration efficiency and the pressure drop increase initially and then decrease. The quality factor of the membrane is highest when the PVDF concentration is 17 wt%. The negative ion release amount is small due to the formation of bead-on-string structure at low solution concentration; then it reaches the highest value of 2026 ions/cm^3 when the PVDF concentration is 17 wt%. However, further increase of the solution concentration intensity the covering of polymeric components on negative ion powder, therefore, the negative ion release amount reduces gradually.

1. Introduction

With rapid urbanization and industrialization, many environmental problems have emerged. Particularly the PM2.5, one of the major sources of air pollution, has posed grave health threats to human extrapulmonary organs and the respiratory tract [1]. According to a report by the United States Environmental Protection Agency, more than 2 million people died worldwide because of the increasing level of PM2.5 [2]. Therefore, high performance filters with excellent air filtration property are in urgent need. Furthermore, air filtration materials with the function of providing healthcare have gained more and more attention. Negative ions can cure asthma, chronic bronchitis, hypertension, and boost the immune system [3]. Herein, artificial generation of negative ions in air is a very useful approach for protecting human health. Electrospinning has gained more and more appeal as an efficient and versatile technique that manufactures sub-micron to nanoscale fibers through an electrically charged jet of polymer solution [4]. Electrospun nanofibrous membranes exhibit attractive features such as ultrathin fiber diameter, high porosity and the ease of functionalization, making them suitable candidate for advanced filtration material production [5].

For the reasons above, electrospun polyvinylidene fluoride (PVDF) nanofibrous membranes doped with negative ion powder, which were capable of releasing negative ions for healthcare and efficiently capturing PM2.5, were introduced in this study. Fiber morphologies were subtlely controlled by tuning the solution concentration. The effect of solution concentration on filtration efficiency, pressure drop and negative ion release amount of the PVDF membranes was systematically analyzed.

2. Experimental

2.1. Materials
Industrial-grade polyvinylidene fluoride (PVDF, M_w=660000) was supplied by Wuxi united Hengzhou chemical Co., Ltd., China. Negative ion powder (with an average particle size of 50 nm) was supplied by Zhengzhou Shengyu chemical products Co., Ltd., China. N,N-Dimethylformamide (DMF, analytical pure) was supplied by Jinan Hengrui chemical Co., Ltd., China. The polypropylene nonwoven substrate (with negligible fitration efficiency of 3% and pressure drop of 2 Pa under the airflow velocity of 32 L/min) was provided by Hebei Hongrun new fabrics Co., Ltd., China.

2.2. Preparation of polymer solutions
Negative ion powder was dissolved in DMF solvent by keeping the negative ion powder concentration of 9 wt%; then PVDF was added to the mixture with vigorous stirring for 16 h at 80 °C. The resultant PVDF concentrations were 15, 17, 19 and 21 wt%, respectively.

2.3. Electrospinning of PVDF nanofibrous membranes
The nanofibrous membranes were fabricated on the DT-1003 electrospinning equipment (Dingtong science and technology Co., Ltd., China). The polymer solution was loaded into a syringe and ejected through the metal needle with a feed rate of 1 mL/h. A direct voltage of 20 kV was applied at the tip of the needle. The nanofibrous membranes deposited on the tumbing barrel covered by the nonwoven substrate, which rotated at a velocity of 60 r/min. The tip-to-collector distance was 15 cm. The temperature and the relative humidity during electrospinning were maintained at 25 °C and 45%.

2.4. Characterization
The morphology of the PVDF membranes was studied by the scanning electron microscope (SEM, TM3000, Hitachi Ltd., Japan). The fiber diameter was measured by image processing software (Image J 1.40G). The negative ion release amount was analyzed by a negative ion detector (KEC-900+, Shanghai IBS instrument equipment Co., Ltd., China). The TSI8530 automated filter tester (TSI, Inc., MN, USA) was used to test the filtration efficiency and the pressure drop of the membranes. Neutralized NaCl aerosol particles of 300 nm mass median diameter were delivered from TSI8530 with a fixed airflow velocity of 32 L/min. To comprehensively evaluate the filtration property of the nanofibrous menbranes, the quality factor (QF) [6] was calculated according to equation (1).

$$QF = -(\Delta p)^{-1}\ln(1-\eta) \quad (1)$$

Where η and Δp were the filtration efficiency and the pressure drop across the filter, respectively. A large quality factor value indicated good filtration performance.

3. Results and discussion

3.1. Effect of solution concentration on morphology of the PVDF membranes
The SEM images of the PVDF membranes obtained by varying the PVDF concentration were shown in figure 1, and the average fiber diameters of the corresponding PVDF membranes were displayed in figure 2. It can be seen in figure 1 that all the nanofibrous membranes had randomly oriented three-dimensional structures. Figure 1 (a) showed that bead-on-string fibers were electrospun from the 15 wt% PVDF solution. The fibers were composed of thin fibers with an average diameter of 163 nm and micro-sized spindle-shaped beads along the fiber axis. This could both be ascribed to the low solution viscosity at low solution concentration. When the PVDF concentration was 17 wt%, the bead-on-string structure disappeared and the average fiber diameter increased slightly to 197 nm. Further increasing the PVDF concentrations to 19 and 21 wt%, no beads could be observed in figure 1 (c)-(d), and the average fiber diameters were 406 and 623 nm, respectively. These changes in fiber morphology could be attributed to the increase of the solution viscosity caused by the increased solution concentration. A higher viscoelastic force helped suppress the surface tension of the polymer solution and prevent the jet from collapsing into droplets before the solvent evaporated, thus bead-free fibers with larger diameters occurred [7].

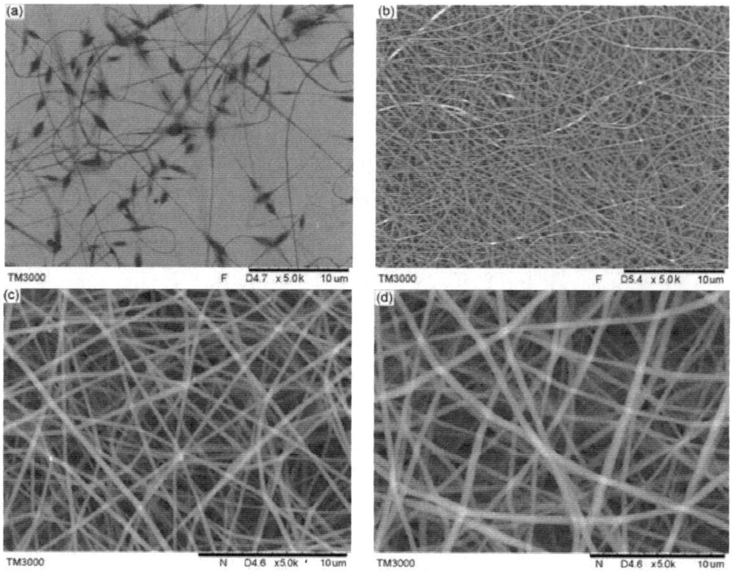

Figure 1. SEM images of the PVDF membranes formed at different solution concentrations of (a) 15 wt%, (b) 17 wt%, (c) 19 wt% and (d) 21 wt%.

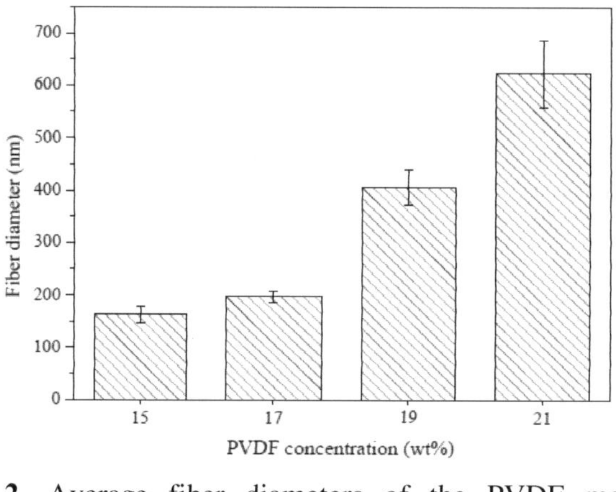

Figure 2. Average fiber diameters of the PVDF membranes fabricated at different PVDF concentrations.

3.2. Effect of solution concentration on air filtration property of the PVDF membranes
The filtration efficiencies of the PVDF membranes (basis weight of ~1.5 g/m^2) fabricated from 15, 17, 19 and 21 wt% solutions were 67.7%, 97.5%, 92.9% and 83.5%, and their pressure drops were 76, 108, 101 and 90 Pa, respectively, as presented in figure 3. Both the lowest filtration efficiency and the pressure drop of the membrane from 15 wt% PVDF solution could be attributed to the wide distribution of beads, which was in favour of air permeability but hindered the fine airborne particles capture. With the aid of continuous and thin fibers, the membrane from 17 wt% PVDF solution had the highest filtration efficiency and relatively low pressure drop. Further increasing the PVDF concentrations to 19 and 21 wt%, the filtration efficiencies dropped rapidly with the increased fiber diameter, meanwhile, the pressure drops decreased rapidly owing to the thicker fibers without beads. The qulitiy factor values of the PVDF membranes were 0.0149, 0.0342, 0.0262 and 0.0200 Pa^{-1} when the solution concentrations were 15, 17, 19 and 21 wt%, respectively, as shown in figure 4. It could be seen that the membrane from 17 wt% PVDF solution possessed the highest qulitiy factor, further

implying the important role of fine fiber diameter and continuous fiber morphology in the construction of nanofiber filter materials.

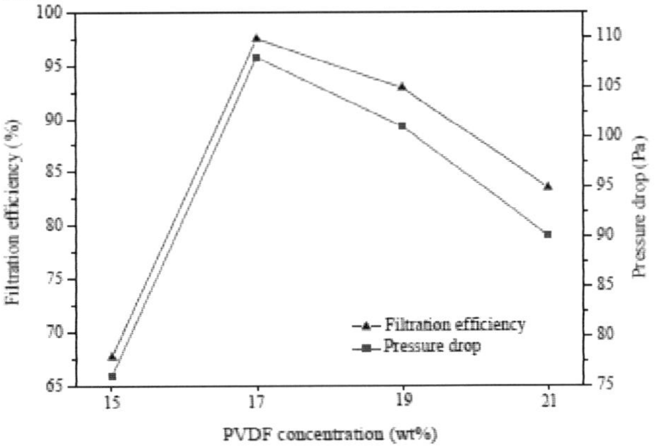

Figure 3. Filtration efficiency and pressure drop of the membranes fabricated at different PVDF concentrations.

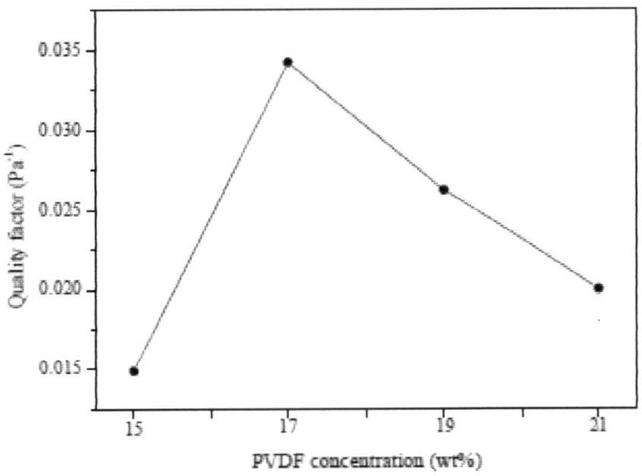

Figure 4. Quality factor of the membranes fabricated at different PVDF concentrations.

3.3. Effect of solution concentration on negative ion release amount of the PVDF membranes

Figure 5 presented the curve between PVDF solution concentration and the negative ion release amount of the relevant membrane. It can be found that the negative ion release amount decreased constantly from 2026 to 1329 ions/cm^3 as the PVDF concentration increased from 17 to 21 wt%. Since less negative ion powder could be naked on the surface of the fibers to aggravate the shielding effect of polymeric components from the electric potential difference between the particles of the negative ion powder, as a result, the negative ion release amount showed a decreasing trend. However, further decrease of the PVDF concentration to 15 wt% led to a significant reduction of the negative ion release amount to 1072 ions/cm^3. This was mainly because the formation of the bead-on-string structure, which stored a lot of negative ion powder in the beads.

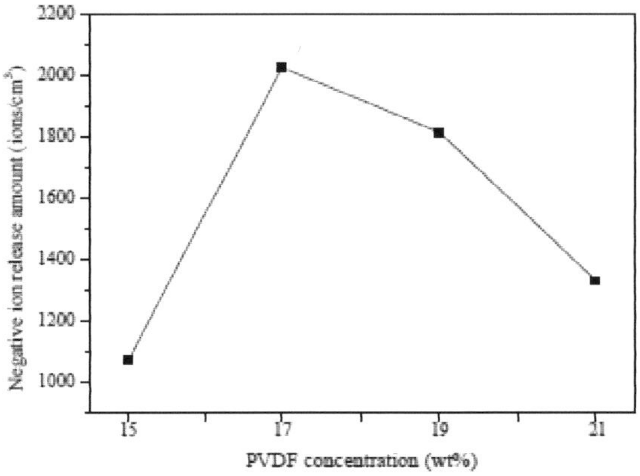

Figure 5. Negative ion release amount of the membranes fabricated at different PVDF concentrations.

4. Conclusion

PVDF nanofibrous membranes capable of releasing negative ions and effectively capturing PM2.5 have been successfully prepared. The morphologies of the PVDF nanofibers are controllable by regulating the solution concentration, which are highly important to air filtration and negative ion release. Results indicate that bead-on-string fibers are formed when the PVDF concentration is 15 wt%, and then smooth fibers with larger diameters are obtained as the soluiton concentration increases. The filtration efficiency and the pressure drop increase initially and decrease gradually. When the PVDF concentration is 17 wt%, the quality factor of the membrane reaches its highest value. Thin fiber diameter is conducive to the improvement of negative ion release amount. On the contrary, thick fiber intensities the covering of polymeric components on negative ion powder, leading to the reduction of the negative ion release amount. Ultimately, the nanofibrous membrane possesses a high PM2.5 purification efficiency of 97.5%, low pressure drop of 108 Pa and good negative ion release amount of 2026 ions/cm^3 when the PVDF concentration is 17 wt%.

5. References

[1] Lee C J, Martin R V, Henze D K, Brauer M, Cohen A and Donkelaar A 2015 *Environ. Sci. Technol.* **49** 4335-44

[2] Chow J C, et al 2006 *J. Air Waste Manage. Assoc.* **56** 1368-80

[3] Nimmerichter A, Holdhaus J, Mehnen L, Vidotto C, Loidl M and Barker A R, 2014 *Int. J. Biometeorol.* **58** 1503-12

[4] Greiner A and Wendorff J H 2007 *Angew. Chem. Int. Edit.* **46** 5670-5703

[5] Wang Z and Pan Z J 2015 *Appl. Surf. Sci.* **356** 1168-79

[6] Wang Z, Zhao C C and Pan Z J 2015 *J. Colloid Interf. Sci.* **441** 121-9

[7] Fong H, Chun I and Reneker D 1999 *Polymer.* **40** 4585-92

Acknowledgments

This research was financially supported by the Science and Technology Research Project for Universities in Hebei Province (ZD2018225), the National Natural Science Foundation of China (51703028), and the Natural Science Foundation of Hebei Province (B2018208095).

IOP Conf. Series: Materials Science and Engineering **409** (2018) 012011 doi:10.1088/1757-899X/409/1/012011

Resonant optical transmission through the sub-wavelength air-hole arrays in a gold thin film for sensing applications

Yuan-Fong Chou Chau*, Chee Ming Lim, Nyuk Yoong Voo, Noormariah Muslim and Muhammad Nur Syafi'ie Md Idris

Centre for Advanced Material and Energy Sciences Universiti Brunei Darussalam, Tungku Link, Gadong, BE1410, Negara Brunei Darussalam

E-mail: a0920146302@gmail.com

Abstract. We have investigated numerically the optical spectral properties of resonant transmission of light through sub-wavelength air-holes in a gold thin film by using the finite element method. The enhanced transmission can be comprehended in the light of interfering surface-wave-like modes propagating in the sub-wavelength air-holes. The effects of geometrical and material properties of the sub-wavelength air-holes in the transmission spectrum were characterized. The proposed structure can be operated as refractive index (RI) sensor with a sensitivity of 524.8 nm/RIU (RIU = refractive index unit) ranging from the visible to the near-infrared. This feature provides a remarkable scheme of nanoscale sensing for the RI variations of the environmental media such as applications in liquid and gas sensing.

1. Introduction

Nanostructured metallic films and their transmission properties are potential for both fundamental studies of light-matter interaction and for developing nanophotonic devices based on surface plasmon resonances (SPRs) [1]. Recently, the SPR effect on metal nanoparticles (MNPs) has been followed out in the related fields of nanophotonics [2]. The optical properties of sub-wavelength air-holes in metallic thin films have been investigated extensively because of the discovery of enhanced optical transmission through such a structure [3]. The enhanced transmission in arrays of sub-wavelength air-holes can be realized to be since the surface plasmon polaritons (SPP) propagating on the metal film [4], or the coupling of incident EM wave into sub-wavelength air-holes waveguide modes [5], or the diffraction theory [6]. SPRs have shown great potential for applications in sub-wavelength optics with components such as the sensitivity of SPRs on the surrounding environment has been exploited in various sensors [7] and nanophotonics device applications [8]. The resonances have also been made use of in surface-enhanced spectroscopy [9].

In this paper, the transmission spectrum of *x*-polarized (transverse magnetic) light through a two-dimensional (2D) periodic sub-wavelength air-holes array in an Au thin film by using three-dimensional (3-D) finite element method (FEM) is investigated and analyzed. The transmission resonance of a single sub-wavelength air-hole as a sensor is proposed to use for detecting small changes in the refractive index (RI) of the surrounding medium. The sensor could be used, for example, to detect biological molecules. By using progressive techniques of nanofabrication [10], the proposed sensor could be integrated in various lab-on-chip applications.

2. Simulation Method And Models

Numerical simulations in this work are performed by using 3-D FEM based on Maxwell's equations. Figure 1 shows the schematic diagram of the truncate view of a 2-D periodic array of air-holes in a gold (Au) thin film with thickness *t* (Fig. 1 (a)) and its unit cell (Fig. 1 (b)). To imitate an infinite periodic array, the simulation model is obtained from the unit cell by performing a periodic boundary condition (PBC) along both the *x*- and *y*-axes (the lateral directions of the unit cell) and the anisotropic perfectly matched layers (PML) condition on the top and bottom surfaces of the unit cell to diminish the influence of light reflection. A 3-D meshing with a high density tetrahedral mesh is used in the simulation structure and the surrounding matrix. The value of the incident electromagnetic (EM) wave is fixed at $|E_0|$=1V/m to be normally incident from the top surface of the structure. The parameters used in this work are Λ (period along the *x* and *y* axes), *t* (thickness of the Au thin film) and R (radius of the circular air-hole in the Au thin film) and the environmental medium is set to be 1.0 for air. The transmittance (T) can be obtained from the scattering parameters (S-parameter), i.e., S_{21}= ((power from the reflected light plane)/(power from the incident light plane))$^{1/2}$. Transmittance is calculated by $|S_{21}|^2$.

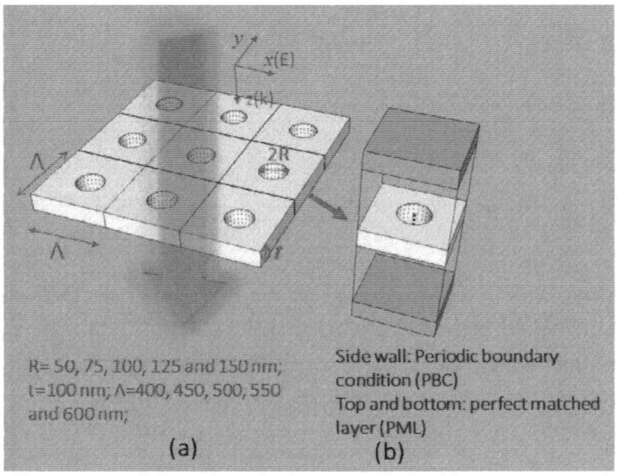

Figure 1. Schematic diagram of the simulation model, (a) truncate view of a periodic array of circular air-holes in an Au thin film, and (b) schematic plot of a unit cell of circular air-hole in an Au thin film. The parameters used in the simulations are listed in the inset of this figure.

3. Results And Discussion

3.1. Transmittance spectrum

It is necessary to investigate the transmission spectrum of a periodic air-holes array with different radius (R) and period (Λ) in an Au thin film to demonstrate which transmission peaks correspond to the resonant localized waveguide (RLW) modes [9], and to know their dependencies. Figures 1 (a)-(e) show the calculated transmittance spectrum of different radius (R=50, 75, 100, 125 and 150 nm) and period (Λ=400, 450, 500, 550 and 600 nm) in an Au thin film with a thickness of *t*=100 nm. Several groups of resonance peaks of the transmittance spectrum are observed in Figs. 2(a)-(e). These transmission peaks are the RLW modes and their magnitudes can be tuned by changing their radii and periods. In all transmission spectra, the smaller ones are located at the ultra-violet (UV) wavelengths while the larger ones are in the range of the visible spectral region, which become stronger as R is increased and Λ is decreased. The period Λ indicates the density of the air-hole arrays in the Au thin film and has a significant contribution to the transmission spectra. The enhanced transmission magnitude with respected to a lager R and a smaller Λ can be intuitively understood since more flux of light (i.e., increasing the R) and number of holes (i.e., decreasing the Λ) per unit area (i.e., decreasing the Λ) lead to more light being transmitted through the air holes. Note that the period Λ denotes the density of the air-hole arrays in the Au thin film and has a significant contribution to the transmittance spectra. It can be demonstrated from Figs. 2 (a)-(e) that the transmittance increases as the value of the period is reduced. This is due to the coupling effect of the shorter period among the air-hole arrays in the Au thin film, as oppose to weaker coupling effect for a longer period. Next, we consider the effect of the thickness of the Au film on the transmittance spectrum. Figure 2(f) depicts the transmittance spectra for different thickness of an Au film (t=100, 150, 200, 300 and 400 nm) with a period Λ and a radius R fixed at 400 nm and 100 nm. The result indicates that as the t is decreased, the resonance wavelength (λres) peaks become significantly higher. Note that the transmittance increases and that the λres peak shifts to the red with decreasing t, whereas the transmission spectra become weaker as t is increased. As the RLW becomes shorter (i.e., smaller t), the real part of the propagation constant of the RLW mode increases. This corresponds to a higher real part of the effective index of refraction, and thus to a shorter optical length of the cavity in the air-holes, which clarifies the red shift of the λres peak.

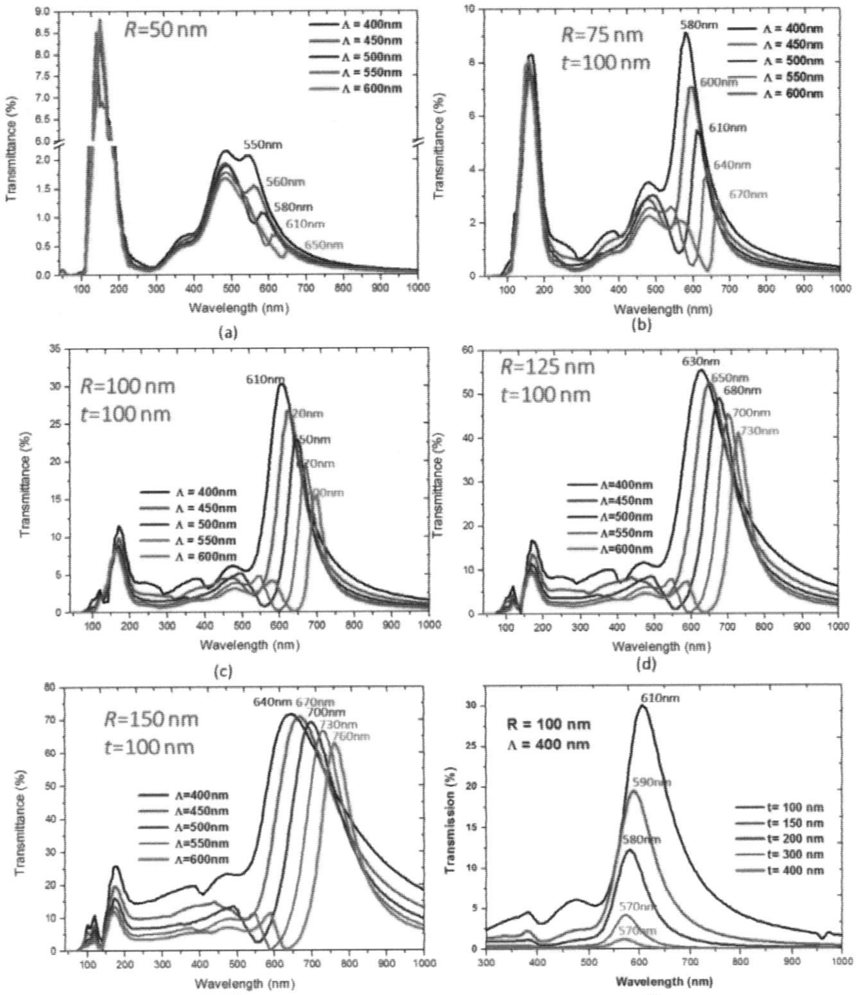

Figure 2. (a)-(e) Transmittance spectrum of different radii (R=50, 75, 100, 125 and 150 nm) and period (Λ=400, 450, 500, 550 and 600 nm) in the Au thin film with a thickness of t=100 nm. (f) Transmittance spectra for different thickness of the Au film (t=100, 150, 200, 300 and 400 nm) with a periodΛand a radius R fixed at 400 nm and 100 nm.

3.2. Near-field intensity and surface charge distributions

The above analysis could be further identified with the near-field intensity and surface charge distributions of air-holes in the Au thin film with different R and Λ with respect to their corresponding λres (results not shown here). We find that surface plasmons can be excited surrounding their surfaces and rims. The *x*-component of the electric field (E*x*) shows two distinct petals of distribution on the top and bottom sides, showing no electric fields in the air-holes. The *y*-component of the electric field (E*y*) is distributed symmetrically along the rim aperture of the air-holes, showing four crossing petals of distribution. In contrast to E*x*, the *z*-component of the electric field (E_z) shows two distinct petals of distribution on the left and right sides, showing more electric fields in the air-hole. The total field distributions (Etotal) at their corresponding resonances show a hybridized plasmon mode on the metallic rim surfaces and in the air-holes, and the strongest field intensity is found in the central part of the air-holes. Owing to different in the number of free electrons, the induced positive-negative surface charge pairs on the rim aperture of the air-holes enhanced each other by induced electromotive force. This behavior could be interpreted by the fact that the increased number of free electrons along the rim aperture of the air-holes affected not only the magnitude of induced surface current but also the full-charging time on the edge area [10].

3.3. Surface plasmon resonance sensing application

To realize the detailed behavior of the field intensity distribution around the air-holes arrays in the Au thin film under SPR excitations, Fig. 3(a) shows the near field intensity corresponding to resonant wavelength versus different radius of air-holes with R= (50, 75, 100, 125, 150) nm along the *x*-axis measured at the middle plane of the Au thin film in the range of *x*= [-200, 200] nm. It is shown that the higher near field intensity can be found at R=75 and 100 nm, while the lower near field intensity occurred at R=50, 125 and 150 nm, respectively. A distinct edge enhancement of the field intensity can be found among the rim aperture of the air-holes. The positive-negative charge pairs on the Au thin film surface can be considered as many dipoles at the symmetry positions around the circumference of the rim aperture air-holes. The *x*-polarization field intensity pattern can be understood as a strong dipolar excitation that involves the entire ring perimeter. Note that the localized electric field enhancement at the edge along the air-hole is due to the edge enhancement effect [11].

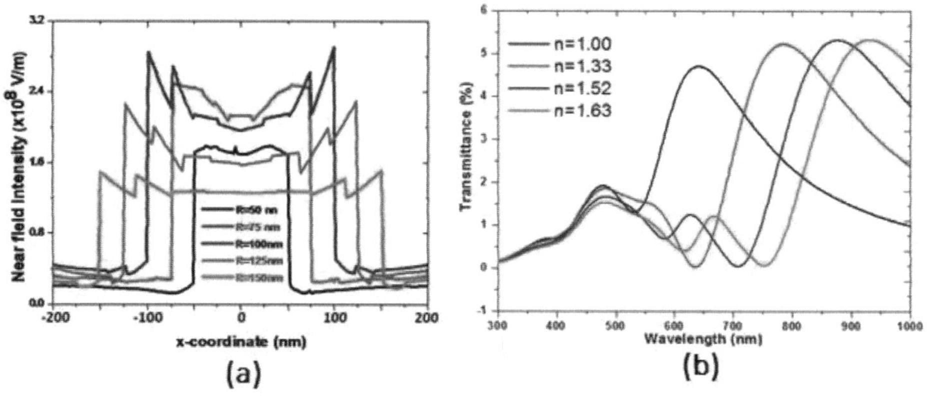

Figure 3. Near field intensity corresponding to resonant wavelength versus different radious of air-hole with R= {50, 75, 100, 125, 150} nm along the *x*- axis measured at the middle plane of the Au thin film in the range of *x*= [-200, 200] nm. (b) Transmittance spectra for fixing *t*=100 nm, Λ=400 nm and R=75 nm as the different material within the air-holes of Au thin film.

In order to verify the dependence of the transmittance resonance on the material surrounding the Au air-hole thin film, the transmittance spectra for air (*n* = 1.0), water (*n* = 1.33), glass (*n* = 1.52), and optical oil (n = 1.63) were investigated (Fig. 3 (b)). The geometrical parameters are fixed at t=100 nm, Λ=500 nm and R=75 nm. As the RI of the air-hole medium increases, the λ_{res} has a significant red shift. The transmittance peak position made a red-shift with a slight increase in magnitude whenever a fluid of higher RI was introduced into the Au thin film with the air-holes arrays. With the increase of RI, a red-shift of the transmittance peak varied from 610 to 940 nm. The sensitivity ($S=\Delta\lambda/\Delta n$) raised to a value of 523.8 nm/RIU, where Δλ is the corresponding central wavelength shift of the resonant dips, and Δn is the difference of the refractive index. When the fluid (or molecular) system is resonant with the Au thin film with the air-holes arrays, the transmittance spectrum is strongly modified, showing characteristics of strong exciton-plasmon coupling and leading to a bonding resonant mode [12]. This observation can be understood based on the analysis of the planar waveguide propagation constant, i.e., a small change in the λ_{res} leads to a large change in the propagation constant. The significant shift of the transmission resonance can be used to monitor changes in the surrounding media [13-15]. This feature provides a remarkable scheme of nanoscale sensing for the RI variations of the environmental medium such as applications in liquid and gas sensing.

4. Conclusion

The optical properties of a 2-D periodic array of sub-wavelength air-holes array in an Au thin film are investigated by using 3-D FEM. The analysis shows that there exists a strong resonance peak in the transmission spectrum of the structure ranging from the visible to near-infrared spectral region. The calculations show that the transmission resonance is a geometric effect resulting from the constructive interference of RLW modes propagating within the air-holes array in an Au thin film. Results show that the geometrical parameters of the structure considerably affect the transmission spectrum. By decreasing the thickness of the air-holes array in the Au thin film, the λres can be shifted from 100 nm to 400 nm. Thus the transmission spectrum can be elastically tailored for various applications. We have shown that the position of the transmission resonance depends strongly on the RI of the medium within the circumference of the air-holes in the Au thin film. This effect can be used in applications including sensors, imaging, and the detection of, e.g. biomolecules and nanophotonic devices.

Acknowledgment

This work was supported by the University Research Grant of Universiti Brunei Darussalam (Grant No. Grant No. UBD/OAVCRI/CRGWG (004)/170101)) and research block funding from Universiti Brunei Darussalam.

References

[1] L. Meng, D. Zhao, Z. Ruan, Q. Li, Y. Yang, M. Qiu, "Optimized grating as an ultra-narrow band absorber or plasmonic sensor," Opt. Lett., 2014, 39, 1137-1140.
[2] Y. F. Chau and C. Y. Jheng, "Buried Effects of Surface Plasmon Resonance Modes for Periodic Metal–Dielectric Nanostructures Consisting of Coupled Spherical Metal Nanoparticles with Cylindrical Pore Filled with a Dielectric," Plasmonics, vol. 9, 2014, pp.1-9.
[3] L. Cong, S. Tan, R. Yahiaoui, F. Yan, W. Zhang, R. Singh, "Experimental demonstration of ultrasensitive sensing with terahertz metamaterial absorbers: A comparison with the metasurfaces," Appl. Phys. Lett., 2015, 106, 031107.
[4] W. Yang, Y. F. Chou Chau, S. C. Jheng, "Analysis of transmittance properties of surface plasmon modes on periodic solid/outline bowtie nanoantenna arrays," Physics of Plasmas, vol. 20, 2013, 064503.
[5] X. Li, J. Zhu, B. Wei, "Hybrid nanostructures of metal/two-dimensional nanomaterials for plasmon-enhanced applications," Chem. Soc. Rev., 2016, 45, pp.3145-3187.
[6] T. Cao, C. Wei, R. E. Simpson, L. Zhang, M. J. Cryan, "Rapid phase transition of a phase-

change metamaterial perfect absorber," Opt. Mater. Express, 2013, 3, 1101
[7] N. T. R. N. Kumara, Y. F. Chou Chau, J. W. Huang, H. J. Huang, C. T. Lin and H. P. Chiang, "Plasmonic spectrum on 1D and 2D periodic arrays of rod-shape metal nanoparticle pairs with different core patterns for biosensor and solar cell applications," J. Opt., vol.18, 2016, 115003.
[8] Y. Z. Ho, et al. "Tunable plasmonic resonance arising from broken-symmetric silver nanobeads with dielectric cores," J. Optics, vol. 14, 2012, 114010.
[9] S. Li, J. Gao, X. Cao, Z. Zhang, Y. Zheng, C. Zhang, "Multiband and broadband polarization - insensitive perfect absorber devices based on a tunable and thin double split-ring metamaterial," Opt. Express, **2015**, 23, pp.3523-3533.
[10] Lai C H et al.," Near infrared surface-enhanced Raman scattering based on starshaped gold/silver nanoparticles and hyperbolic metamaterial, Sci. Rep., **2017**, **7** 5446.
[11] Z. Yong, S. Zhang, C. Gong, S. He, "Narrow band perfect absorber for maximum localized magnetic and electric field enhancement and sensing applications," Sci. Rep. 2016, 6, 24063.
[12] Y. F. Chou Chau, "Investigation of plasmonic effects on the metal nanoparticle arrays for biosensor applications," IOP Conf. Series: Materials Science and Engineering, 191, 2017, 012016, doi:10.1088/1757-899X/191/1/012016
[13] Y. F. Chou Chau, C. K. Wang, L. Shen, C. M. Lim, H. P. Chiang, C. T. Chou Chao, H. J. Huang, C. T. Lin, N. T. R. N. Kumara and N. Y. Voo, "Simultaneous realization of high sensing sensitivity and tunability in plasmonic nanostructures arrays," Sci. Rep., vol. 7, 2017, 16871, doi:0.1038/s41598-017-17024-7
[14] M. S. Kwon, B. Ku, Y. Kim, "Plasmofluidic Disk Resonators," Sci. Rep. 2016, 6, 23149.
[15] H. J. Huang, B. H. Liu, J. Su, C. T. Lin, H. P. Chiang, T. S. Kao, Y. F. C. Chau, C. C. Kai. C. H. Hwang, "Light energy transformation over a few nanometers," Journal of Physics D: Applied Physics, vol. 50, 2017, pp.375601.

A Comparative Study of Visible Raman Scattering of Ceria Prepared by Sol-gel and Hydrothermal Techniques with Gold Nanoparticles

Muhammad Nur Syafi'ie Md Idris[1,*], Hai-Pang Chiang[2,3], Noormariah Muslim[1], Yuan-Fong Chou Chau[1], Abdul Hanif Mahadi[1], Nyuk Yoong Voo[1], and Chee Ming Lim[1]

[1]Centre for Advanced Material and Energy Sciences, Universiti Brunei Darussalam, Jalan Tungku Link, Gadong, BE 1410, Brunei Darussalam
[2]Institute of Optoelectronic Sciences, National Taiwan Ocean University, No. 2 Pei-Ning Rd., 202, Keelung, Taiwan
[3]Institute of Physics, Academia Silica, Taipei 11529, Taiwan

*E-mail: 14h8703@ubd.edu.bn / nursyafiee@hotmail.com

Abstract. Nano-structured ceria was synthesized by using sol-gel and hydrothermal techniques. Prior to measurements by Raman spectroscopy, the prepared ceria was dispersed onto glass substrates by spin-coating, and this was followed by the deposition of gold nano-particles onto the surface of the ceria by direct current magnetron sputtering. The F_{2g} Raman peak, oxygen vacancy and longitudinal optical modes of ceria were successfully observed with visible Raman laser excitation wavelength of 532 nm at non-resonance Raman conditions. The surface oxygen adsorption on ceria surface was also detected. The F_{2g} Raman peaks of ceria for all of the samples were red-shifted because of the nano-crystalline structure and strain effect on the samples. In the case of the hydrothermally processed gold coated ceria, the F_{2g} Raman peak was further red-shifted to 451 cm^{-1} because of temperature effect and anharmonicity, and the peak intensity was also found to be enhanced by at least one order of magnitude (x13). This increase in peak intensity was attributed to the electrical field enhancement via localized surface plasmon resonance of the gold nano-particles.

1. Introduction

Cerium dioxide (ceria, CeO_2) has been widely known to have variations of defects in its lattice structure. The energy calculations show that these defects can be allocated into four particular groups and these are; (a) oxygen vacancy site, (b) anion Frenkel pairs, (c) interstitial sites, and (d) Schottky disorders [1-3]. Amongst these defects, the most attractive to our work is the oxygen vacancy site (i.e. one of the lattice position experience an oxygen atom lost) which are essential for oxygen storage [4, 5], oxygen reduction[6], water splitting [7-10], thermal conductivity [11] and carbon monoxide oxidation [12, 13] applications.

UV Raman spectroscopy has been extensively used to identify oxygen vacancy site in ceria because of the strong absorption of UV light and the effect of Raman resonance which provides high probing sensitivity [14-16]. The UV excitation activates overtone bands of second-order longitudinal optical (2LO) mode of ceria which are caused by multi-phonon excitation at resonance. The observations of oxygen vacancy and 2LO mode are rarely reported in visible Raman spectroscopy because of the weak absorption and non-resonance Raman conditions.

However, in this study, by dispersing ceria on glass substrates, prominent Raman spectra of ceria oxygen vacancy and 2LO were observed. In addition, ceria was synthesized by using two different techniques, i.e. sol-gel (SG) and hydrothermal (HT) for comparison purposes. Besides, in order to study the effect of gold on the Raman spectra of ceria, gold nano-particles (NPs) were deposited on top of ceria by sputtering method. A finite element method was performed to demonstrate the electrical field intensity distribution of ceria with gold NPs. Crystal structure, surface morphology, and optical properties of the samples were also investigated.

2. Experimental Section

2.1. Preparation of Ceria by Sol-Gel (SG) Method

Cerium (III) nitrate hexahydrate ($Ce(NO_3)_3.6H_2O$, 98%) and ethanol (C_2H_5OH, absolute, 99.9%) from Merck Millipore were used in the sol-gel method for the synthesizations of ceria. Initially, an aqueous solution was prepared by dissolving 5 g of $Ce(NO_3)_3.6H_2O$ in 10 ml of ethanol inside a beaker. Then, the beaker was covered with aluminium foil. The solution was continuously stirred for 20 h in ambient atmosphere. A small portion of the solution was dropped onto a flat glass substrate and heated in air at 100 °C for 1 h and 150 °C for 20 min on a hot plate. As a result, chalky foam was produced which was then dried at 150 °C for 1 h and 250 °C for 10 min. In order to obtain the powder, the dried foam was crushed by using an agate mortar and pestle. Finally, the powder was calcined at 500 °C for 1 h (2 °C/min). The powder was labelled as ceria-SG.

2.2. Preparation of Ceria by Hydrothermal (HT) Method

In this method, cerium (III) nitrate hexahydrate ($Ce(NO_3)_3.6H_2O$, 98%) and sodium hydroxide (NaOH, 93.0%, Merck Millipore) were used for the ceria production. At first, 7.282 g of NaOH was dissolved in 30 ml of distilled water. Then, 0.746 g of $Ce(NO_3)_3.6H_2O$ was added into the solution and stirred for 30 minutes. The solution with the precipitate was then transferred to an autoclave and heat-treated at 100 °C for 24 h. As the autoclave was cooled down to room temperature, the excess solution was poured off to retrieve the solid. It was filtered and washed with distilled water until neutral pH was reached, and finally dried at 100 °C for 24 h. The solid was in the form of powder and labelled as ceria-HT.

2.3. Deposition of Gold NPs onto Ceria Surfaces

The prepared ceria-SG (50 mg) was ultrasonicated in ethanol (5 ml) for 30 min (step 1). The suspension (0.1 ml) was then dropped onto cleaned soda lime glass substrate (step 2) and spin-coated for 1 min at spinning rate of 2000 r.p.m (step 3). The dispersed ceria was left to dry for 1 min (step 4). These steps (2-4) were repeated for 3 times on the same surface. The same procedure was also employed for ceria-HT. All of the dispersed ceria on the glass substrates was dried in an oven at 100 °C for 4 h. Gold was then deposited onto the surface of ceria via direct current (DC) magnetron sputtering technique (SPI Module, Sputter Coater) at DC supply of about 7.0 – 9.0 mA and pressure of 10^{-1} Torr at room temperature with deposition time of 90 s. The samples were named as gold-ceria-SG and gold-ceria-HT, respectively.

2.4. Sample Characterization

X-ray diffraction (XRD-7000, Shimadzu) was used to analyze the crystal structure of the samples with monochromatic high-intensity CuKα (λ=1.54056 Å) radiation operated at 40 kV and 30 mA. The samples were scanned at a rate of 2°/min in the range of 2θ = 10–80° and at incident angle of 1.0°. A JEOL JSM-7610F field emission scanning electron microscopy (FE-SEM) was used to obtain the image of the surface of the sample. The images were captured at magnifications of x50,000 with an accelerating voltage of 5.0 kV and 7.0 kV. The measurements of optical absorbance of the samples were conducted by using a UV-Vis-NIR spectrophotometer (CARY 5000) in wavelength range of 300 nm to 800 nm. Moreover, a confocal micro-Raman spectrometer (CRM200, WITec) with a semiconductor laser excitation of 532 nm at 30 mW was used to record the Raman spectra of the samples at room temperature. A computational simulation by using finite element method (FEM) was conducted [17-20].

3. Results and Discussion

3.1. XRD and FE-SEM Analysis of Ceria and Gold NPs

The XRD patterns of dispersed ceria coated with gold NPs on glass substrates are presented in Fig.1. All samples indexed to the CeO_2 with cubic fluorite structure, as certified with the standard ICDD: card no. 00-043-1002. The diffraction peaks of ceria at 2θ: 28.6° (111), 32.6° (200), 47.6° (220) and 56.3° (311) were identified for ceria-SG and gold-ceria-SG. In the case of ceria-HT and gold-ceria-HT, the diffraction peaks were at 2θ: 28.6° (111), 32.7° (200), 47.4° (220) and 56.4° (311). The diffraction peaks of gold appeared at 2θ: 37.9° (111), 44.1° (200) and 64.5° (220) for gold-ceria-SG and at 2θ: 38.1° (111) for gold-ceria-HT (standard ICDD: card no. 00-004-0784). Based on Scherrer equation and line broadening, the average crystallite size of ceria for ceria-SG, gold-ceria-SG, ceria-HT and gold-ceria-HT were approximated to be 13.2 nm, 13.5 nm, 12.9 nm and 12.7 nm, respectively. The estimated average crystallite size of gold was 32.3 nm and 64.6 nm for gold-ceria-SG and gold-ceria-HT, respectively.

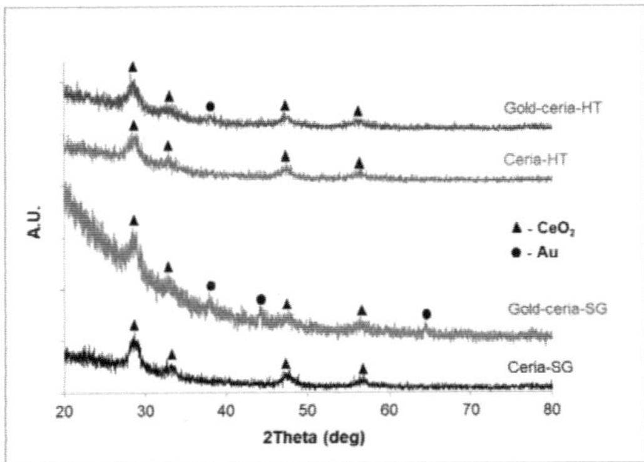

Figure 1. X-Ray Diffraction patterns of: ceria-SG (black); gold-ceria-SG (red); ceria-HT (green); and gold-ceria-HT (blue).

Figure 2. FE-SEM images of: [A] ceria-SG; [B] gold-ceria-SG; [C] ceria-HT; and [D] gold-ceria-HT.

The FE-SEM images of the samples are displayed in Fig.2. The images show the surface morphology of the dispersed ceria. The ceria-SG (Fig.2(A)) has different surface morphology than the ceria-HT (Fig.2(C)). No distinct profile was observed for the surface of ceria-SG, while ceria-HT was found to be in cauliflower-like surface. In general, metal was found to nucleate in the oxygen vacancy sites in fluorite structure [21]. Moreover, high concentration of oxygen vacancies contributes to the high dispersion of gold [22] that leads to the formation of smaller gold NPs. Thus, in this work, the average crystallite size of gold for gold-ceria-SG (32.3 nm) was smaller than that of gold-ceria-HT (64.6 nm).

3.2. UV-Vis Absorbance and Raman Spectra Analysis of Ceria and Gold NPs

The absorption spectra of the samples in the range of 300 nm to 800 nm are presented in Fig.3. Ceria coated with gold NPs showed an increase in the absorption compared to un-coated ceria. Furthermore, the absorption of ceria-SG and gold-ceria-SG (Fig.3(A)) was higher than the ceria-HT and gold-ceria-HT (Fig.3(B)). Despite of the low absorption of ceria-HT, it showed an increase in absorption in the visible range (grey region) due to gold-ceria interactions [23].

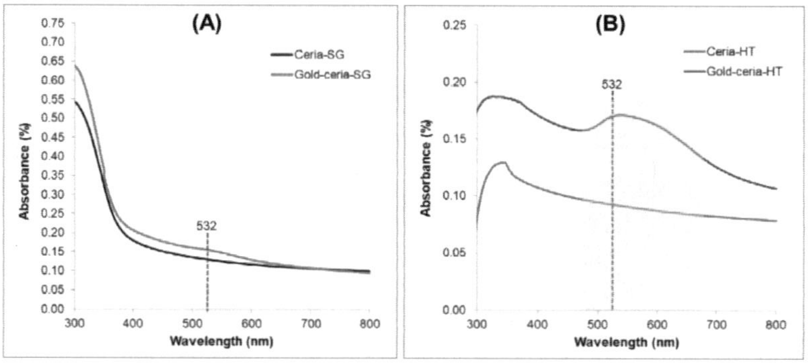

Figure 3. UV-Vis absorption spectra of: (A) ceria-SG (black); gold-ceria-SG (red); and (B) ceria-HT (green); gold-ceria-HT (blue).

Fig.4 shows the Raman spectra of the samples measured at laser excitation wavelength of 532 nm with power of 30 mW. In Fig.4(A), the Raman spectra of ceria-SG showed several Raman peaks centered at 461 cm^{-1}, 558 cm^{-1}, 783 cm^{-1} and 1098 cm^{-1}, whereas peaks for gold-ceria-SG were at 457 cm^{-1}, 557 cm^{-1}, 781 cm^{-1} and 1100 cm^{-1}. These Raman peaks have been corresponded to the first-order F_{2g} Raman mode for cubic ceria [24], oxygen vacancy defects (D) [25], surface oxygen adsorption [15] and second-order longitudinal optical (2LO) mode [14, 26], respectively. In the case of ceria-HT as shown in Fig.4(B), the peaks were at 457 cm^{-1}, 555 cm^{-1}, 785 cm^{-1}, and 1098 cm^{-1}, while peaks for gold-ceria-HT were at 451 cm^{-1}, 567 cm^{-1}, 763 cm^{-1} and 1101 cm^{-1}. In this study, visible Raman laser was used to reveal the oxygen vacancy and 2LO spectra. This could be made possible because of the dispersion of ceria on the glass substrate which increased the exposure of the oxygen vacancy site and allowed the activation of the 2LO Raman mode. Hence, this method could improve the detection sensitivity of visible Raman spectroscopy of ceria.

The F_{2g} Raman peak, full-width half maximum (FWHM), intensity, as well as, the defects (D) peak (oxygen vacancy) of the samples are tabulated in Table 1. In general, the F_{2g} Raman peak of bulk ceria has been reported to be at 464 cm^{-1}[27]. Nonetheless, all of the samples showed a decrease (red-shift) in the F_{2g} Raman peak of ceria. This has been attributed to the nano-structured ceria and strain effects [28]. In addition, the further red-shift of the F_{2g} Raman peak of ceria to 451 cm^{-1} for gold-ceria-HT (Fig.4(B)) was caused by temperature effects and anharmonicity [29], which originated from plasmon heating by localized surface plasmon resonance (LSPR) decay of the gold NPs [30]. Apart from that, there was a significant increase in the intensity of the F_{2g} Raman peak of gold-ceria-HT when compared to ceria-HT, i.e. from 259 a.u. to 3542 a.u. which was 13 times higher than ceria without gold NPs. The gold NPs amplified the Raman signals as a result of surface-enhanced Raman scattering (SERS) from the LSPR of gold NPs that occurred at the laser excitation wavelength of 532 nm. Furthermore, an increase in the absorption of visible spectra was observed, as shown in the grey region of Fig.3(B), where the Raman laser excitation wavelength of 532 nm was located. In contrast, there was no intensification of the Raman intensity of the F_{2g} Raman peak for gold-ceria-SG. This might be due to the size of the gold NPs which were not favorable for LSPR to occur.

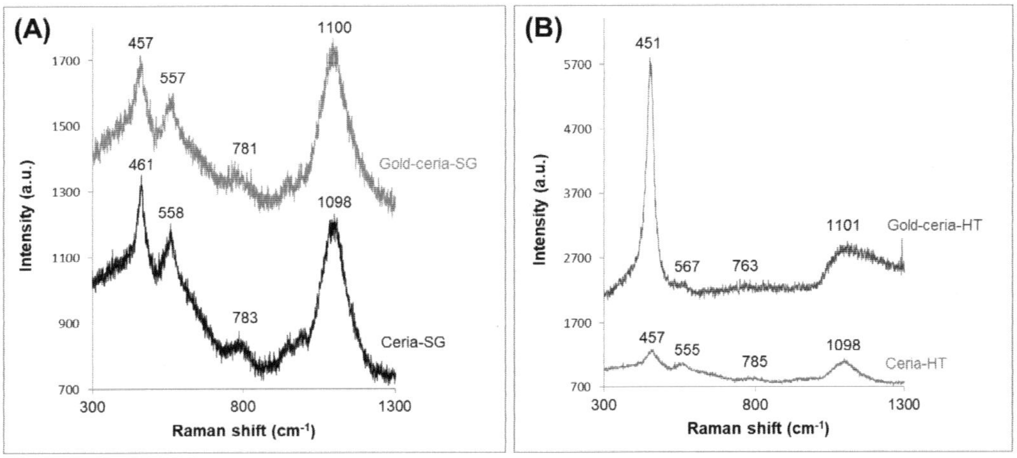

Figure 4. Raman spectra of: (A) ceria-SG (black); gold-ceria-SG (red); and (B) ceria-HT (green); gold-ceria-HT (blue).

Table 1. Average Crystallite Size, F_{2g} Raman Peaks and Intensities, D Raman Peaks and Intensities, and Estimated FWHM Of The Samples.

Sample	CeO_2 Crystallite size (nm)	Au Crystallite size (nm)	F_{2g} Raman Peak (cm^{-1})	F_{2g} FWHM (cm^{-1})	F_{2g} Raman intensity (a.u.)	D Raman Peak (cm^{-1})	D Raman intensity (a.u.)
Ceria [SG]	13.2	-	461	25.3	296	558	188
Gold-ceria [SG]	13.5	32.3	457	36.7	241	557	176
Ceria [HT]	12.9	-	457	36.7	259	555	132
Gold-ceria [HT]	12.7	64.6	451	29.5	3542	567	-

3.3. Finite element method simulation

In order to analyze the existence of the enhanced electrical field (E-field) intensity distributions (Fig.5(A)), and the surface charge densities (Fig.5(B)) contributed by the gold NPs from gold-ceria-SG (32.3 nm) and gold NPs from gold-ceria-HT (64.6 nm) at the excitation wavelength (λ) of 532 nm with x-polarization, computational simulations were performed by using the finite element method (FEM) with perfect matched layers on the top and bottom surface of the unit cell. The imaginary dielectric constant of gold will give a negative damping effect to the LSPR and the gold permittivity data cited in [31] was used. It could be seen that the enhanced E-fields and surface charges could be excited around the gold NPs and the interface between gold NPs and ceria, which is caused by the LSPR effects [18, 32, 33]. Moreover, the E-field enhancement could also be attributed to gap plasmon resonance (GPR) [19, 34, 35]. Note that the different in the charge pair configuration (Fig.5(B)) is due to different resonant mode arising from different incident resonant wavelength. In addition, the E-field intensity distribution obtained from larger gold NPs (Fig.5(A,ii)) was higher than that of the smaller gold NPs (Fig.5.(A,i)). This is because of the existence of more charge pairs and hence their higher density distribution on the surface of the larger gold NPs (Fig.5(B,ii)). Thus, this E-field enhancement leads to the surface-enhanced Raman scattering (SERS) which is responsible for the intensification of ceria F_{2g} Raman peak of gold-ceria-HT sample.

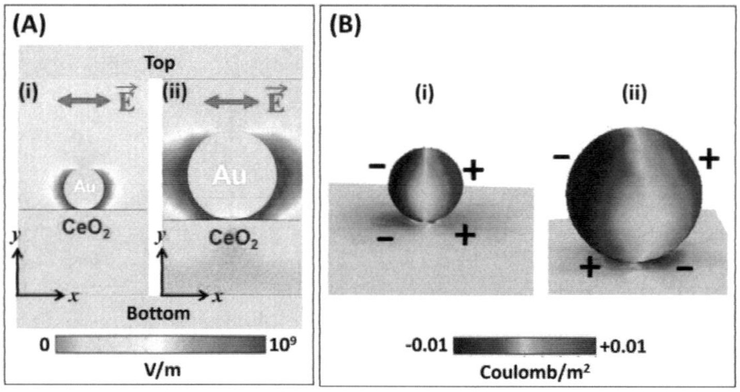

Figure 5. Calculated E-field intensity distributions of gold NPs on ceria for gold-ceria-SG (A,i) and gold-ceria-HT (A,ii); surface charge densities of gold NPs on ceria for gold-ceria-SG (B,i) and gold-ceria-HT (B,ii), excited by an incident light with linearly polarized plane wave (λ= 532 nm).

4. Conclusion

In this study, visible Raman spectroscopy with laser excitation wavelength of 532 nm was demonstrated to detect the F_{2g} Raman peak, oxygen vacancy, oxygen adsorption as well as longitudinal optical mode of ceria prepared by sol-gel and hydrothermal techniques. The dispersion of ceria onto glass substrates assisted in revealing these peaks in non-resonance Raman conditions. Furthermore, the F_{2g} Raman peaks of ceria were red-shifted because of the nano-structured ceria and strain effect. The additional red-shift of ceria prepared by hydrothermal method demonstrated temperature effect and anharmonicity originated from plasmon heating via LSPR decay of gold NPs. Besides, the observation of E-field enhancement attributed to LSPR effect which caused SERS produced a significant amplification of the F_{2g} Raman peak of ceria.

Acknowledgment

Universiti Brunei Darussalam (Grant No. UBD-ORI-URC-RG331-U01 and UBD/OVACRI/CRGWG(004)/170101), and Ministry of Science and Technology of Taiwan (MOST 106-2112-M-019-005-MY3). We also would like to thank Prof. Hai-Pang Chiang from Institute of Optoelectronic Sciences, National Taiwan Ocean University, Keelung 202, Taiwan, for his assistance with the Raman measurement.

References

[1] L. Minervini, M.O. Zacate, and R.W. Grimes, "Defect cluster formation in M_2O_3-doped CeO_2," Solid State Ionics, vol. 116, Jan. 1999, pp. 339-349, doi: https://doi.org/10.1016/S0167-2738(98)00359-2.

[2] A. Walsh, S.M. Woodley, C.R.A. Catlow, and A.A. Sokol, "Potential energy landscapes for anion Frenkel-pair formation in ceria and india," Solid State Ionics, vol. 184, Mar. 2011, pp. 52-56, doi: https://doi.org/10.1016/j.ssi.2010.08.010.

[3] T.X.T. Sayle, M. Molinari, S. Das, U.M. Bhatta, G. Möbus, S.C. Parker, S. Seal, and D.C. Sayle, "Environment-mediated structure, surface redox activity and reactivity of ceria nanoparticles," Nanoscale, vol. 5, May 2013, pp. 6063-6073, doi: https://doi.org/10.1039/C3NR00917C.

[4] E. Mamontov, T. Egami, R. Brezny, M. Koranne, and S. Tyagi, "Lattice Defects and Oxygen

Storage Capacity of Nanocrystalline Ceria and Ceria-Zirconia," The Journal of Physical Chemistry B, vol. 104, Oct. 2000, pp. 11110-11116, doi: https://doi.org/10.1021/jp0023011.
[5] H. He, H.X. Dai, and C.T. Au, "Defective structure, oxygen mobility, oxygen storage capacity, and redox properties of RE-based (RE = Ce, Pr) solid solutions," Catalysis Today, vol. 90, July 2004, pp. 245-254, doi: https://doi.org/10.1016/j.cattod.2004.04.033.
[6] K. Sudarshan, S.K. Sharma, R. Gupta, S.K. Gupta, F.N. Sayed, and P.K. Pujari, "Role of surface defects in catalytic properties of CeO_2 nanoparticles towards oxygen reduction reaction," Materials Chemistry and Physics, vol. 200, Oct. 2017, pp. 99-106, doi: https://doi.org/10.1016/j.matchemphys.2017.07.064.
[7] J.A. Rodriguez, S. Ma, P. Liu, J. Hrbek, J. Evans, and M. Pérez, "Activity of CeO_x and TiO_x nanoparticles grown on Au(111) in the water-gas shift reaction," Science, vol. 318, Dec. 2007, pp. 1757-1760, doi: https://doi.org/10.1126/science.1150038.
[8] J. Shi, A. Schaefer, A. Wichmann, M.M. Murshed, T.M. Gesing, A. Wittstock, and M. Bäumer, "Nanoporous gold-supported ceria for the water–gas shift reaction: UHV Inspired Design for Applied Catalysis," The Journal of Physical Chemistry C, vol. 118, July 2014, pp. 29270-29277, doi: https://doi.org/10.1021/jp505433a.
[9] M. Molinari, S.C. Parker, D.C. Sayle, and M.S. Islam, "Water adsorption and its effect on the stability of low index stoichiometric and reduced surfaces of ceria," The Journal of Physical Chemistry C, vol. 116, Feb. 2012, pp. 7073-7082, doi: https://doi.org/ 10.1021/jp300576b.
[10] S. Gangopadhyay, D.D. Frolov, A.E. Masunov, and S. Seal, "Structure and properties of cerium oxides in bulk and nanoparticulate forms," Journal of Alloys and Compounds, vol. 584, Jan. 2014, pp. 199-208, doi: https://doi.org/10.1016/j.jallcom.2013.09.013.
[11] H. Muta, H. Kado, Y. Ohishi, K. Kurosaki, and S. Yamanaka, "Effect of oxygen defects on thermal conductivity of thorium-cerium dioxide solid solutions," Journal of Nuclear Materials, vol. 483, Jan. 2017, pp. 192-198, doi: https://doi.org/10.1016/j.jnucmat.2016.10.043.
[12] M. Wilklow-Marnell, and W.D. Jones, "Catalytic oxidation of carbon monoxide by α-alumina supported 3nm cerium dioxide nanoparticles," Molecular Catalysis, vol. 439, Oct. 2017, pp. 9-14, doi: https://doi.org/10.1016/j.mcat.2017.06.015.
[13] M. Lohrenscheit, and C. Hess, "Direct evidence for the participation of oxygen vacancies in the oxidation of carbon monoxide over ceria-supported gold catalysts by using operando Raman spectroscopy," ChemCatChem, vol. 8, Feb. 2016, pp. 523-526, doi: https://doi.org/10.1002/cctc.201501129.
[14] T. Taniguchi, T. Watanabe, N. Sugiyama, A.K. Subramani, H. Wagata, N. Matsushita, and M. Yoshimura, "Identifying defects in ceria-based nanocrystals by UV resonance Raman spectroscopy," The Journal of Physical Chemistry C, vol. 113, Oct. 2009, pp. 19789-19793, doi: https://doi.org/10.1021/jp9049457.
[15] Z. Wu, M. Li, J. Howe, H.M. Meyer III, and S.H. Overbury, "Probing defect sites on CeO_2 nanocrystals with well-defined surface planes by Raman spectroscopy and O_2 adsorption," Langmuir, vol. 26, July 2010, pp. 16595-16606, doi: https://doi.org/10.1021/la101723w.
[16] S. Agarwal, X. Zhu, E.J.M. Hensen, L. Lefferts, and B.L. Mojet, "Defect chemistry of ceria nanorods," The Journal of Physical Chemistry C, vol. 118, Feb. 2014, pp. 4131-4142, doi: https://doi.org/10.1021/jp409989y.
[17] Y.F. Chou Chau, C.M. Lim, C. Lee, H.J. Huang, C.T. Lin, N.T.R.N. Kumara, V.N. Yoong, and H.P. Chiang, "Tailoring surface plasmon resonance and dipole cavity plasmon modes of scattering cross section spectra on the single solid-gold/gold-shell nanorod," Journal of Applied Physics, vol. 120, Sept. 2016, pp. 093110, doi: https://doi.org/10.1063/1.4962175.
[18] M.W. Chen, Y.F. Chau, and D.P. Tsai, "Three-dimensional analysis of scattering field interactions and surface plasmon resonance in coupled silver nanospheres," Plasmonics, vol. 3, Dec. 2008, pp. 157-164, doi: https://doi.org/10.1007/s11468-008-9069-8.
[19] Y.F. Chau, and Z.H. Jiang, "Plasmonics effects of nanometal embedded in a dielectric

substrate," Plasmonics, vol. 6, Sept. 2011, pp. 581-589, doi: https://doi.org/10.1007/s11468-011-9238-z.
[20] K.Y. Yang, Y.F. Chau, Y.W. Huang, H.Y. Yeh, and D.P. Tsai, "Design of high birefringence and low confinement loss photonic crystal fibers with five rings hexagonal and octagonal symmetry air-holes in fiber cladding," Journal of Applied Physics, vol. 109, May 2011, pp. 093103, doi: https://doi.org/10.1063/1.3583560.
[21] M.G. Sanchez, and J.L. Gazquez, "Oxygen vacancy model in strong metal-support interaction," Journal of Catalysis, vol. 104, Mar. 1987, pp. 120-135, doi: https://doi.org/10.1016/0021-9517(87)90342-3.
[22] W.Y. Hernández, F. Romero-Sarria, M.A. Centeno, and J.A. Odriozola, "In situ characterization of the dynamic gold–support interaction over ceria modified Eu3+. Influence of the oxygen vacancies on the CO oxidation reaction," The Journal of Physical Chemistry C, vol. 114, Jun. 2010, pp. 10857-10865, doi: https://doi.org/10.1021/jp1013225.
[23] S. Bhagat, N.V.S. Vallabani, V. Shutthanandan, M. Bowden, A.S. Karakoti, and S. Singh, "Gold core/ceria shell-based redox active nanozyme mimicking the biological multienzyme complex phenomenon," Journal of Colloid and Interface Science, vol. 513, Mar. 2018, pp. 831-842, doi: https://doi.org/10.1016/j.jcis.2017.11.064.
[24] Y. Lee, G. He, A.J. Akey, R. Si, M. Flytzani-Stephanopoulos, and I.P. Herman, "Raman analysis of mode softening in nanoparticle $CeO_{2-\delta}$ and Au-$CeO_{2-\delta}$ during CO oxidation," Journal of the American Chemical Society, vol. 133, July 2011, pp. 12952-12955, doi: https://doi.org/10.1021/ja204479j.
[25] K. Sambasivudu, Y.B. Reddy, J.S. Yadav, G. Sabitha, and D. Shailaja, "Ceria-supported vinylpyridine polymers: synthesis, characterization and application in catalysis," International Journal of Polymeric Materials and Polymeric Biomaterials, vol. 57, July 2008, pp. 891-903, doi: https://doi.org/10.1080/00914030802153199.
[26] W.H. Weber, K.C. Hass, and J.R. McBride, "Raman study of CeO_2: Second-order scattering, lattice dynamics, and particle-size effects," Physical Review B, vol. 48, July 1993, pp. 178-185, doi: https://doi.org/10.1103/PhysRevB.48.178.
[27] J.E. Spanier, R.D. Robinson, F. Zhang, S.W. Chan, and I.P. Herman, "Size-dependent properties of CeO_{2-y} nanoparticles as studied by Raman scanttering," Physical Review B, vol. 64, Nov. 2001, pp. 245407, doi: https://doi.org/10.1103/PhysRevB.64.245407.
[28] Z.V. Popović, Z. Dohčević-Mitrović, M. Šćepanović, M. Grujić-Brojčin, and S. Aškrabić, "Raman scattering on nanomaterials and nanostructures," Annalen der Physik, vol. 523, Jan. 2011, pp. 62-74, doi: https://doi.org/10.1002/andp.201000094.
[29] Z.V. Popović, Z. Dohčević-Mitrović, A. Cros, and A. Cantarero, "Raman scattering study of the anharmonic effects in CeO_{2-y} nanocrystals," Journal of Physics: Condensed Matter, vol. 19, Nov. 2007, pp. 496209, doi: https://doi.org/10.1088/0953-8984/19/49/496209.
[30] J.R. Adleman, D.A. Boyd, D.G. Goodwin, and D. Psaltis, "Heterogenous catalysis mediated by plasmon heating," Nano Letters, vol. 9, Nov. 2009, pp. 4417-4423, doi: https://doi.org/10.1021/nl902711n.
[31] P. B. Johnson, and R. W. Christy, "Optical constants of the noble metals," Physical Review B, vol. 6, Dec. 1972, pp. 4370-4379, doi: https://doi.org/10.1103/PhysRevB.6.4370.
[32] Y.F. Chau, Z.H. Jiang, H.Y. Li, G.M. Lin, F.L. Wu, and W.H. Lin, "Localized resonance of composite core-shell nanospheres, nanobars and nanospherical chains," Progress In Electromagnetics Research B, vol. 28, 2011, pp. 183-199, doi: https://doi.org/10.2528/PIERB10102705.
[33] Y.F. Chau, H.H. Yeh, and D.P. Tsai, "Surface plasmon resonances effects on different patterns of solid-silver and silver-shell nanocylindrical pairs," Journal of Electromagnetic Waves and Applications, vol. 24, 2010, pp. 1005-1014, doi: https://doi.org/10.1163/156939310791586098.
[34] Y.F. Chau, and H.H. Yeh, "A comparative study of solid-silver and silver-shell nanodimers on

surface plasmon resonances," Journal of Nanoparticle Research, vol. 13, Feb. 2011, pp. 637-644, doi: https://doi.org/10.1007/s11051-010-0058-4.
[35] Y.F. Chau, "Surface plasmon effects excited by the dielectric hole in a silver-shell nanospherical pair," Plasmonics, vol. 4, Dec. 2009, pp. 253, doi: https://doi.org/10.1007/s11468-009-9100-8.

IOP Conf. Series: Materials Science and Engineering **409** (2018) 012013 doi:10.1088/1757-899X/409/1/012013

Shock Wave Effect on Micromechanics Properties of Sn-Ag-Cu solder (SAC 387) via Nanoindentation

Nur Shafiqa Safee[1], Wan Yusmawati Wan Yusoff[1*], Ariffin Ismail[1], Norliza Ismail[2], Emee Marina Salleh[2], Maria Abu Bakar[3], Azman Jalar[2,3]

[1]Center for Defence Foundation Studies, Universiti Pertahanan Nasional Malaysia, Kem Sg Besi,57000 Kuala Lumpur, Malaysia
[2]School of Applied Physics, Faculty of Science and Technology,
[3]Institute of Microengineering and Nanoelectronics, Universiti Kebangsaan Malaysia, 43600 Bangi, Selangor, Malaysia,

E-mails: nshafiqasafee@gmail.com; yusmawati@upnm.edu.my*; ariffin@upnm.edu.my; iza_nza@yahoo.com; emme@ukm.edu.my; mabmarya@gmail.com; azmn@ukm.my

Abstract. A variety of Sn-Ag-Cu solder alloys have been proposed and recommended for use in industrial organizations for example Sn-95.5Ag3.8Cu-0.7 (SAC 387) by the European Consortium Brite-Euram in the European Union. In this study an effect of shock wave on micromechanics properties of SAC 387 solder was investigated via nanoindentation approach. Solder joint of SAC 387 on copper substrate was prepared by hand soldering techniques. Then soldered samples were exposed to shock wave at different distance of 1 m, 2 m and 3 m. The shock wave test was conducted using 500 g of trinitrotoluene (TNT). Hardness value of SAC 387 decreased from 300 MPa to 130 MPa as well as reduced modulus drop from 115 GPa to 43 GPa after exposure. However the hardness and reduced modulus increased with increasing distance of the shock wave. The phenomenon revealed that shock wave deteriorated the micromechanics properties of soldered samples.

1. Introduction

Lead-free solder has been spread extensively in manufacturing many electronic applications due to the toxicity of lead for human health. A Sn-Ag-Cu ternary alloy is recommended to be a replacement of a Sn-Pb eutectic solder which has been used in electronic device. It's due to Sn-Ag-Cu ternary solder has superb mechanical properties compared to Sn-Pb eutectic solder which are creep resistance, fatigue resistance, thermal resistance and tensile strength [1].

The mechanical stability of solder interconnects during dynamic conditions is crucial for high reliability electronic device especially in military in order to fulfill a vital role in the success of a mission [2]. The demand for complexity of electronic equipment such as low volume and highly sophisticated especially in military has resulted in an increase need to understand the failure mechanism dynamic load including drop, vibration and shock event.

A shock wave within a distance of about ten mean-free-path length was characterized by the very rapid increase, of all the physical properties, of the ambient gas, namely, the temperature, entropy density, particle velocity, and hydrostatic pressure. Based on previous research study by Wang et. al, there are many advantages of high pressure in reducing atom diffusion coefficient, to refine of grain and crystal structure, facilitate nucleus formation, and resulting in restrain in grain growth. Today, high pressure has been appealing researchers' huge attention because it can enhance microstructure of alloys [3-4]. The shock wave expands spherically and achieves pressure peaks for forming processes in the range of some mega Pascal [5]. Research conducted by Trdan et. al [6] revealed that shock wave refined the grain size of Al-Mg-Si alloy. Meanwhile Aydogan et. al [7], found shock loading increase the dislocation density and indirectly increase the hardness of HT-9 ferritic/martensitic steels.

The development of machine using nanoindentation method gained popularity that were capable by recording very small loads, displacements to a high level of accuracy and precision [8]. Nanoindentation employs an indirect measurement of the contact area between indenter and specimen in order to determine the material properties, namely, hardness and reduced modulus. Due to the miniaturization of device and needed smaller size of solder, conventional mechanical test such as tensile test, shear test and macro hardness test is not relevant to be used. One of the promising and power tools for determining the mechanical properties of solder is nanoindentation. This method gave an advantage of localized micromechanical properties in addition to other mechanical properties in conventional stress-strain data [9].

Therefore in this paper, effect of shock wave to micro mechanical properties of SAC 387 lead free solder were determined by using nanoindentation approach. There is only a little literature reported on lead free solder reliability due to the shock wave impact. This study becomes a great interest due to application of solder in military electronic devices and importance of their reliability properties.

2. Experimental Procedure

2.1. Sample Preparation

The lead-free solder wire Sn-95.5Ag-3.8Cu-0.7 (SAC 387) with diameter size of 0.6 mm supplied by Redring Solder (M) Sdn. Bhd. and a copper substrate with 2 cm × 1 cm × 0.3 cm dimensions had been used in this experiment. The copper substrates were cleaned with acetone in ultrasonic bath for 5 min before starting the soldering process. The solder wire was soldered on a substrate at 100°C on electronic hot plate and cool down at room temperature [10]. Then the samples were place at distance 1 m, 2 m and 3 m from shock wave using 500 g of TNT equivalent and the non-exposure sample was used as control. The samples were cross sectioned and go through metallographic techniques. The samples were then cold mounted using epoxy resin followed by grinding using silicon carbide abrasive paper from 400 until 1200 grits and lastly polished with DP Nap polishing cloth, liquid lubricant and 1μm of diamond spray for nanoindentation test [11].

2.2. Nanoindentation

The nanoindenter machine (Micro Materials, NanoTest^{TM}) with Berkovich tip was used to perform the nanoindentation test at room temperature as shown in figure 1. Berkovich tip is preferred because a three-sided pyramid is more easily ground to a sharp point, and suitable for small scale indentation. Normally the face angle for Berkovich indenter is 65.27°. The indentation test was run at the centre of solder crossed section. The rate of 0.5 mN/s for a constant of loading and unloading was adapted to the sample surface until achieved at 10 mN (maximum load). At maximum load the dwell time is 30 second followed by the unloading process [11].

The graph from this experiment shows the hysteresis of load versus depth obtaining a means to determine hardness and modulus of the sample material, but also can identify the non linear events for example cracking, phase transformation and delamination of samples. This hysteresis consists of three segments, which are loading, dwell time at target load, and the unloading processes. These types of tests allow the measurement of viscoelastic properties of the specimen material. The thermal drift correction is 60 second hold time was applied at 90% unloading. The unloading curve is conceding as the elastic recovery of the material. The Poisson's ratio value of indenter used in this experiment is 0.07 and Young's modulus of 1140 GPa.

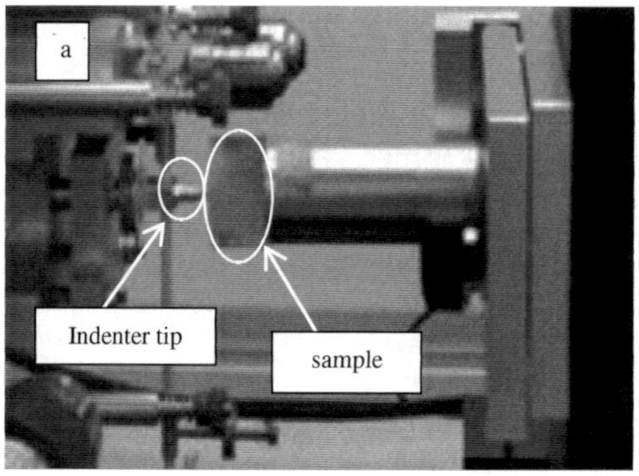

Figure 1. Nanoindentation test

3. Results And Discussion

3.1. Depth, Displacement and Hardness Analysis

Semi- analytical procedures which is the experimental *P-h* response has been proposed by various scientists to measure pseudo properties like characterizing resistance to deformation such as hardness *H*, plastic properties for example yield stress % and strain hardening exponent *n* and elastic properties including elastic modulus *E* [12]. Figure 2 shows the load versus depth of the SAC 387 alloy before (control) and after expose to shock wave with different distances. From this figure it shows that the sample at 2 m and 3 m distance consist of pop in and there is no superposition at 1 m sample indicating that the elastic deformation takes place inside the alloy. During the loading step in the indentation depth, these serrated jumps generally more pronounced Cu_6Sn_5 compare to Cu_3Sn. Pop in event have been discovered in epitaxial ZnSe on GaAs, single-crystal metals (Al, Ni, W), ionic cystals (BaF_2, CaF_2), semiconductors (Si, GaAs, InP), and attributed to the nucleation of dislocation [13].

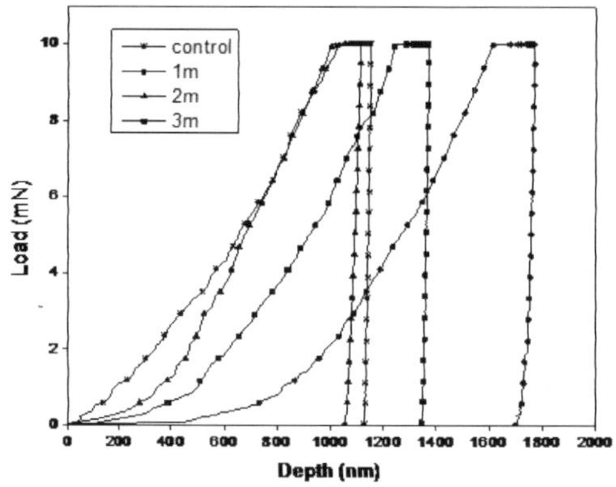

Figure 2. *P-h* profile of SAC 387 solder joint

Figure 3 (a) show the hardness values of SAC 387 after exposed to the shock wave at distance of 1 m, 2 m and 3 m. From the results, hardness of soldered samples obviously decreases from 300 MPa for control sample to 130 MPa, 170 MPa and 210 MPa for 1 m, 2 m and 3 m distance samples respectively. It's revealed that shock wave degrade hardness of soldered samples. Besides, the distance of shock wave also influenced hardness by decreasing the values. Generally hardness *(H)* is a measure of a material's resistance to deformation by surface indentation. The motion of dislocations in the atomic structure of a material will caused plastic deformation. By inhibiting dislocation motion through grain boundaries, imperfections or alloys, the yield strength of material may be changed. Material hardness can be increased by different means, including interstitial or substitution hardening, where atoms are either added between or substituted in the atomic lattice. Ignatova et. al [14] state that by increasing of shock wave intensity (pressure) lead to a decrease the dislocation density. The grain size also increase due to increasing intensity of shock wave and reduction of mechanism of fine grained copper to the level of coarse crystalline copper [14]. In Sn-Ag-Cu solder the primary phase formed is β-Sn phase followed by eutectic phase and intermetallic Ag_3Sn and Cu_6Sn. Therefore coarsening of grain size or sub grain of β-Sn tend to soft and resulted in low hardness value.

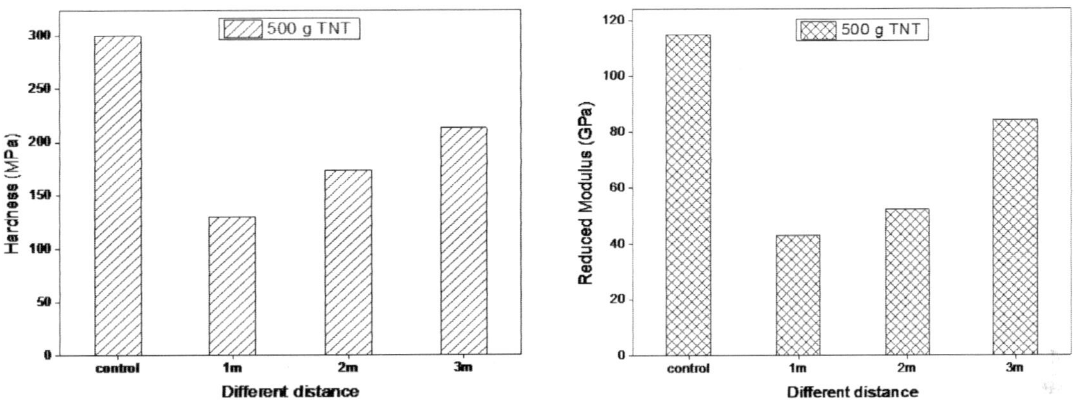

Figure 3.(a) Hardness and (b) Reduced Modulus of SAC 387 before and after shock wave.

3.2. Reduced Modulus Analysis

As reported by Suhling et. al [15], the modulus of elasticity corresponds to the inter atomic forces between adjacent atoms was illustrated by Hertzberg. Figure 3(b) shows the reduced modulus of SAC 387 for control sample which is 115 GPa higher than the samples of different distance from shock wave. On the other hand, the reduced modulus for different distance from shock wave increase rapidly which are 43 GPa for 1 m followed by 52 GPa, and 84 GPa, at 2 m and 3 m distance respectively. Under normal conditions, the atoms exhibit an equilibrium position to balance the attraction and repulsion forces. There is no broken in interatomic bond when an external force in shock wave is applied within the elastic region and only the balance of the attraction and repulsion forces changes [16]. The atoms will return to their original equilibrium positions after the external force is relieved.

3.3. Plastic and Elastic Work Analysis

Figure 4 the plastic work (*Wp*) and elastic work (*We*) data showed that, in all cases, the plastic work was more than the elastic recovery. All the plastic and elastic works were in the range of 0-10 nJ. The value of plastic work before shock wave is 5.28 nJ which is higher compare to the samples of different distance from shock wave which are 5.27 nJ, 4.48 nJ and 4.90 nJ for 1 m, 2 m and 3 m. However the value of elastic work for control sample is 0.09 nJ which is lower than the sample subjected to shock wave which are 0.21 nJ, 0.14 nJ and 0.11 nJ for 1 m, 2 m, and 3 m respectively. *Wp* and *We* can be correlated with the *H* and *E* values of the materials, respectively. Hence, low plastic work and low elastic recovery effectively translate to materials with high hardness and high Young's modulus [17]. According to Khan et. al [18], the results that obtained by Xu and Li showed that changes in plastic work/total work (*Wp/Wt*) are independent of the strain hardening of the material and directly dependent on the *E/rY* ratio of the material.

IOP Conf. Series: Materials Science and Engineering **409** (2018) 012013 doi:10.1088/1757-899X/409/1/012013

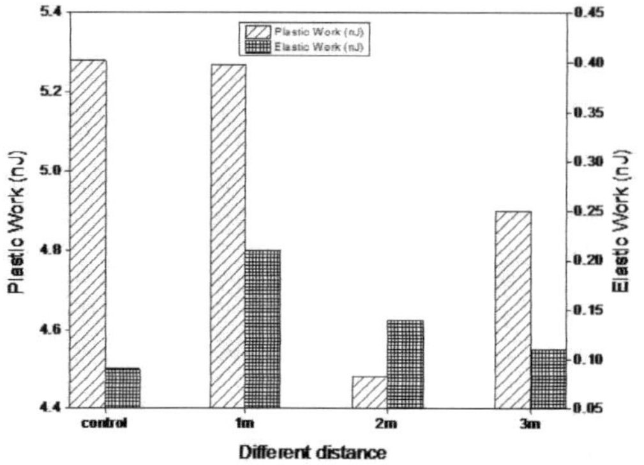

Figure 4. Plastic and Elastic Work before and after shock wave.

4. Conclusion

The micromechanical properties of SAC 387 solder prepared by hand soldering with different distance from shock wave has successfully investigated using nanoindentation technique. The hardness and reduced modulus values increase with increasing distance of shock wave. After being exposed to shock wave the sample becomes softer. Shock wave resulted the microstructure changes of the solder material due to dislocation motion. Future work concerns deeper analysis on microstructure change such as grain boundary effect the failure analysis towards shock wave.

Acknowledgement

The authors are gratefully acknowledged the financial support from FRGS/1/2015/SG06/UPNM/03/1, Universiti Kebangsaan Malaysia (UKM) and Redring Solder (M) Sdn. Bhd. for research materials and collaboration work.

References

[1] I. Shohji, H. Goto, K. Nakamura, & T. Ookubo, "Influence of Surface Finish of Cu Electrode on Shear Strength and Microstructure of Solder Joint with Sn-3Ag-0 . 5Cu," vol. 300, Jan 2005, pp. 2864–2869, doi:10.4028/www.scientific.net/kem.297-300.2864

[2] E. Habtour, B. Werner, A. Hilburn, & H. I. Saraidaridis, "Physics of failure for portable electronic devices in military applications". Proc. Annual Reliability and Maintainability Symposium, January 2015,doi.org/10.1109/RAMS.2013.6517620

[3] Z. X. Wang, J. B. Lu & Y. J. Xi, "Effects of quenching under high pressure on crystallization of $Cu_{60}Zr_{20}Hf_{10}Ti_{10}$ bulk metallic glass" , Mater.Sci. Eng.A vol.499, January 2009, pp.192-194. doi.org/10.1016/j.msea.2007.09.096

[4] Z. X.Wang, F. Y. Li & M. X. Pan,"Effects of high pressure on the nucleation of $Cu_{60}Zr_{20}Hf_{10}Ti_{10}$ bulk metallic glass",J.Alloy.Compd.Vol.388, February 2005, pp.262-265. doi.org/10.1016/j.jallcom.2004.07.025

[5] T. Czotscher, S. Veenaas, F. Vollertsen, "Possibilities to Characterise Laser Induced Shock Waves", Journal for Technology of Plasticity, Vol. 42 2017, Number 1 doi: 10.24867/jtp.2017.42-1.1-7

[6] U. Trdan, M. Skarba, & J. Grum, "Laser shock peening effect on the dislocation transitions and grain refinement of Al–Mg–Si alloy", Materials Characterization, Vol. 9, November 2014, pp. 57-68 doi.org/10.1016/j.matchar.2014.08.020

[7] G. Constantinides, E. C. C. M Silva, G. S. Blackman, & K. J. Van Vliet, "Dealing with Imperfection: quantifying potential length scale Artfact from normally spherical indenter Probes", Nanotechnology pp:18/30 2007 doi.org/10.1088/0957-4484/18/30/305503

[8] E. Aydogan, O. Anderoglu, S. A. Maloy, V. Livescu, G. T. Gray III, S. Perez-Bergquist, D. J. Williams "Effect of shock loading on the microstructure, mechanical properties and grain boundary characteristics of HT-9 ferritic/martensitic steel", Materials Science and Engineering: A vol. 651, 10 January 2016, pp. 75-82, doi.org/10.1016/j.msea.2015.10.102

[9] M. A. Bakar, A. Jalar, A. R. Daud, R. Ismail, N. A. C. Lah, N. S Ibrahim, "Nanoindentation approach on investigating micromechanical properties of joining from green solder materials", Sains Malaysiana vol. 45(8) 2016 pp.1275–1279

[10] N. N. M. Fedzir, W. Y. W. Yusoff, & A. Jalar, "Effect of substrate pre-soldering temperature on wettabillty of SAC387 lead free solder", J. Fundem. Appl. Sci., 2017, 9 (3S), pp.154-166, dx.doi.org/10.4314/jfas.v9i3s.13

[11] M. A. Bakar, A. Jalar, R. Ismail, A. R. Daud, Nanoindentation of Sn3.0Ag0.5Cu/ENIG Solder Joint after High Temperature Storage", Materials Science Forum vol. 857(3) 2016 pp.40–43, doi:10.4028/www.scientific.net/msf.857.40.

[12] G. Constantinides, & F. J. Ulm, & K. Van Vliet, "On the Use of Nanoindentation for Cementitious Materials. Materials and Structures" , vol:36. 2003 pp:191-196,doi:10.1617/14020.

[13] C. A. Schuh, J. K. Masoanda, & C. Lund, "Quantitative insight into dislocation nucleation from high-temperature nanoindentation experiments" , Nat Mater 2005; vol 4 pp:612-617,doi:10.1038/nmat1429.

[14] O. N. Ignatova, I. I. Kaganova, & A. N. Malyshev, "Effect of shock-wave loading on the internal microstructure and mechanical properties of fine- grained copper" , Combustion, Explosion and Shock Waves, vol. 46(6), 2010,pp.719–723.,doi.org/10.1007/s10573-010-0096-3.

[15] H. Ma, & J. C. Suhling, "A review of mechanical properties of lead-free solders for electronic packaging" , J. Mater Science vol. 44(5), March 2009 pp.1141–1158, doi:10.1007/s10853-008-3125-9.

[16] O. Deubzer, H. Hamano & T. Suga, "Lead-free soldering-toxicity, energy and resource consumption" , In: Proc of the 2001 IEEE international symposium on electronics and the environment, May 2001, pp.290–295, doi:10.1109/isee.2001.924542.

[17] A. Dey, & A. K. Mukhopadhyay, "*Nanoindentation of brittle solids*" ,(Boca Raton, FL: CRC Press) 25 June 2014.

[18] M. K. Khan, M. E. Fitzpatrick, S. V. Hainsworth, & L. Edwards, "Effect of residual stress on the nanoindentation response of aerospace aluminium alloys",*Computational Materials Science*, August-september 2011. vol*50*(10), pp.2967-2976. doi.org/10.1016/j.commatsci.2011.05.015

Monitoring of Various Minerals of Flash Furnace Charge and their Impacts on Furnace Operation

Ali Manoochehri, Mahmoud Khodadadi, Mehdi balochi, Roohallah Abbasi, Hossin Mahmoudi, Ali Karimi far, Peyman Jamali, Ahmad Karami

National Iranian Copper Industries Company, Khatoon Abad Copper Smelter, Kerman , Iran

E-mail: Ali_manoo111@yahoo.com

Abstract. One of the most common copper concentrate smelting furnaces is Flash smelting furnace which smelts over the half of the world's copper concentrates. The copper concentrate, air or oxygen enriched air and auxiliary fuel oil are charged into this furnace. Due to pyrolysis of copper concentrate, and reaction the decomposed products with oxygen intake, a large amount of heat release that causes the melting of copper concentrate. Since the copper concentrate, contains various minerals, and each of their reactions with atmospheric oxygen in the furnace creates certain conditions; by varying the percentage of minerals in the concentrate charged into the furnace, many problems are raised. In order to fix the problems, and also recognizing the proper actions at the time of changing minerals; within 6 months, inlet concentrate were mineralogically and chemically analyzed and the effect of concentrate minerals on-air and fuel consumption as well as the dust production were studied and The most appropriate concentrate which creates the best conditions in the Khatoon Abad Flash Furnace is obtained.

1.Introduction

Now a day, the Flash smelting process involves half of the copper production factory in the world. This process, which is an autogenously process, uses copper concentrate that involve sulfide minerals such as chalcopyrite, Covellite and pyrite and the like .Due to the charging sulfide copper concentrate, preheated air with 450 ċ or oxygen-rich air and fuel to flash furnace with a temperature of 1250 ċ; (because of concentrate reaction with oxygen the copper matte, slag, and sulfur dioxide exhaust gas will be produced.. In this process, matte molten droplets and slag are settled at the bottom of the furnace, where create the matte phase consists of a copper-rich phase and the slag phase consists of a silicon-rich phase. The exhaust gas containing sulfur dioxide with a temperature of about 1280 ċ and some of the dust caused by the reaction of the concentrate with air, are transferred to the boiler for dust settling and reduce the off gas temperature and also produced steam in boiler will be send for various uses, oil & air pipes and etc.

However, it is important that how much(%) of each minerals must be in the concentrate until the furnace reach to the optimal conditions, because the combustion temperature of some minerals is low (Chalcopyrite and pyrite) and some minerals (Chalcocite) are very high and close to the furnaces operating temperature. This case Further can cause some problem in the melting process of the concentrate. Since using of different concentrate is one of the main advantages of the flash furnace, attended to the change concentrate minerals changing its necessary to adjust the furnace input in order to keep the furnace stable. Therefore, by performing chemical and mineralogical analysis of the various concentrate and studying the furnace conditions, the most stable conditions were determined and by applying this information arrive to optimizing the combustion process in the furnace, improving productivity, reducing costs and protecting the environment.

2.Theory

According to table 1, the copper concentrate with 100% chalcopyrite, has a copper content of 35%, iron 30% and sulfur content of 35%. If the percentage of pyrite in the concentrate is increased, the percentage of copper in the concentrate decreases and the amount of iron and sulfur is increased.

Table 1. Comparison Of Concentrates Contain With Various Percent Of Mineral Chalcopyrite And Pyrite; And Its Effect On Air Consumption, Matte, Slag And Off Gas.(NM3: Normal Cubic Meter). (Ton: Meteric Ton, Equivalent 1000 Kilogram)

Table Head	Concentrate				Matte(Cu2S,FeS)				Air consumption	Slag (FeO)2SiO2		SO2 Gas	
	Ton	Cu	Fe	S	Ton	Cu: 51.6%	Fe: 22.6%	S: 25.8%	NM3	Ton	Fe	SiO2	NM3
CuFeS2 (100%)	100	35	30	35	67	35	15	17	76049	28	15	8	12168
CuFeS2 (80%) FeS2 (20%)	80 20	28	34	38	54	28	12	14	107536	39	22	12	17206
CuFeS2 (50%) FeS2 (50%)	50 50	17	39	44	34	17	8	9	154710	56	31	17	24754
FeS2 (100%)	100	0	47	53					233333	85	47	25	37333

For example, the concentrate include 50% Chalcopyrite and 50% pyrite have a copper content of 17%, iron 39% and sulfur 44%.

If desired, by concentrates with different minerals in the table (1) want to produce matte with constant copper grade (copper 51.6%, iron 22.6% and sulfur 25.8%); 100 tons of copper concentrate contain 100% Chalcopyrite needed 76049 Nm3 air that produced 67 tons of copper matte(It contain 35 tons of copper, 15 tons of iron and 17 tons of sulfur). If the copper concentrate contains 20% pyrite and 80% Chalcopyrite, then for the production of copper matte according to the table (1), required 107536 Nm3 air and produced 54 ton of copper matte(which produces 28 tons of copper, 12 tons of iron and 14 tons of sulfur). As it is known, when the amount of Chalcopyrite in the concentrate decrease and instead the amount of pyrite in the concentrate increase, the percentage of copper will decrease and the percentage of iron and sulfur in the concentrate will increase, which brings the following results:

A) for the production of matte with a specific grade (according to the table (1)) iron and sulfur burning need more combustion air (air consumption is indicated in the preceding paragraph and shown in table).

B) Sulfur burning of concentrate will increase with increasing of combustion air in flash furnace, and more SO2 gas will be produced (according to the table (1), proportionally to increased pyrite, the SO2 gas produced will be increased).

C) Because of dust mechanical conveying from flash furnace to boiler the rate of output dust from the furnace increase proportion to increase output gas,(increasing losses and reducing productivity).

D) Proportion to increase the furnace air consumption, increases the combustion iron of concentrate, and more silica(SiO_2) should be consumed in order to carry out slag treatment (Fe/ SiO_2 ratio in slag is about 1.2), which results in more slag production (according to table (1) produced slag increases-increasing costs).

E) Since the slag is blocked some of the copper and exhausted from the furnace, with increasing slag production, the amount of copper losses in the slag will also increase (increasing losses and decreasing recovery).

3.Investigation Steps

According to the above, for investigate in effect of parameters such as concentrate copper grade and concentrate mineralogical composition on air consumption and dust composition; is evaluated the actual performance of the flash smelting furnace(which uses a preheated air of 450Ċ for concentrate combustion) on a monthly basis. Its complementary resources are as follows:

According to Figure (1) (a,b), proportional to increasing the percentage of chalcopyrite mineral in the concentrate , the percentage of copper in the concentrate increased and also proportional to increasing of pyrite mineral in the concentrate, the percentage of copper in the concentrate has decreased (copper content has a direct relation with chalcopyrite and has a negative relation with the amount of pyrite in the concentrate).

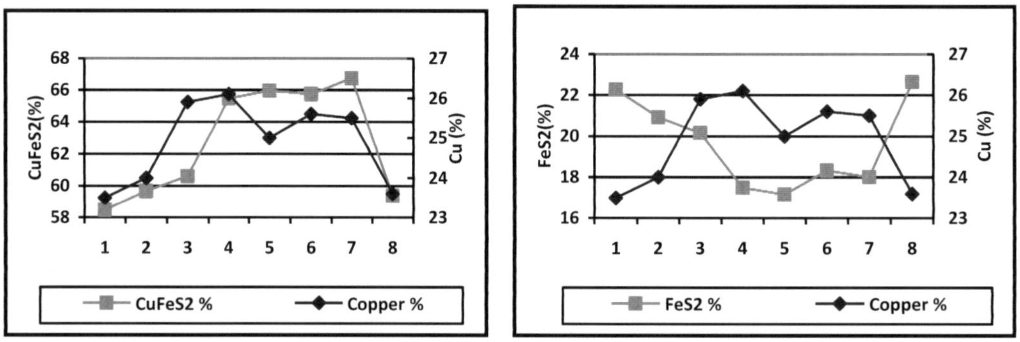

Figure 1. (a)Compar percentage of copper with chalcopyrite mineral in concentate. (b)Compar percentage of copper with pyrite mineral in concentate.

According to Figure (2a), proportional to increasing the percentage of chalcopyrite mineral in the concentrate, the amount of air consumed per tone of concentrate by the furnace decreased(to product a certain grade of copper matte, air consumed with the percentage of chalcopyrite had a inverse ratio). Also, according to Figure (2b), proportional to decreasing the percentage of pyrite, the amount of air consumed has decreased.

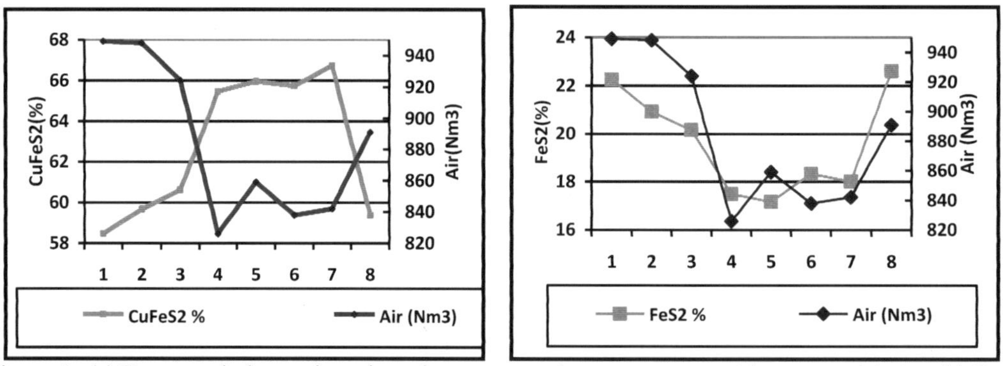

Figure 2. (a)Compar chalcopyrite mineral percentage in concentrate with consumed air. (b)Compar pyrite mineral percentage in concentrate with consumed air.

According to Figure (3), at the months that the concentrate had the highest copper percentage, the consumption air of flash furnace per tone of concentrate was minimal, and when the concentrate copper percentage was reduced, the air consumption has increased.

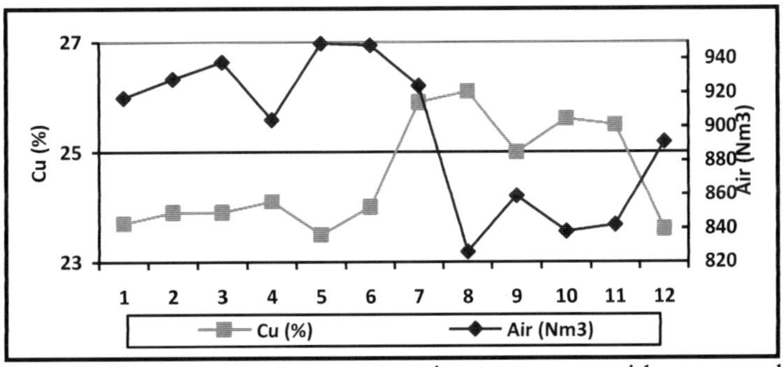

Figure 3. Compar copper percentage in concentrate with consumed air.

According to Figure (4), the dust production in months of flash furnace when low-grade copper concentrates are charged to the furnace is more than months when higher-grade copper concentrate are charged. so that is shown, when the concentrate with the highest copper grade(26.06%) is charging in furnace, The dust production was minimum by at least 12.6% and When the concentrate has the lowest grade (23.6%), the dust production is maximum (14.7%).

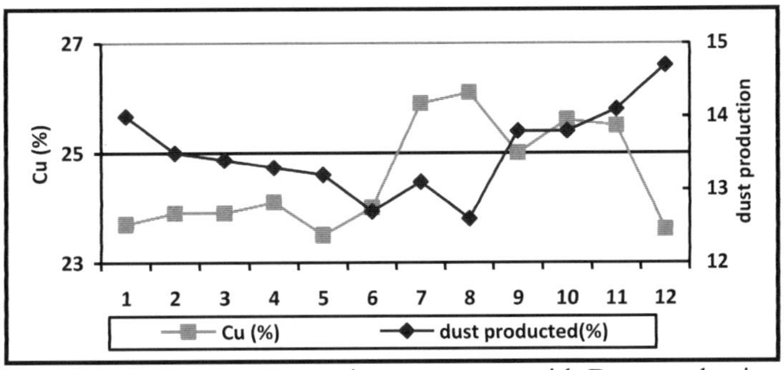

Figure 4. Compar copper percentage in concentrate with Dust production percentage.

According to figure (5), the dust production of the flash furnace associated with the air consumption.

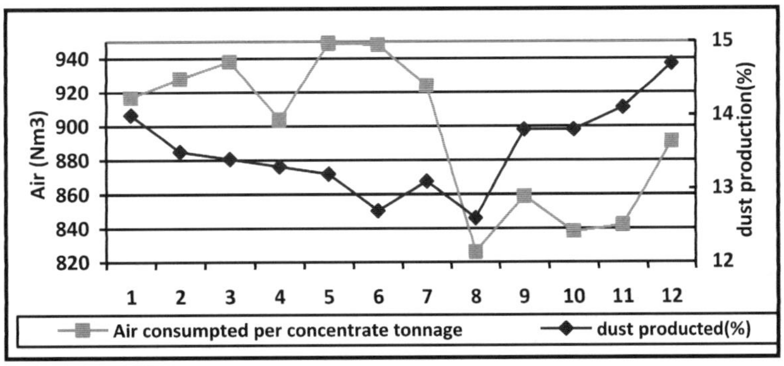

Figure 5. Compar air consumpted per concenterate tonnage with Dust production percentage.

The amount of iron in the copper concentrate increases With increasing the pyrite, which requires iron more oxidation of the copper concentrate to produce the specific matte content. As a result of more oxidation of iron, more magnetite is produced in the furnace. With producing more magnetite created many problem such as separation between copper matte and slag in the settler of flash furnace.

4.Conclusion

The best performance of a Flash smelting furnace in addition to continuous and uniform smelting operations of copper concentrate, suitable separation of the matte and slag produced, is the time when, and also have the least amount of dust exhaust from the furnace, and the furnace output temperature (matte, slag and gas) is also appropriate. This matter is possible that a number of input and output factors of the furnace are controlled. According to studies and diagrams, the best performance of the Flash furnace is when the lowest consumption of the combustion air per ton of concentrate is entered into the furnace.

The lowest consumption combustion air was due to concentrates with higher chalcopyrite and lower pyrite, that way the concentrate have higher grade of copper. So that the concentrate with higher copper grade consumes less air and produces less dust.

Chalcopyrite and pyrite minerals are major minerals in copper concentrate, and higher percentage of copper at concentrate have more chalcopyrite and less pyrite. At the same temperatures, Pyrite releases more heat energy than chalcopyrite.

Therefore, to produce the matte with a specific percentage of copper, if the higher copper grade concentrate (due to the lower pyrite) are used, therefore the lower heat released. In this case, for complete the smelting operation in the furnace, it is necessary that when charging a concentrate with a higher copper grade, the heavy oil was increased (proportional to the heat released is lower) until there was no disturbance in the smelting operation of the concentrate.

Therefore, since the amount of heat released from the combustion of the heavy oil in the reaction shaft of Flash furnace is very high (compared to the heat released from the reaction of the air with the concentrate), the increase of the heavy oil in the reaction shaft of Flash furnace causes to accelerates the particle temperature of the concentrates, and smelting to be faster and also makes easier agglomerate of the molten concentrate particles. As a results, settling of the formation melting particles in the floor zone of the furnace increases, and the production of dust decreases.

Since a higher grade copper concentrate (than to lower grade copper concentrate) consumes lesser air for produce a copper matte with certain grade, it also consumes less silica and produces less dust; therefore, to reduce costs and better protect from devices and the environment, concentrates with high Percentage of the chalcopyrite is more appropriate.

References

[1] Miettinen Elli. " Thermal Conductivity and Characteristics of Copper Flash Ssmelting Flue Dust Accretions" December 2008, Finland.

[2] Ts.Markova, BiBoyanov, S.Pironkov and N.Shopov. " Investigation of Dusts from Waste-Heat Boiler and Electrostatic Precipitators after Flash Smelting Furnace for Copper Concentrates" May 2000, Bulgaria.

[3] Yang, Y. "Computer Simulation of Gas Flow and Heat Transfer in Waste-Heat Boilers of the Outokumpu Copper Flash Smelting Process". Acta Polytechnica Scandinavica Chemical Technology Series 242. Doctoral thesis Helsinki University of Technology 1996.

[4] Samuelsson, C. and Björkman, " B. Dust Forming Mechanisms in the Gas Cleaning System after the Copper Converting Process " Scandinavian Journal of metallurgy Vol 27 (1998).

[5] Mark E. Schlesinger, Matthew J. King, Kathryn C. Sole, William G. Davenport "Extractive Metallurgy of Copper" Fifth Edition 2011.

[6] Davenport w, Jones D, King M. J, & Partelpoeg E. H. "Flash smelting, analysis, control and optimization". Warrendale, PA: TMS. (2001).

[7] Chen S. L, & Mansikkaviita H. "The beneficial effects of feeding dry copper concentrate to smelting furnaces and development of the dryers". In R. T. Jones (Ed.), Southern African pyrometallurgy 2006 (pp. 265e272). http://www. saimm.co.za/Conferences/Pyro2006/265_Kumera.pdf. Johannesburg: South African Institute of Mining and Metallurgy. (2006).

[8] Walker P, Coleman M, & Money G. "Energy efficient and reliable pneumatic conveying solutions at Aurubis". In Copper 2010, Vol. 6: Economics, process control, automatization and optimization (pp. 2517e2526). Clausthal-Zellerfeld, Germany (2010).

[9] Reed M. E, Jones, C. U, Snowdon B, & Coleman M. "flash furnace feed system": the development cycle. Clyde-WorleyParsons. In Copper 2010, Vol. 3: Pyrometallurgy II (pp. 1079e1094). Clausthal-Zellerfeld, Germany. (2010).

[10] Outotec. "Minerals and metals, copper, flash smelting and flash converting".http://www.outotec.com/pages/Page38082.aspx?epslanguage ¼ EN. Finland(2010).

[11] Nagai K, Kawanaka K, Yamamoto K, & Sasai S. "Development of Sumitomo concentrate burner". In Copper 2010, Vol. 3: Pyrometallurgy II (pp. 1025e1034). Clausthal-Zellerfeld, Germany: GDMB. (2010).

[12] Miettinen E, Ahokainen T, & Eklund K. "Management of copper flash smelting off-gas line gas flow and oxygen potentia"l. In Copper 2010, Vol. 3: Pyrometallurgy II (pp. 1003e1012). Clausthal-Zellerfeld,Germany: GDMB. (2010).

[13] McKenna K, Newman C, Voermann N, Veenstra R, King M, & Bryant, J. "Extending copper smelting and converting furnace campaign life through technology". In Copper 2010, Vol. 3: Pyrometallurgy II (pp. 971e986). Clausthal-Zellerfeld, Germany: GDMB. (2010).

[14] Kojo I.V & Lahtinen M. "Outokumpu blister smelting processes e clean technology standards". In A. E. M. Warner, C. J. Newman, A. Vahed, D. B. George, P. J. Mackey & A. Warczok (Eds.), Cu2007,Vol. III, Book 1: The Carlos Dı´az symposium on pyrometallurgy (pp. 183e190). Montreal (2007).

IOP Conf. Series: Materials Science and Engineering **409** (2018) 012015 doi:10.1088/1757-899X/409/1/012015

Productivity Improvement Through Line Balancing at Electronic Company – Case Study

Fansuri, A.F.H, Rose, A.N.M, Ab Rashid, M.F.F, Nik Mohamed, N.M.Z, Ahmad, H

Faculty of Mechanical Engineering, Universiti Malaysia Pahang, 26600 Pekan, Pahang, Malaysia
TATI University College, 24000 Kemaman, Terengganu, Malaysia

Abstract. A good manufacturing practice for production assembly can result in better productivity for the organization. This study focusses on productivity improvement of Electronics Manufacturing Service Company (EMSC) production assembly. It also aims to reduce waste and to propose a leaner line balancing with the elimination of non-value-added activities during the assembly. This study also aims to utilize current workstation to carter for the needs of the workstation to achieve 39 seconds takt time. Time study has been conducted for all workstations and found that all of the cycle time for each workstation exceeds the required takt time of 39 seconds. The bottleneck is at workstation number two with a recorded standard time of 54.1 seconds to complete the assembly thus causes high work in progress at workstation number two and excessive operator idle time to subsequent workstations. The main issue of EMSC is high volume demands from the customer and committing to the shipment plan. By adding one additional operator to workstation number two and a proper distribution of tasks may help EMSC to increase their production output. Time study is recorded in standard work form for future improvement.

1. Introduction

The electronics industry has been facing fierce competition among competitors on product varieties, high production volume, product quality and cost. Due to stiff competition, all electronic industries have to come out with a solution to stay at their top performance.

In this paper, work standardization is implemented as to improve the current process, reduce cycle time, increase output to meet customer's demand and add discipline to train new operators. Standardised work also includes the documentation of the current process. Later it will be used for further Kaizen activities based on takt time. It is to ensure all operators are performing the task within takt time. Work standardization is part of lean manufacturing tools which could increase motivations among workers to deliver higher efficiency and quality at a minimum cost [1].

The studied organization in this paper is facing high volume with low part number varieties. The standard procedure of product assembly in this company is called Work Instructions (WI) and Temporary Process Change Notice (TPCN) when there is a need to change the procedure at short notice. Currently, the company produces an electronic part version 1.2 (V.12). The newly developed and improved design product with additional features and new functions is called version 1.2+ (V1.2+). This version needs to be assembled and shipped out according to the shipment plan. During the study, the production assembly sequence and manpower utilization for a new version are the same as old version V1.2. However, due to additional parts number to be assembled and other functional test need to be carried out, the production line was imbalance thus increases cycle time, work in progress (WIP), the idle time of operator and overall output decreases by 26.3%. In summary, this company has facing high cycle time, work in progress and inefficient resource utilization. Therefore, this paper is aim to balance the distribution of workload of each workstation through line balancing technique. Any non-value-added activities will be identified and eliminated as to achieve optimum cycle time.

This paper is structured into five sections, with section one highlight brief introduction regarding the paper, followed by literature study regarding the concept related to the case study, the method to carry out research study and data analysis. The fifth and final section will describe the conclusion and future work.

2. Literature Review

Many companies had turned to lean manufacturing to improve their performance [2-4]. Lean manufacturing is a system that focuses on elimination of wastes such as time reduction. Through this implementation, the company could sustain competitive advantages which gained from Kaizen and will improve stakeholder value [5, 6]. Apart from that, the operations efficiency also could be improved tremendously [7]. Due to electronics products has a short product cycle and evolves rapidly from various customer needs and unpredictable had caused many companies to adopt lean manufacturing philosophy [8]. One of the critical lean tools is standard work as part of continuous improvement. Standard work is the standardization of procedures in production operation. As cited by Masaki Imai, *"there can be no Kaizen without standardization"* [9].

2.1. Standardized work

Standardisation is the most powerful lean tools but least used in the production [1]. It is by documenting current production best practice to form the fundamental baseline for Kaizen [1]. It provides a routine for consistency of an operation or a process and a basis for improvement. According to Johansson [9], there are eight steps in developing standardized work [9] such as forming an improvement team, identifying production takt time, identifying cycle time, identifying production assembly sequence, identifying standard quantity for work in progress (WIP), providing standardised workflow, providing standard operations sheet and continuing to improve standard operations.

The goal of production standardization is to implement all processes right at the first time with zero waste. There are five characteristics need to be followed for standardization. Firstly, work instruction has to contain the only necessary process for the operator to follow. Work instruction has to be simple and provide visualization for easy understanding. Next is the possibility of immediate changes in process parameter. Fourthly, to ensure that every assembly operator has relevant activities which tie to the process and finally able to monitor operations standard and the impact it has on process area. [1]

2.2. Time Study

Time study is identifying required time by a qualified assembly operator working at a normal pace to the assembly operation. Time study is also described as an assembly measuring technique where the times are recorded and the rate of the specific assembly is carried out under a specified condition. The data is then analyzed to obtain the optimum time to carry out the production assembly. The result obtained from time study is called standard time [10]. Line balancing is a way to minimise imbalance workloads between workers as to achieve the desired output. Workstation assembly line should be analyzed in terms of the assembly process, layout and also cycle time [11].

3. Methodology

Firstly, the research area is identified and agreed based on preliminary observation of production yield. During the visit, work instruction is studied to understand the upgraded process from V1.2 to V1.2+. Cycle time data is then collected by using a stopwatch for ten times for each workstation to get the accurate average value based on the company standard work format. Then, data evaluation is evaluated to perform line balancing for each of the workstations. The company considered the performance rating for the operator is 100% at its normal pace with 15% allowed allowances. A prior study, the operator was explained clearly details of the process need to be carried out. This operator is well trained and has good experience in this process and training records had been provided by human resource department to signify that the operators are qualified for the task. Apart from that, a team was formed to study the current process and propose possible action should be carried out. This team was led by senior engineer and four members. All activities were recorded for reviewing process. Interview also had been conducted to the workers for data clarification. Table 1 shows the average of cycle time for each of workstation. Then, after four months, new cycle time data is collected, recorded and analysed in standard work to compare before and after improvement implementation to the line cycle time. Next, normal time and standard time are calculated as Eq. 1 Normal Time (s)=Average Time ×Rating Factor and Eq. 2 Standard time (s)=(Normal Time)/(1-Allowance Factor) [12]. Table1shows the averages of assembly time study for 7 workstations.

Table 1. Summary Of Standard Time For All Assembly Workstations

Assembly Station	Average Cycle Time (s)	Operator's Efficiency (%)	Standard Time (s)
1	42.0	100	49.4
2	46.0	100	54.1
3	42.0	100	49.4
4	40.0	100	47.1
In-Line potting	23.0	100	27.1
5	36.0	100	42.4
6	34.0	100	40.0
Total	263	100	309.5

In this study, the takt time is calculated based on Rohani & Zahraee as in Eq. 3 Takt Time=(Available time for production)/(Required unit of production) which produces takt time of 39.6s [13]. The highest standard time is 54.1s at assembly workstation 2 which equivalent of producing 737 devices per cell per day. This performance shows the achievement of workstation number 2 is about 73.7% of required target per day. The cycle time is imbalance and most of the assembly workstations are above than takt time except in-line potting. Workstation 2 shows the highest cycle time which causes high work in progress (WIP) of 28% and affects production flow.

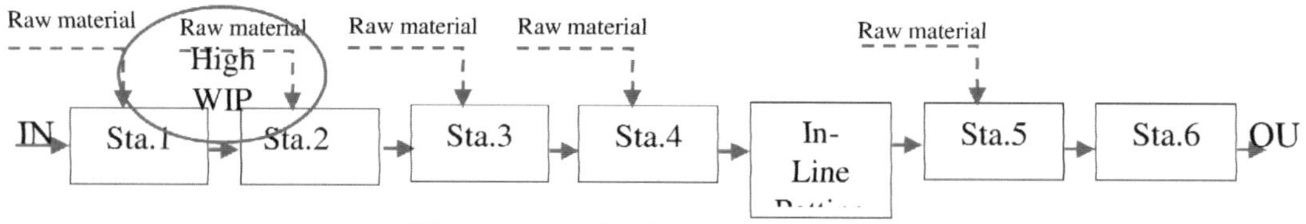

Figure 1. Production Process Flow

The assembly flow is shown as in

. From the illustration, it is clear to see that the existence of WIP at workstation 2. The reason for high WIP at workstation 2 is due to the additional vibration motor soldering and heater wire soldering onto the heater PCBA. The previous process of version 1.2 is no vibration motor needed to be soldered to the PCBA board and the design of heater frame is different and need more soldering time to solder heater wire to the area.

Below is the calculation of production efficiency by using Eq. 4 Efficiency=(∑ Task time)/(Actual number of workstation ×Largest assigned cycle time) [11]. The calculated line efficiency is 81.7%. Although the efficiency is quite good, the company still want to improve line efficiency as to increase production output. There are 42 electronic device assembly workstations to carter estimated annual demand of 12 million devices per year. Each cell is required to produce 1000 devices per day. Currently, there are seven assembly workstations per cell including the automated machine for in-line potting. From time study, the highest standard time of 54.1s is used to calculate the total output. Daily production target is to assemble 1000 electronic devices. Based on the highest standard time 54.1s, it takes 15.03 hours to assemble 1000 devices. Current workers used for assembly are two workers. Thus, the current productivity is calculated by using Eq. 5 Labour hour per unit= (15.03 hours)/ (1000 devices) and Eq. 6 Labour productivity= 1/ (labour hour per unit). By substituting Eq. 5 and Eq. 6, Eq. 7 is produced Labour productivity= 1/ (labour hour per unit) which shows a productivity of 67 devices/hr. Thus, in one day, EMSC is only able to produce 737 assembled electronic devices per cell in one day.

4. Discussion

The desired cycle time was calculated based on maximum availability capacity 17,688 assembled devices per month per cell and monthly target output 24,000 devices per cell per month. This was based on 24 working days. Desired cycle time is calculated by using Eq. 8 C_d= (Productivity time available)/ (Desired units of output) which then resulted in 39s desired cycle time. Based on the desired cycle time and measured cycle time, a minimum number of workstation can be determined by using Eq. 9 $N=(\sum_{i=1}^{n}\times t_i)/C_d$. The calculation shows the minimum number of workstation should be 8.

Table 2. Work Element And Precedence Step.

Assembly Station	Element No.	Work Element	Precedence	Standard Time (s)
1	1	Contact PCBA Soldering Heater + Press Fit Jig	-	49.4
2	2	Heater wire to Heater PCBA Soldering	1	54.1
	3	Vibration Motor to Heater PCBA Soldering + Insert to Frame	2	
	4	Pneumatic Press Fit Jig	3	
	5	Heater PCBA to Frame Soldering	4	
3	6	Heater PCBA 6 Points Soldering	5	49.4
	7	Battery to Frame Soldering	6	
4	8	MT 4	7	47.1
	9	Frame to Middle Part Insertion	8	
	10	Middle Part Press Fit Jig	9	
In Line potting	11	In Line Potting	10	27.1
5	12	Ring Insertion	11	42.4
	13	Button Insertion	12	
	14	Staple Insertion	13	
	15	Rear Housing Press Fit Jig	14	
	16	Front Housing	15	
6	17	Vibration Test	16	40
	18	MT 5	17	
	19	TSH Test Using MU	18	

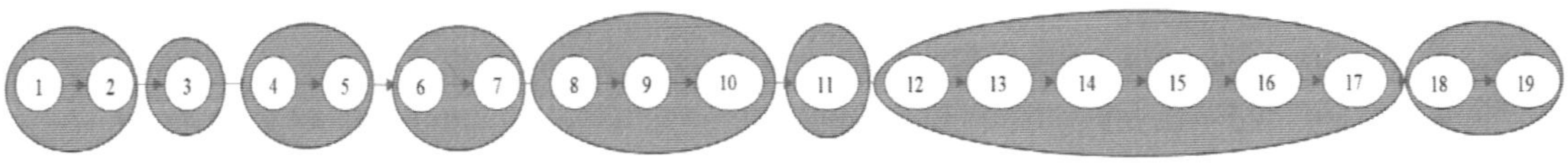

Figure 2. Line Balancing.

Figure shows the group of eight workstations. The blue coloured region is indicating one assembly workstation. The difference in this assembly workstation compared with the old assembly workstation is an additional one operator based on line balancing calculation to do the assembly at workstation number 2. Soldering of heater wire to heater PCBA process is moved to assembly workstation number 1 to balance out the process. At workstation number 2, the operators have to insert vibration motor to the heater PCBA and solder it in place. While at workstation number 3, the operator has to peel off vibration sticker, add the heater PCBA to frame and solder 5 points on the heater PCBA. As for operator at workstation number 4, one-point soldering to connect heater PCBA and contact PCBA is carried out at this station. There is also a combination of workstation 7 and part of assembly procedure from assembly station 8 (Vibration test), a 5s manufacturing test that was moved to workstation 7.

It is identified that at assembly workstation 4 and 6 are longer cycle time due to the server has reached its maximum capacity, thus causes the scanning process and testing time become higher. In the midst of server migration, it is expected to cause longer cycle time. Once the migration is completed, the cycle time at that station will be twice faster than recorded. Four months later after an additional operator, the new cycle time is measured. The time had significantly improved over the past months after improvement. With server's migration, the scanning process at station 4 and 6 also improves. On average, the cycle time has improved by 45.5%. The new standard time after improvements were calculated as shown in Table 3.

Table 3. Summary Of Standard Time For After Improvement.

Assembly work station	Standard time (s)
Station 1	37.13
Station 2	29.53
Station 3	32.75
Station 4	30.20
Station 5	32.20
In-Line potting	30.87
Station 6	36.82
Station 7	26.49
TOTAL	255.99

Line efficiency is calculated using Eq. 4. New production efficiency is 86.2% compared to 81.7% before the improvement. This show new production efficiency is increased by 4.5%. The increment may seem quite small, but based on labour productivity as shown below, the impact has increased the production. Using the highest standard time which is 37.13s at assembly station 1, it will take 10.31 hours to complete 1000 devices per day for one cell. The improved productivity calculation calculated as in Eq. 7. Therefore, by improving line balancing, EMSC productivity is 97 devices per hour. Previously, before improvement, EMSC productivity is 67 devices per hour. Based on this improvement, EMSC productivity has increased from 67 devices per hour to 97 devices per hour with an increment of 44.8% in terms of productivity that is more than required production goal of 1000 devices per day. The excess of 67 devices will be accounted as a buffer which may be used when there is a part shortage or production line down or production off day.

5. Conclusion

In this study, the aim was to improve the productivity of an electronic device produced by EMSC. The real problem is identified and it is found to be associated with the line balancing and standardization. Time study was conducted and process flow is analysed thoroughly. The study shows some processes were above and below takt time. Improvement on the production line is applied through line balancing techniques. This study also focused on the improvement of high work-in-progress of 28% which disrupted the production flow. The additional operator in workstation number 2 and segregate assembly flow accordingly to each assembly workstation have increased the productivity. In addition, this improvement also has eliminated WIP at assembly workstation number 2 by balancing the work segregation to operator 3. Time study is recorded in standard work form for future research in reducing the number of operators through the integration of automation to the assembly line. After improvement, the production efficiency has increased from 81.7% to 86.2%. The productivity also has increased from 67 units per hour to 97 units per hour. As overall the line balancing technique in this study has improved the efficiency and productivity of this company.

References

[1] Mĺkva, M., et al., *Standardization – One of the Tools of Continuous Improvement.* Procedia Engineering, 2016. **149**(Supplement C): p. 329-332.

[2] Lyonnet, B. and R. Toscano, *Towards an adapted lean system – a push-pull manufacturing strategy.* Production Planning & Control, 2014. **25**(4): p. 346-354.

[3] Gonçalves, M.T. and K. Salonitis, *Lean Assessment Tool for Workstation Design of Assembly Lines.* Procedia CIRP, 2017. **60**(Supplement C): p. 386-391.

[4] Rose A.N.M, A.R.M.F.F., Nik Mohamed N.M.z, Ahmad H, *Similarities of lean manufacturing approaches implementation in SMEs towards the success: Case study in the automotive industry.* MATEC Web of Conferences, 2015. **87**.

[5] Brown, C.B., T.R. Collins, and E.L. McCombs, *Transformation From Batch to Lean Manufacturing: The Performance Issues.* Engineering Management Journal, 2006. **18**(2): p. 3-14.

[6] Veza, I., et al., *Development of Assembly Systems in Lean Learning Factory at the University of Split.* Procedia Manufacturing, 2017. **9**(Supplement C): p. 49-56.

[7] Venkataraman, K., et al., *Application of Value Stream Mapping for Reduction of Cycle Time in a Machining Process.* Procedia Materials Science, 2014. **6**(Supplement C): p. 1187-1196.

[8] Nguyen, M.-N. and N.-H. Do, *Re-engineering Assembly Line with Lean Techniques.* Procedia CIRP, 2016. **40**(Supplement C): p. 590-595.

[9] Johansson, P.E.C., et al., Current State of Standardized Work in Automotive Industry in Sweden. Procedia CIRP, 2013. 7(Supplement C): p. 151-156.

[10] Hartanti, L.P.S., *Work Measurement Approach to Determine Standard Time in Assembly Line* International Journal of Management and Applied Science (IJMAS), 2016. **2**(10): p. 192-195.

[11] Lam, N.T., et al., *Lean Line Balancing for an Electronics Assembly Line.* Procedia CIRP, 2016. **40**(Supplement C): p. 437-442.

[12] Chueprasert, M. and P. Ongkunaruk, *Productivity improvement based line balancing: A case study of pasteurized milk manufacturer*. Vol. 22. 2015. 2313-2317.

[13] Rohani, J.M. and S.M. Zahraee, Production Line Analysis via Value Stream Mapping: A Lean Manufacturing Process of Color Industry. Procedia Manufacturing, 2015. **2**(Supplement C): p. 6-10.

Developments of Thermochromic VO_2–Based Thin Films For Architectural Glazing

Noormariah Muslim[1,*], Ying Woan Soon[2], Chee Ming Lim[1], Nyuk Yoong Voo[1]

[1]Centre for Advanced Material and Energy Sciences, Universiti Brunei Darussalam,
[2]Physical and Geological Sciences, Faculty of Science, Universiti Brunei Darussalam, Jalan Tungku Link, Gadong BE 1410, Brunei Darussalam

*E-mail: 14h8702@ubd.edu.bn, noormariah.muslim@hotmail.com

Abstract. Investigation on properties of vanadium dioxide (VO_2) thin films has received significant attention as it is a thermochromic material potentially for applications on architectural glazing in order to reduce energy consumption from usage of air-conditioning system. Most studies have focused on the hysteresis behavior of thermochromic VO_2 thin films that occurs at a certain phase transition temperature. Although large area coating techniques have recently been developed, there are still many obstacles in trying to deploy energy-efficient windows. Hence, this work concisely reviews on VO_2 thin films as thermochromic materials, and developments on preparation methodologies in order to enhance the thermochromic characteristics of VO_2 thin films.

1. Introduction

VO_2 thin films as thermochromic materials are widely known for its well-defined critical or transition temperature (Tt) whereby no electrical power supply is needed. A simple illustration for thermochromic VO_2 thin films in architectural glazing is shown in Fig.1 (a). Thermochromic VO_2 thin films have been studied on glass substrates by several deposition techniques, for instances chemical vapor deposition [1], reactive pulsed laser deposition [2], sputtering [3], and solution-based methods [4].

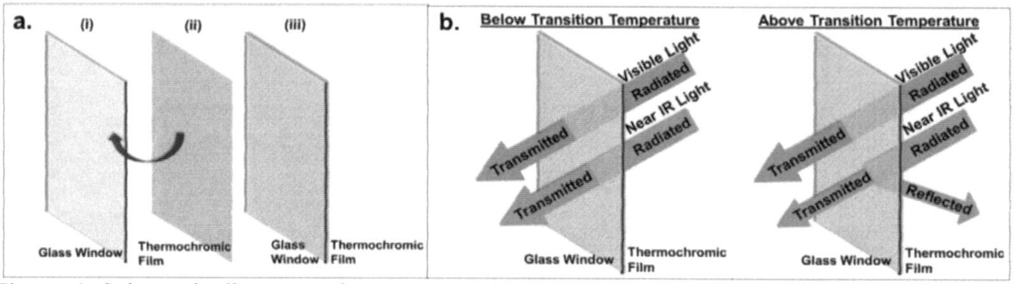

Figure 1. Schematic diagrams of (a) an example of glass window coating where (i) a glass window (ii) a thermochromic VO_2 film and (iii) thermochromic VO_2 thin film deposited on glass window, and (b) thermochromic VO_2 film deposited on glass window with temperature below and above Tt.

Fig.1 (b) shows schematic diagrams of thermochromic VO_2 films operating as intelligent window coatings. The illustration was suggested by Kiri et al. [5] in their study on thermochromic materials.

Theoretically, at first, the thermochromic VO_2 film transmits incident solar radiation quite well while acting as a semiconductor in its monoclinic state at low temperature (which preferably at room temperature) i.e. below the Tt. Then, as the VO_2 film heats up which means its temperature increases and becomes higher than the Tt due to high solar radiation and/or high outdoor temperatures, the thin film switches to a metal in its tetragonal state (or metallic state/ rutile state) and reflects infrared (IR) solar radiation. Therefore, this phenomenon is called semiconductor to metal transition (SMT) of thermochromic films and has been illustrated in Fig.2 (a) by Budai J.D. et al. [6]. This structural transformation can be first- or second-order in nature and may be reversible or irreversible, depending on the thermodynamics of the system [7].

Eventually, VO_2 is the only metal oxide that exhibits Tt close to room temperature i.e. 68 °C [7,8]. The thermochromic properties can been seen in Fig.2 (b-c) which show the transmittance spectra of VO_2 films with heating effect. Fig.2 (b) represents a change in solar transmittance that can be influenced by increasing the temperature from 25 °C (below Tt (which was also denoted as Tc)) to 90 °C (above Tt/Tc). Fig.2 (c) shows a hysteresis loop representing the presence of phase transition of VO_2 films at 68 °C. However, it is difficult to deposit pure, stoichiometric single crystal of VO_2 films that has a distinct thermochromic effect due to the existence of many other vanadium oxides [7,9,10]. VO_2(M) and VO_2(B) are two different VO_2 structures i.e. monoclinic and metastable respectively, whereby high vanadium contents are required to get VO_2(M) which has thermochromic property [11,12]. Besides that, VO_2(M) films have to be crystallized to exhibit thermochromism, thus, substrate heating is needed. The substrate temperature was 300–500 °C during depositions without further annealing. On the other hand, some studies have annealed VO_2 films at 400–550 °C with or without argon (Ar) gas. Rapid thermal annealing in air has also been used as a possible alternative method to crystallize VO_2 films [13].

Four major obstacles for practical uses of thermochromic VO_2 thin films in architectural glazing are (i) Tt is still high, (ii) visible light transmittance is too low at high temperature (i.e. above the Tt), (iii) wide hysteresis width, and (iv) VO_2 is not a thermodynamically stable oxide which means its properties may alter with time. Hence, the following sections would concisely review several literatures on the developments of thermochromic VO_2-based thin films with dopants and structured film systems.

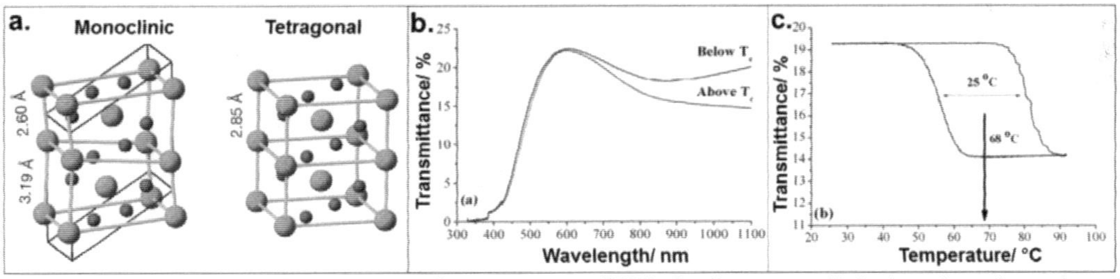

Figure 2. (a) Illustrations on atomic structures of VO_2 in the monoclinic (semiconductor) and tetragonal (metallic) phases, with V and O atoms in green and red colours, respectively [6]. (b) Transmittance spectra of VO_2 thin films at 25 °C (below Tt (or Tc); semiconductor state) and 90 °C (above Tt (or Tc); metallic state), and (c) Transmittance at 1100 nm against heating temperature of the VO_2 thin films [1].

2. Elemental Dopants

Doped-VO_2 thin films can be prepared by embarking dopants during depositions of VO_2 thin films. Dopants were found to corporate with VO_2 and have a significant potential in changing phase Tt of VO_2 films. Hence, as a result, this makes doped-VO_2 thin films commercially viable. Atomic radius of dopants plays an important role i.e. when it is larger than V^{4+} ion (of VO_2), it results in a reduction in Tt because of a charge-transfer mechanism taking place by extra electrons from the dopant into the vanadium d-bands [14,15]. Several dopants that have been applied in synthetizations of doped-VO_2

thin films are aluminium (Al^{3+}), cerium (Ce^{4+}, Ce^{3+}), fluorine (F), molybdenum (Mo^{5+}, Mo^{6+}), silicon (Si^{4+}), titanium (Ti^{4+}), and tungsten (W^{4+}, W^{6+}). Some of these dopants were reviewed particularly in the following sub-sections.

2.1. Aluminium Dopant

Al-doped VO_2 thin films have been deposited on silicon (100) wafers by means of RF magnetron co-sputtering with V and Al targets at 490 °C [16]. Then, the Al-doped VO_2 films were in-situ annealed. The prepared thin films exhibited almost the same Tt as undoped thermochromic VO_2 thin films i.e. 67.9 °C as the films lost its special discontinuous change in electrical resistance as the Al doping increased.

2.2. Cerium Dopant

Ce-doped VO_2 thin films have been prepared on muscovite substrates by spin coating method with a precursor i.e. a mixture powder of cerium (III) nitrate hexahydrate and V_2O_5 heated at 850 °C for 30 mins and then stirred in 150 ml deionized water at room temperature for 2 h, forming a brown sol [17]. The films showed a single orientated (011) plane of monoclinic (VO_2) at $2\theta = 27.87°$ and 27.85° for 0.84 % and 1.68 % of Ce doping, respectively, whereby the diffraction peaks were actually shifted from $2\theta = 27.88°$ diffraction peak for undoped VO_2 films. In addition, phase Tt of the films decreased as Ce doping concentration increased i.e. the Tt was 64 °C and 60 °C for 0.84 % and 1.68 % of Ce doping, respectively.

2.3. Fluorine Dopant

Depositions of F-doped VO_2 films on SiO_2 coated float glass have been performed by aerosol assisted chemical vapour deposition with different vanadyl acetylacetonate, ethanol and F doping concentrations [18]. Phase Tt of the films was ~60 °C. RF-sputtered F-doped VO_2 films on quartz glass have also been prepared at 327 °C [19]. As the F doping concentration increased from 0 % to 2.1 %, the Tt of the films decreased from 58 °C to 20 °C.

2.4. Molybdenum Dopant

Mo-doped VO_2 films have been prepared on soda lime glass slides by dip coating method with a precursor i.e. a mixture powder of ammonium molybdate tetrahydrate and V_2O_5 stirred in 500 ml deionized water at room temperature, and heated at 500 °C [20]. The Tt of the films decreased as Mo doping concentration increased i.e. the Tt was 62.5 °C, 47.5 °C, 34.0 °C and 24.0 °C for 0 %, 1 %, 4 % and 7 % of Mo doping concentrations, respectively. Optical reflectance of the films was higher than the undoped VO_2 films in its metallic state, and increased with doping concentration in its semiconductor state.

2.5. Silicon Dopant

Si-doped VO_2 thin films have been synthesized by DC magnetron sputtering on ITO-coated glass at 100 °C and in-situ annealing at 255 °C [21]. Optical transmittance and phase Tt of the films have been investigated to increase and decrease respectively with increasing Si doping concentration. When the Si/V ratio was 0.17, the Tt of the films was 46.1 °C.

2.6. Titanium Dopant

Ti-doped VO_2 thin films have been RF magnetron sputtered on a TiO_2 buffer layer that was deposited on silicon or fused quartz (SiO_2) substrates [22]. The prepared films showed a decrease in Tt and a transmittance change with increasing doping concentration due to the thermochromism effect.

2.7. Tungsten Dopant

W-doped VO_2 thin films have been deposited on silicon (100) wafers by means of RF magnetron co-sputtering with V and W targets at 490 °C [23]. Tt of the deposited films was 68.2 °C, 61.4 °C,

IOP Conf. Series: Materials Science and Engineering **409** (2018) 012016 doi:10.1088/1757-899X/409/1/012016

54.0 °C and 49.5 °C for sputtering power of 25 W, 75 W, 88 W and 100 W, correspondingly. By means of DC magnetron sputtering, W-doped VO_2 thin films have been prepared on Corning 7059 glass, standard glass and beryllium foil substrates at substrate temperature of 400 °C [24]. The deposited films showed lower Tt than the undoped films of 65 °C. The Tt was 32 °C with 0.0032 W/V ratio.

3. Structured Films

Another way to alter thermochromic characteristics of VO_2 films is by adding another layer(s) of anti-reflection and/or anti-oxidation and/or photocatalytic self-cleaning coatings into VO_2 film(s), hence making bilayer or multilayer structured films. Deposition techniques employed could be the same or different for all films in one structured film system. Several examples of the depositions were reviewed in the following sub-sections for bilayer and multilayer structured–based VO_2 thin films.

3.1. Bilayer structured films

Al oxide films with 30 nm thickness have been deposited on 80 nm thickness thermochromic VO_2 films by DC magnetron sputtering with V and Al targets at 380 °C and annealed at 300 °C [25]. The prepared films exhibited a slight increase in visible transmittance with same IR transmittance in metallic state of VO_2. The Al oxide also showed good protection and delayed oxidation for more than one day. Furthermore, 85 nm thickness VO_2 films have been prepared by RF magnetron sputtering on different substrates i.e. glass coated with ZnON (as a buffer layer), glass coated with SnO_2 (as a buffer layer), silica float glass and silicon [26]. The deposited VO_2 films at 300 °C and 400 °C showed thermochromic characteristics with highest visible transmittance of ~40 % and lowest Tt of 43 °C obtained for glass/ZnON/VO_2 films deposited at 400 °C. Depositions of VO_2 films by DC magnetron sputtering at 300 °C on RF sputtered buffer layers i.e. TiO_2, SnO_2 and SiO_2 have also been carried out [27]. All deposited films exhibited thermochromism except for glass/TiO_2/VO_2 films because no change in optical transmittance was observed in both semiconductor and metallic states. The highest visible transmittance of ~40 % with lowest IR transmittance of ~8 % in metallic state and lowest Tt of 55 °C were obtained for glass/SnO_2/VO_2 films. This was due to the deposited VO_2 film on glass/SnO_2 being looser and more porous causing high visible transmittance. Moreover, SiO_2 can be used as an anti-reflection coating when deposited on VO_2 film by RF magnetron sputtering [28,29]. With 100 nm thickness of SiO_2 coating on the VO_2 film, an increase in luminous transmittance from 42 % to 55 % at semiconductor state and from 40 % to 50 % at metallic state was observed. The study also reported that thermochromism of the films was enhanced due to an improvement in film crystallinity during deposition of the anti-reflection coating, however the Tt was constant at 70 °C. Vanadium oxide films have also been prepared on V, W, Fe, Ni, Ti and Pt metal buffer layers by RF magnetron sputtering at substrate temperature of 400 °C with a V target [30]. Nevertheless, thermochromic VO_2 films were obtained on V and W metal buffer layers of 10 nm thickness. Both bilayer structured films exhibited Tt of 68 °C and 53 °C for silica glass/V/VO_2 and silica glass/W/VO_2 films, respectively. Thermochromic VO_2 film with 120 nm thickness have been overlaid with nanostructured CeO_2 film deposited by RF magnetron sputtering at room temperature [31]. Optical transmittance of the deposited films decreased from its semiconductor to metallic state with Tt of 68 °C. 56 nm thick zirconium oxide (ZrO_2) have been deposited on 50 nm thick VO_2 film by RF magnetron sputtering at 500 °C and room temperature for VO_2 and ZrO_2 thin films, respectively [32]. The structured films showed Tt at 52 °C. Thermochromic VO_2 films have been layered with ZnO films that deposited by RF magnetron sputtering at room temperature [33]. The films exhibited visible transmittance of more than 40 % in semiconductor and metallic states, and IR transmittance of less than 20 % in its metallic state.

3.2. Multilayer Structured Films

Multilayer structured glass/TiO_2/VO_2/TiO_2 and glass/TiO_2/VO_2/TiO_2/VO_2/TiO_2 thin films have been prepared by DC magnetron sputtering at substrate temperature of 450 °C [34]. For glass/TiO_2/VO_2/TiO_2 films, the VO_2 and TiO_2 films had 50 nm and 75 nm film thickness, respectively, with refractive index of 2.2, whereas for glass/TiO_2/VO_2/TiO_2/VO_2/TiO_2 films, the VO_2 and TiO_2 films had 50 nm and 130 nm film thickness, respectively. Luminous transmittance of the films was 62.8 % and 58 % for three-layer films, and 45 % and 42.3% for five-layer films, in its semiconductor and metallic states, respectively. In addition, Tt of the films was 60 °C. A thermochromic VO_2 thin film deposited between two F-doped SnO_2 (FTO) films by DC magnetron sputtering whereby the FTO films were fabricated using photolithography and a chemical etching process, hence, this resulted in a multilayer structured films of conductive glass/FTO/VO_2/FTO [35]. The maximum optical transmittance for the structured films was 50 % and 35 % for its semiconductor and metallic states, respectively. A multilayer structured silicon substrate/Ag/HfO_2/VO_2 thin films where HfO_2 is hafnium oxide, has been prepared by DC magnetron sputtering, electron beam evaporation and RF magnetron sputtering for Ag, HfO_2 and VO_2 films, respectively [36]. The films exhibited a decrease in optical reflectance in its metallic state with Tt of ~58 °C. The films were further studied by using W-doped VO_2 film. The films showed decreasing Tt as the W doping concentration increased. Moreover, depositions of float glass/TiO_2(1)/VO_2:W/TiO_2(2)/Si_xN_y structured films have been conducted with substrate temperature of 270 °C, 210 °C and 230 °C for W-doped VO_2, TiO_2 and Si_xN_y films, respectively [37]. The Tt of the films were reduced to 33 °C with an increase in visible transmittance of ~58 % in both states. Furthermore, a thermochromic VO_2 film has also been stacked between two silicon suboxide (a-SiO_x) films by RF magnetron sputtering with substrate temperature of 600 °C [38]. The fused quartz/a-SiO_x/VO_2/a-SiO_x films with film thickness of 100 nm and 25 nm for VO_2 and a-SiO_x, respectively, showed same visible transmittance of ~48 % in both states with lower IR transmittance of ~20 % in its metallic state while the thermochromism preserved. Multi-functions window coatings of glass/TiO_2(R)/VO_2(M)/TiO_2(A) consisting of energy-saving, anti-fogging and self-cleaning effects have been obtained by medium frequency reactive magnetron sputtering [39]. The films exhibited Tt of 61.5 °C with visible transmittance of ~40 % in both states.

4. Conclusion

Depositions of VO_2-based thin films were developed accordingly with dopants and structured film systems to improve the thermochromic properties of VO_2 films. Dopants corporate with VO_2 very well i.e. as atomic radii of the dopants are larger than V^{4+} ions (of VO_2), Tt of VO_2 reduced, and vice versa. Meanwhile, structured films were also intensively studied by stacking films of different properties such as anti-reflection, anti-oxidation, photocatalytic self-cleaning, and anti-fogging, with thermochromic VO_2 films. Nevertheless, major challenges with preparing thermochromic VO_2-based thin films for architectural glazing applications still remain. For instance, when Tt of the VO_2 films was lower than 68 °C, only slight changes were detected in the visible transmittance, and broad hysteresis width was obtained. Despite that, this research area is still of interest to many researchers which progresses at a healthy pace to achieve commercial application of intelligent window coatings.

Acknowledgment
This research was supported in part by the Brunei Research Council under Grant No. S&T 17.

References

[1] D. Louloudakis, D. Vernardou, E. Spanakis, N. Katsarakis, and E. Koudoumas, "Thermochromic vanadium oxide coatings grown by APCVD at low temperatures," Physics Procedia, vol. 46, July 2013, pp. 137-141, doi: https://doi.org/10.1016/j.phpro.2013.07.055.

[2] A. Hendaoui, N. Émond, S. Dorval, M. Chaker, and E. Haddad, "Enhancement of the positive emittance-switching performance of thermochromic VO_2 films deposited on Al substrate for an efficient passive thermal control of spacecrafts," Current Applied Physics, vol. 13, Jan. 2013, pp. 875-879, doi: https://doi.org/10.1016/j.cap.2012.12.028.

[3] X. Dong, Z. Wu, X. Xu, T. Wang, and Y. Jiang, "Effects of duty cycle and oxygen flow rate on the formation and properties of vanadium oxide films deposited by pulsed reactive

sputtering," Vacuum, vol. 104, Jun. 2014, pp. 97-104, doi: https://doi.org/10.1016/j.vacuum.2014.01.020.

[4] H. Zhang, X. Xiao, X. Lu, G. Chai, Y. Sun, Y. Zhan, and G. Xu, "A cost-effective method to fabricate VO_2 (M) nanoparticles and films with excellent thermochromic properties," Journal of Alloys and Compounds, vol. 636, Jul. 2015, pp. 106-112, doi: https://doi.org/10.1016/j.jallcom.2015.01.277.

[5] P. Kiri, G. Hyett, and R. Binions, "Solid state thermochromic materials," Advanced Material Letters, vol. 1, 2010, pp. 86-105, doi: https://dx.doi.org/10.5185/amlett.2010.8147.

[6] J.D. Budai, J. Hong, M.E. Manley, E.D. Specht, C.W. Li, J.Z. Tischler, D.L. Abernathy, A.H. Said, B.M. Leu, L.A. Boatner, R.J. McQueeney, and O. Delaire, "Metallization of vanadium dioxide driven by large phonon entrophy," Nature, vol. 515, Nov. 2014, pp. 535-539, doi: https://doi.org/10.1038/nature13865.

[7] N.F. Mott, "Metal-insulator transitions," in Physics and Chemistry of Electrons and Ions in Condensed Matter, vol. 130, J.V. Acrivos, N.F. Mott, A.D. Yoffe, Eds. Dordrecht: Springer, 1984, pp. 287-296.

[8] J.B. Goodenough, "The two components of the crystallographic transition in VO_2," Journal of Solid State Chemistry, vol. 3, Nov. 1971, pp. 490-500, doi: https://doi.org/10.1016/0022-4596(71)90091-0.

[9] H. Zhang, Z. Wu, W. Yang, and Y. Jiang, "Large phase-transition hysteresis for nanostructured VOx film prepared on ITO conductive glass by DC reactive magnetron sputtering," Vacuum, vol. 94, Aug. 2013, pp. 84-86, doi: https://doi.org/10.1016/j.vacuum.2013.01.027.

[10] X. Wei, S. Li, J. Gou, X. Dong, X. Yang, W. Li, T. Wang, Z. Wu, Y. Jiang, and Z. Chen, "Preparation and characteristics of vanadium oxide thin films by controlling the sputtering voltage," Optical Materials, vol. 36, Jun. 2014, pp. 1419-1423, doi: https://doi.org/10.1016/j.optmat.2014.02.021.

[11] C. Batista, V. Teixeira, and J. Carneiro, "Structural and morphological characterization of magnetron sputtered nanocrystalline vanadium oxide films for thermochromic smart surfaces," Journal of Nano Research, vol. 2, Aug. 2008, pp. 21-30, doi: https://doi.org/10.4028/www.scientific.net/JNanoR.2.21.

[12] S. Saitzek, F. Guinneton, G. Guirleo, L. Sauques, K. Aguir, and J.R. Gavarri, "VO_2 thin films deposited on silicon substrates from V_2O_5 target: limits in optical switching properties and modeling," Thin Solid Films, vol. 516, Jan. 2008, pp. 891-897, doi: https://doi.org/10.1016/j.tsf.2007.04.129.

[13] Y.K. Dou, J.B. Li, M.S. Cao, D.Z. Su, F. Rehman, J.S. Zhang, and H.B. Jin, "Oxidizing annealing effects on VO_2 films with different microstructures," Applied Surface Science, vol. 345, Aug. 2015, pp. 232-237, doi: https://doi.org/10.1016/j.apsusc.2015.03.044.

[14] M.H. Lee, M.G. Kim, and H.K. Song, "Thermochromism of rapid thermal annealed VO_2 and Sn-doped VO_2 thin films," Thin Solid Films, vol. 290-291, Dec. 1996, pp. 30-33, doi: https://doi.org/10.1016/S0040-6090(96)09201-2.

[15] P. Jin, S. Nakao, and S. Tanemura, "Tungsten doping into vanadium dioxide thermochromic films by high-energy ion implantation and thermal annealing," Thin Solid Films, vol. 324, Jul. 1998, pp. 151-158, doi: https://doi.org/10.1016/S0040-6090(98)00362-9.

[16] A. Paone, R. Sanjines, P. Jeanneret, H.J. Whitlow, E. Guibert, G. Guibert, F. Bussy, J.L. Scartezzini, and A. Schüler, "Influence of doping in thermochromic $V_{1-x}W_xO_2$ and $V_{1-x}Al_xO_2$ thin films: twice improved doping efficiency in $V_{1-x}W_xO_2$," Journal of Alloys and Compounds, vol. 621, Feb. 2015, pp. 206-211, doi: https://doi.org/10.1016/j.jallcom.2014.08.264.

[17] L. Song, Y. Zhang, W. Huang, Q. Shi, D. Li, Y. Zhang, and Y. Xu, "Preparation and thermochromic properties of Ce-doped VO_2 films," Materials Research Bulletin, vol. 48, Jun. 2013, pp. 2268-2271, doi: https://doi.org/10.1016/j.materresbull.2013.02.016.

[18] P. Kiri, M.E.A. Warwick, I. Ridley, and R. Binions, "Fluorine doped vanadium dioxide thin

films for smart windows," Thin Solid Films, vol. 520, Dec. 2011, pp. 1363-1366, doi: https://doi.org/10.1016/j.tsf.2011.01.401.
[19] W. Burkhardt, T. Christmann, S. Franke, W. Kriegseis, D. Meister, B.K. Meyer, W. Niessner, D. Schalch, and A. Scharmann, "Tungsten and fluorine co-doping of VO_2 films," Thin Solid Films, vol. 402, Jan. 2002, pp. 226-231, doi: https://doi.org/10.1016/S0040-6090(01)01603-0.
[20] T.J. Hanlon, J.A. Coath, and M.A. Richardson, "Molybdenum-doped vanadium dioxide coatings on glass produced by the aqueous sol–gel method," Thin Solid Films, vol. 436, Jul. 2003, pp. 269-272, doi: https://doi.org/10.1016/S0040-6090(03)00602-3.
[21] X. Wu, Z. Wu, H. Zhang, R. Niu, Q. He, C. Ji, J. Wang, and Y. Jiang, "Enhancement of VO_2 thermochromic properties by Si doping," Surface and Coatings Technology, vol. 276, Aug. 2015, pp. 248-253, doi: https://doi.org/10.1016/j.surfcoat.2015.07.007.
[22] H. Kakiuchida, P. Jin, and M. Tazawa, "Optical characterization of vanadium–titanium oxide films," Thin Solid Films, vol. 516, May 2008, pp. 4563-4567, doi: https://doi.org/10.1016/j.tsf.2007.05.096.
[23] A. Paone, R. Sanjines, P. Jeanneret, and A. Schüler, "Temperature-dependent multiangle FTIR NIR–MIR ellipsometry of thermochromic VO_2 and $V_{1-x}W_xO_2$ films," Solar Energy, vol. 118, Aug. 2015, pp. 107-116, doi: https://doi.org/10.1016/j.solener.2015.05.020.
[24] M.A. Sobhan, R.T. Kivaisi, B. Stjerna, and C.G. Granqvist, "Thermochromism of sputter deposited $W_xV_{1-x}O_2$ films," Solar Energy Materials and Solar Cells, vol. 44, Dec. 1996, pp. 451-455, doi: https://doi.org/10.1016/S0927-0248(95)00051-8.
[25] Y.X. Ji, S.Y. Li, G.A. Niklasson, and C.G. Granqvist, "Durability of thermochromic VO_2 thin films under heating and humidity: Effect of Al oxide top coatings," Thin Solid Films, vol. 562, Jul. 2014, pp. 568-573, doi: https://doi.org/10.1016/j.tsf.2014.03.043.
[26] M. Panagopoulou, E. Gagaoudakis, E. Aperathitis, I. Michail, G. Kiriakidis, D. Tsoukalas, and Y.S. Raptis, "The effect of buffer layer on the thermochromic properties of undoped radio frequency sputtered VO_2 thin films," Thin Solid Films, vol. 594, Nov. 2015, pp. 310-315, doi: https://doi.org/10.1016/j.tsf.2015.06.010.
[27] B. Zhu, H. Tao, and X. Zhao, "Effect of buffer layer on thermochromic performances of VO_2 films fabricated by magnetron sputtering," Infrared Physics & Technology, vol. 75, Mar. 2016, pp. 22-25, doi: https://doi.org/10.1016/j.infrared.2016.01.004.
[28] M.H. Lee, and J.S. Cho, "Better thermochromic glazing of windows with anti-reflection coating," Thin Solid Films, vol. 365, Apr. 2000, pp. 5-6, doi: https://doi.org/10.1016/S0040-6090(99)01112-8.
[29] M.H. Lee, "Thermochromic glazing of windows with better luminous solar transmittance," Solar Energy Materials and Solar Cells, vol. 71, Mar. 2002, pp. 537-540, doi: https://doi.org/10.1016/S0927-0248(01)00135-0.
[30] H. Miyazaki, and I. Yasui, "Effect of buffer layer on VO_x film fabrication by reactive RF sputtering," Applied Surface Science, vol.252, Sept. 2006, pp. 8367-8370, doi: https://doi.org/10.1016/j.apsusc.2005.11.040.
[31] S. Saitzek, G. Guirleo, F. Guinneton, L. Sauques, S. Villain, K. Aguir, C. Leroux, and J.R. Gavarri, "New thermochromic bilayers for optical or electronic switching systems," Thin Solid Films, vol. 449, Feb. 2004, pp. 166-172, doi: https://doi.org/10.1016/j.tsf.2003.10.013.
[32] G. Xu, P. Jin, M. Tazawa, and K. Yoshimura, "Optimization of antireflection coating for VO_2-based energy efficient window," Solar Energy Materials and Solar Cells, vol. 83, Jun. 2004, pp. 29-37, doi: https://doi.org/10.1016/j.solmat.2004.02.014.
[33] H. Zhou, J. Li, S. Bao, J. Li, X. Liu, and P. Jin, "Use of ZnO as antireflective, protective, antibacterial, and biocompatible multifunction nanolayer of thermochromic VO_2 nanofilm for intelligent windows," Applied Surface Science, vol. 363, Feb. 2016, pp. 532-542, doi: https://doi.org/10.1016/j.apsusc.2015.12.045.
[34] N.R. Mlyuka, G.A. Niklasson, and C.G. Granqvist, "Thermochromic multilayer films of VO_2 and TiO_2 with enhanced transmittance," Solar Energy Materials and Solar Cells, vol. 93,

Sept. 2009, pp. 1685-1687, doi: https://doi.org/10.1016/j.solmat.2009.03.021.

[35] R. Hao, Y. Li, F. Liu, Y. Sun, J. Tang, P. Chen, W. Jiang, Z. Wu, T. Xu, and B. Fang, "Electric field induced metal–insulator transition in VO_2 thin film based on FTO/VO_2/FTO structure," Infrared Physics & Technology, vol. 75, Mar. 2016, pp. 82-86, doi: https://doi.org/10.1016/j.infrared.2015.12.012.

[36] X. Wang, Y. Cao,Y. Zhang, L. Yan, and Y. Li, "Fabrication of VO_2-based multilayer structure with variable emittance," Applied Surface Science, vol. 344, Jul. 2015, pp. 230-235, doi: https://doi.org/10.1016/j.apsusc.2015.03.116.

[37] G. Fu, A. Polity, N. Volbers, and B.K. Meyer, "Annealing effects on VO_2 thin films deposited by reactive sputtering," Thin Solid Films, vol. 515, Dec. 2006, pp. 2519-2522, doi: https://doi.org/10.1016/j.tsf.2006.04.025.

[38] H. Kakiuchida, P. Jin, and M. Tazawa, "Control of thermochromic spectrum in vanadium dioxide by amorphous silicon suboxide layer," Solar Energy Materials and Solar Cells, vol. 92, Oct. 2008, pp. 1279-1284, doi: https://doi.org/10.1016/j.solmat.2008.04.025.

[39] J. Zheng, S. Bao, and P. Jin, "TiO_2(R)/VO_2(M)/TiO_2(A) multilayer film as smart window: combination of energy-saving, antifogging and self-cleaning functions," Nano Energy, vol. 11, Jan. 2015, pp. 136-145, doi: https://doi.org/10.1016/j.nanoen.2014.09.023.

IOP Conf. Series: Materials Science and Engineering **409** (2018) 012017 doi:10.1088/1757-899X/409/1/012017

Analysis of Deformation and Texture Gradient in Shot Peened Pure Titanium

H P Yang and Y M Wang

School of Metallurgical Engineering, Xi'an University of Architecture and Technology, Xi'an 710055, China
E-mail: yanghp409@163.com

Abstract. The plastic deformation of shot peened pure titanium and deformed texture along the depth were simulated by finite element method(FEM) and visco-plasticity self-consistent(VPSC) method respectively. The results show that the stress state is compressive stress in three principle directions from subsurface to the depth at 600μm. The strain state is a complicated strain state and exhibits strain gradient along the depth. Strain gradient results in texture gradient in shot peened pure titanium. The texture in pure titanium with initial typical recrystallized texture is strengthened after shot peening, with strongest texture at the depth where the maximum absolute values of strain components exist.

1. Introduction

Pure titanium and its alloys have been widely used in many key fields such as aerospace, chemical, energy and ocean industries due to their high specific strength and excellent corrosion resistance [1]. During the manufacturing process of titanium engineering parts, shot peening is usually employed to improve fatigue strength and other mechanical properties. It is well known that compressive residual stress and work hardening effects which are induced by shot peening contribute to the property improvement [2, 3]. In addition, it has been verified that distinct texture evolution occurs during the surface plastic deformation [4-6]. Chen et al. [4] investigated texture evolution in pure iron subjected to surface mechanical attrition treatment by using steel balls in diameter of 8 mm impacting target and found that the density of texture was strengthened and reached the maximum intensity in the micro-sized regime as increased the depth from the top surface. As for the shot peening deformation, Man and Zhai [5] have reported that texture was strengthened after shot peening in Ti-6Al-4V alloy and a basal fiber texture was formed in the surface layer, whereas Maawad et.al [6] have shown that the texture in hot rolled Ti-2.5Cu was weakened after shot peening and the texture intensity was weakest at the top surface. Extensive research has indicated that texture influence fatigue properties and other mechanical properties obviously [5, 7]. Therefore, in order to improve the performance of pure titanium and its alloy, further investigation should be conducted to clarify the texture evolution mechanism caused by shot peening.

In the present study, computer simulations by combining FEM and VPSC were carried out to investigate the deformation and texture evolution in shot peened pure titanium.

2. Model description

In this work, a 3D numerical simulation was carried out by using commercial finite element code ABAQUS/Explicit to calculate the stress and strain distribution in pure titanium after shot peening. A symmetry-cell model with single impacting ball was used to simulate shot peening process. The target with dimension of 2mm×2mm×4mm was used. A C3D8R 8-node linear brick element with reduced integral and hourglass control was used for the target. A quarter of the steel ball with a diameter of 0.8 mm was modelled. A C3D4 4-node linear tetrahedron element was used for the ball. Additionally, the refinement mesh near the treated surface with 1 mm in depth has been adopted to improve the calculation accuracy. As for boundary conditions, the bottom and two side faces were fixed, and symmetry displacement conditions were applied on the two faces which were in the impacted area. A general contact with an isotropic friction coefficient of 0.1 was used.

Plasticity deformation behavior of pure titanium was represented by Hollomon's equation as equation (1):

$$\sigma = K\varepsilon_P{}^{n} \tag{1}$$

where, σ is true stress, ε_p is plastic strain, K is a material constant and n is strain hardening exponent. In this study, K=780.95 MPa and n=0.159 were used. Steel shot was assumed to be elastic during the impact. The material parameters used in the simulation are summarized in Table 1.

Table 1. Material parameters used in FEM calculation.

Objects \ Parameters	Density [kg/m^3]	Elastic modulus [GPa]	Poisson's ratio
Target	4510	114.8	0.32
Steel shot	7800	200.0	0.26

In the FEM simulation, an initial velocity of 80 m/s was imposed on the steel ball. The velocity was decreasing during impacting. When the velocity reached zero, it began to rebound. Strain along the depth at this moment calculated by FEM was treated as load for VPSC calculation.

The shot peened texture along the depth was simulated using VPSC code. The detailed VPSC model and the predominant twin reorientation scheme are described in the work of Tomé and Lebensohn et al. [8, 9]. The Voce law, shown by equation (2), is used to describe hardening behavior on different slip and twin systems.

$$\tau^s(\Gamma) = \tau_0 + (\tau_1 + \theta_1\Gamma)\left[1 - \exp\left(\frac{-\theta_0\Gamma}{\tau_1}\right)\right] \tag{2}$$

$$\Gamma = \int_0^t \sum_s \left|\dot{\gamma}^s\right| dt \tag{3}$$

where Γ in equation (3) is the accumulated shear within time t on all slip systems, and $\dot{\gamma}^s$ is the shear rate on slip system s. In the simulation, the prismatic <a> slip $\{10\bar{1}0\}\langle\bar{1}2\bar{1}0\rangle$, the basal <a> slip $\{0001\}\langle\bar{1}210\rangle$, the first-order pyramidal <c+a> slip $\{10\bar{1}1\}\langle 11\bar{2}3\rangle$and two twinning modes, namely the tensile twinning $\{10\bar{1}2\}\langle\bar{1}011\rangle$ and compressive twinning $\{11\bar{2}2\}\langle\bar{1}\bar{1}23\rangle$, were taken into account. τ_0, τ_1, θ_0 and θ_1 are hardening parameters for each deformation system. The hardening behaviors representing the four parameters of the five deformation modes that lead to the best texture predictions are illustrated in Figure 1.

The texture evolution was characterized by the orientation distribution function (ODF) in the Euler space of (φ_1, ϕ, φ_2) following Bunge convention. When describing a specific orientation, the relationship between crystal and sample coordinate system is defined as $[\bar{1}010]$, $[\bar{1}2\bar{1}0]$ and [0001] corresponding to 1, 2 and 3 respectively. The texture was calculated using MTEX toolbox [10] by imposing the orthorhombic sample symmetry for shot peening process.

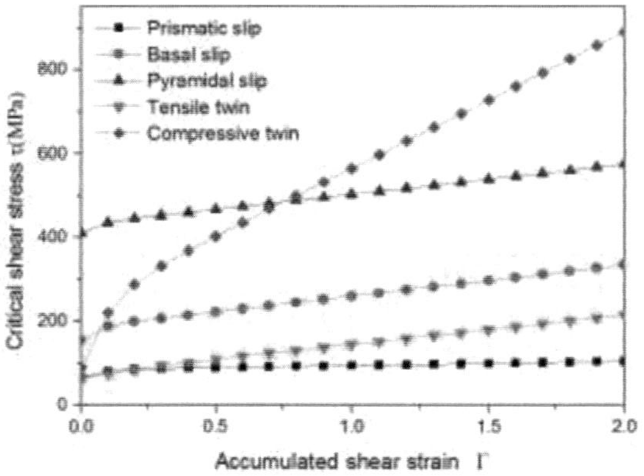

Figure 1. Hardening behaviors of the slip/twin systems.

3. Results and discussion

The schematic diagram of shot peening modeled by a single ball impacting and the calculated stress components by FEM are shown in figure 2(a) and figure 2(b) respectively. It is noted that three shear stress components can be negligible compared with the values of three normal stress components. That is to say, the stress state can be approximated three normal stress components as three principle stresses. At the top surface, the stress state can be described as compressive stress in two directions due to small value of σ_{33}. With the depth increased to about 600 μm, the stress state is compressive stress in three directions, as illustrated in figure 2(b). As the depth further increased towards interior, the stress state is changed from tensile stress in two directions and compressive stress in one direction to tensile stress in three directions, but the value of three normal stresses is relatively small.

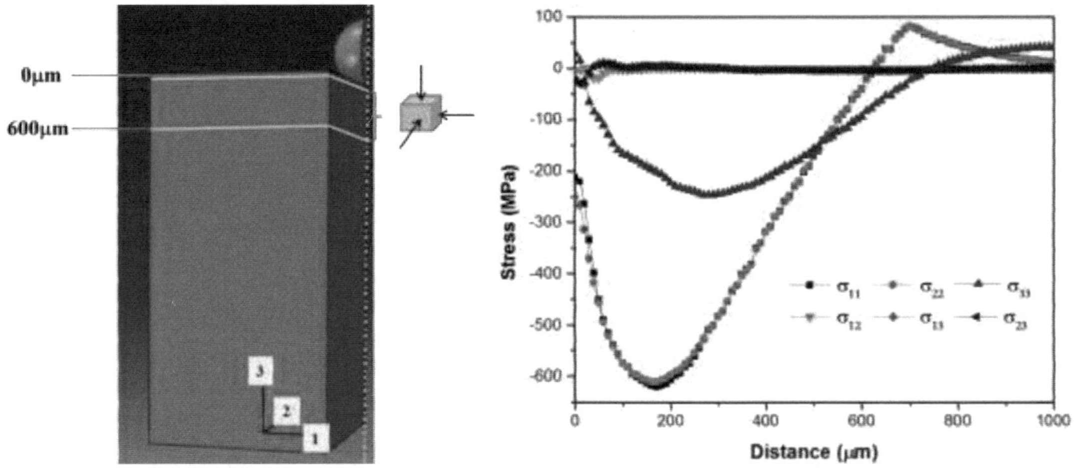

(a) (b)

Figure 2. The schematic diagram of shot peening (a) and the calculated stress components (b).

The simulated equivalent plastic strain (PEEQ) and the strain components are given in figure 3. The PEEQ along the depth from the top surface after shot peening is increased firstly to maximum value and then decreased continuously to zero. The distribution of PEEQ is qualitatively good agreement with the work of Frija et al. [11]. The corresponding strain components are shown in figure 3(b). It can be seen that the strain state shows a gradient distribution along the depth and the strain affected range is between the top surface and the depth at 600μm. Among those strain components, the normal and shear strain components are nonzero except for shear strain ε_{12}, which means that shot peened deformation is a complicated strain state. And those nonzero strain components exhibit similar

IOP Conf. Series: Materials Science and Engineering **409** (2018) 012017 doi:10.1088/1757-899X/409/1/012017

trend, i.e., with increasing the depth, the magnitude of strain is increased firstly to maximum value at depth around 60 μm and then decreased to zero at the depth near 600 μm.

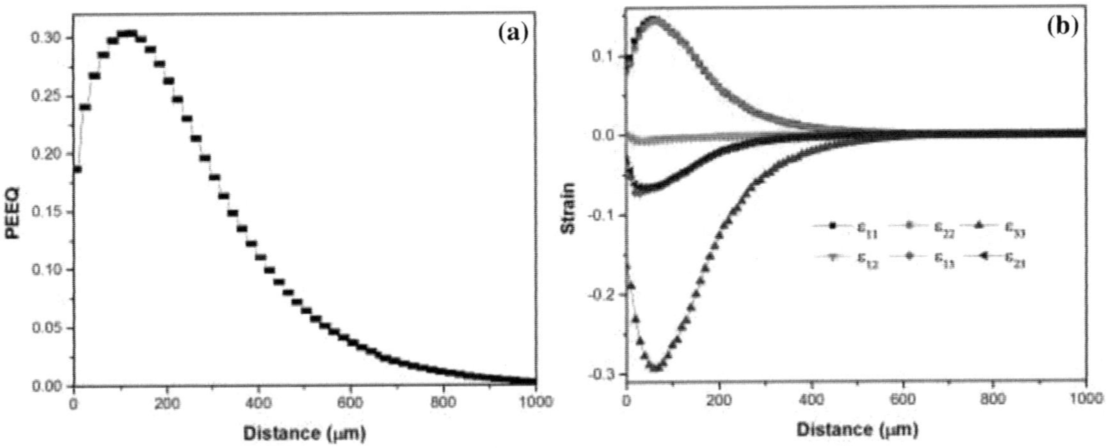

Figure 3. The distribution of PEEQ (a) and strain components (b).

The initial texture of pure titanium is illustrated by figure 4. It exhibits a tilt-model distribution of the basal poles and the $(\bar{2}110)$ pole is distributed along the rolling direction, which is typical texture commonly found in recrystallized pure titanium and can be characterized by basal poles tilted from the normal direction to transverse direction, i.e., two main components with Euler angles (0°, 40°, 30°) and (0°, 40°, 0°) [12]. Figure 5 shows the VPSC simulated texture at different depths after shot peening. From the (0001) pole figures, it can be seen that the deformed texture at different depths is all strengthened compared with initial texture. Moreover, it is noted that strongest texture is formed at the depth where the maximum absolute values of strain components exist, corresponding to the depth of 60 μm. The simulated results are contrary to the work of Maawad et.al [6], where the shot peened texture is weakened along the depth and the texture is weakest at the top surface, whereas they are agreement with the experimental result of Man et al. [5], who reported that texture was strengthened by shot peening.

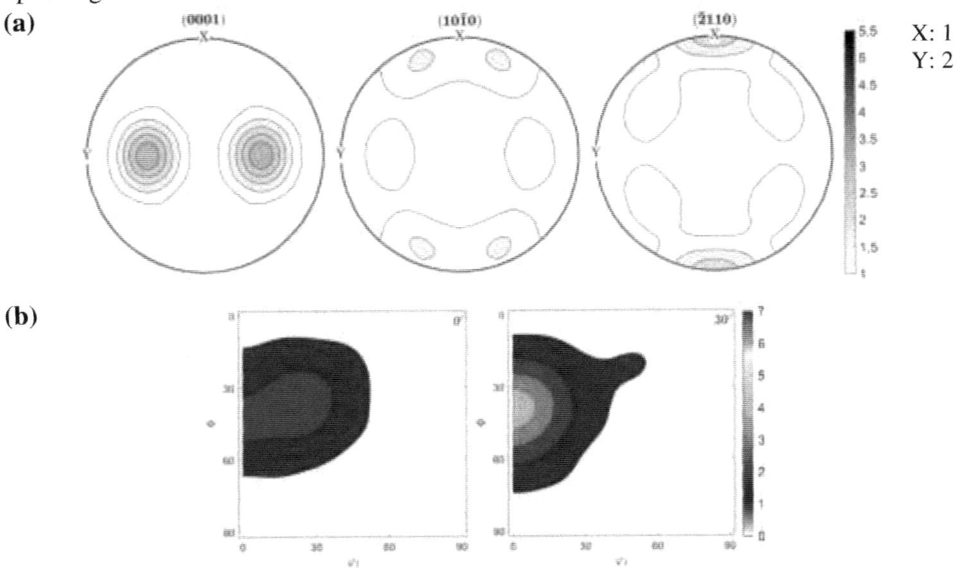

Figure 4. The pole figures of (0001), $(10\bar{1}0)$ and $(\bar{2}110)$ (a) and the ODF constant φ_2=0°and φ_2=30° (b) in initial pure titanium.

IOP Conf. Series: Materials Science and Engineering **409** (2018) 012017 doi:10.1088/1757-899X/409/1/012017

In order to investigate the variation of specific orientation, the ODF constant φ_2=0°and φ_2=30° at different depths is shown in figure 6. It can be seen that the deformed texture under complicated strain state caused by shot peening is quite different from plane strain rolled texture [12], the initial two main components (0°, 40°, 30°) and (0°, 40°, 0°) are strengthened simultaneously, and both are shifted towards little ϕ and it shifts much more at the depth of 60 μm with peak intensity at (0°, 25°, 30°) and (20°, 25°, 0°). Also, the ND//(01$\bar{1}$4) and ND//($\bar{1}$2$\bar{1}$7) fiber textures are formed at depth of 60 μm. And it can be seen the weak basal fiber texture or basal components are formed at different depths. At the depth of 300 μm, the deformed texture is nearly unchanged compared with the initial state.

As known, the deformed texture is determined by the activity of slip/twin system during plastic deformation. The simulated activity of slip/twin system is shown in figure 7. It is noted that prismatic slip, basal slip and both tensile twin and compressive twin are activated in shot peening. And it can be seen the activation of prismatic slip under complicated strain state is decreased compared to plane strain state [13]. However, further research is necessary to elucidate the specific mechanism.

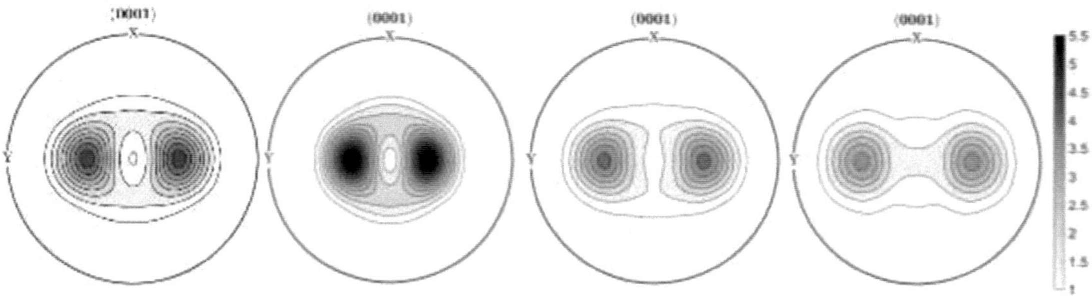

(a) (b) (c) (d)

Figure 5. The simulated (0001) pole figure of shot peened pure titanium at depth of (a) 0 μm, (b) 60 μm, (c) 220 μm and (d) 300 μm from the top surface.

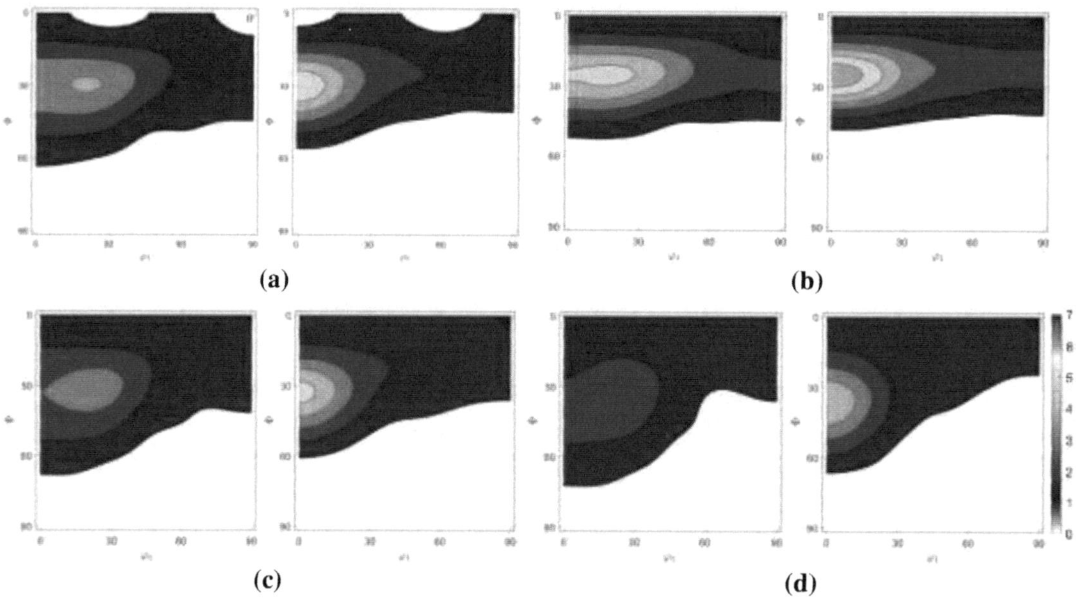

Figure 6. The simulated ODF constant φ_2=0°and φ_2=30° sections of shot peened pure titanium at depth of (a) 0 μm, (b) 60 μm, (c) 220 μm and (d) 300 μm from the top surface.

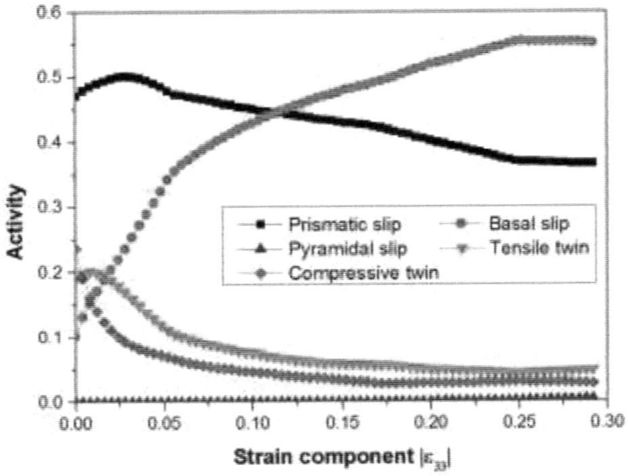

Figure 7. Slip activity for slip/twin systems at depth of 60μm in shot peened pure titanium.

4. Conclusions

The stress state in shot peened pure titanium can be described as compressive stress in three principle directions between subsurface and the depth at 600 μm. The strain state is a complicated strain state and the distribution of strain is in gradient along the depth.

The shot peened pure titanium with initial typical recrystallized texture is strengthened to different degrees at the depth between top surface and depth of 300μm. And the larger the magnitude of strain components, the stronger the deformed texture.

Acknowledgments

Financial support from Science and Technology Department of Shaanxi Province (2017GY-115) and Education Department of Shaanxi Province (16JK1466) are gratefully acknowledged.

References

[1] Lütjering G and Williams J C 2007 *Titanium* (Berlin: Springer-Verlag)
[2] Gao Y K 2016 *Acta Metall. Sin.* **52** 915-22
[3] Soady K A 2013 *Mater. Sci. Technol.* **29** 637-51
[4] Chen W Y, Tong W P, He C S, Zhao X and Zuo L 2012 *Mater. Sci. Forum* **702-703** 920-23
[5] Man C S and Zhai T G 2006 *Final Performance Report*
[6] Maawad E, Brokmeier H G and Wagner L 2010 *Solid State Phenom.* **160** 141-46
[7] Subhasis S and Gurao N P 2017 *Mater. Sci. Eng. A* **691** 100-09
[8] Tomé C N, Lebensohn R A and Kocks U F 1991 *Acta Metall. Mater.* **39** 2667-80
[9] Lebensohn R A and Tomé C N 1994 *Acta Metall. Mater.* **41** 2611-24
[10] Bachmann F, Hielscher R and Schaeben H 2010 *Solid State Phenom.* **160** 63-68
[11] Frija M, Hassine T, Fathallah R, Bouraoui C and Dogui A 2006 *Mater. Sci. Eng. A* **426** 173-80
[12] Zhong Y, Yin F X and Nagai K 2008 *J. Mater. Res.* **23** 2954-66
[13] Bozzolo N, Dewobroto N, Wenk H R and Wagner F 2007 *J. Mater. Sci.* **42** 2405-16

IOP Conf. Series: Materials Science and Engineering **409** (2018) 012018 doi:10.1088/1757-899X/409/1/012018

Application of Taguchi Method in Investigating The Effect of Layer Thickness and Fill Angle on FDM Parts

Siti Maryam Md Nor[1], Mohd Nizam Sudin[2] and Muhamad Farid Bin Razali[3]

Department of Design and Innovation, Faculty of Mechanical Engineering, Universiti Teknikal Malaysia Melaka, Centre for Advanced Research on Energy (CARe) Melaka, Malaysia

E-mails faridrazali27@gmail.com, maryam_3101@yahoo.com, nizamsudin@utem.edu.my

Abstract. This paper studies the effect of Layer Thickness and Fill Angle on tensile stress of Fused Deposition Modeling (FDM) parts for Acrylonitrile Butadiene Styrene (ABS) and Carbon Fiber Reinforced ABS (CFRABS) materials by using Taguchi Method, signal-to-noise (S/N) ratio analysis and Analysis of Variance (ANOVA). From the results, it was found that layer thickness contribute to the tensile stress performance of ABS and fill angle has significant effect on tensile stress of CFRABS. It can be concluded that the optimized parameter obtained in this study for both materials are 0.18mm layer thickness and 90° fill angle.

1. Introduction

Fused Deposition Modelling (FDM) is a popular Additive Manufacturing (AM) machine that able to manufacture parts with complex geometries easily in a minimal time compared to traditional machine [1]. Acrylonitrile butadiene styrene (ABS) has been used as a homopolymer FDM feedstock due to its properties which provides chemical resistance, heat stability, toughness, impact strength, rigidity and easiness of process ability [2]. Reference [3] stated that the range of the application of polymer can be extended further by the reinforcement of carbon fiber, ceramics, metals and fiber glass in polymer matrix. Reference [4] revealed that one of the possible ways to increase the strength of FDM printed thermoplastic parts is by adding reinforced materials such as carbon fibre into plastic materials to form thermoplastic matrix carbon fibre reinforced plastic (CFRP) composites. Nowadays, CFRP are applied in various industries. The examples of application are components in automotive and endoscopic surgery. Taguchi approach provides statistical method of analyzing is the best way to investigate the effect of FDM process parameters such as layer thickness and fill angle on tensile stress of FDM part for ABS and CFRABS where details of the experiment can be discussed accurately through S/N ratio and ANOVA analyses.

2. Methodology

2.1. Samples Preparation and Analysis

The geometrical model of Dog bone specimen ASTM D638 was produced by using CATIA V17 solid modelling software and saved as a stereolithographic file (STL). This .STL file then was loaded into the slicing software or interface platform namely 'Flash Print'. The file was converted to G-file and uploaded into FDM Flashforge Dreamer modeller. The 3D model was sliced into layers at the required

IOP Conf. Series: Materials Science and Engineering **409** (2018) 012018 doi:10.1088/1757-899X/409/1/012018

thickness and fill angle. A molten Acrylonitrile Butadiene Styrene (ABS) thermoplastic and Carbon Fiber Reinforced ABS (CFRABS) with a diameter of 1.75mm were used as FDM feedstock and extruded by the heated nozzle which was deposited on the bed as the nozzle travels along the platform bed. 9 samples of ABS and CFRABS respectively were produced and detached from the heat bed upon completion. The statistical software of Minitab 17 and Microsoft Excel 2010 were utilized for the S/N ratio and ANOVA analysis.

2.2. Taguchi Method

Taguchi technique was applied in this project as it is a powerful statistical tool in simplifying experimental plan where number of experiments, time and cost can be reduced without giving effect to the product's quality [5]. This approach proposes an experimental plan that consist series of L_9 Orthogonal Array as in Table 1 which provides combination of process parameters and their level to the required performance quality of the product [5]. Hence, correlation between process parameters and their levels to the quality response can also be determined using this tool [6]. 'The-larger-the-better' objective function was implemented to meet the purpose of this project that is maximizing the tensile stress of FDM parts. Several steps in parameter design had been proposed by Günay and Yücel [7] to optimize the process parameters was implemented in this research as the following:

- Process parameters and performance characteristics were selected.
- The number of levels for the process parameters and possible interactions between the process parameters were determined.
- Appropriate orthogonal array and assignment of process parameters to the orthogonal array was selected.
- The experiments were carried out based on the arrangement of the orthogonal array.
- Total loss function and the S/N ratio were calculated.
- Experimental results using the S/N ratio and ANOVA were analyzed.
- Optimal levels of process parameters were selected.
- Optimal process parameters were verified through the confirmation experiment.

Table 1. L_9 Orthogonal Array

Experiment No	Process Parameters/Levels	
	A	B
1	1	1
2	1	2
3	1	3
4	2	1
5	2	2
6	2	3
7	3	1
8	3	2
9	3	3

2.3. Selection of Process Parameters

In this experiment, layer thickness and fill angle parameters are variable process parameters while humidity and temperature were kept constant. Layer thickness indicates the height of deposited slice from FDM nozzle whereas fill angle is the method where roads can be deposited to fill the interior part. In this experiment, values of the layer thickness for level 1, level 2 and level 3 are 0.18mm, 0.25mm and 0.31mm respectively whereas 30°, 45° and 90° were used for each level of the fill angle respectively. This experiment used Taguchi method which proposed L_9 Orthogonal Array of three rows and nine columns with three process parameters at three different levels.

2.4. *Tensile Test Experiment*

The research test in this project was tensile strength test. The tensile machine used in testing the Specimens of ABS and CFRABS was Universal Test Machine: INSTRON 8832. The dog bone of both specimens were clamped and stretched until fracture. The data of the tensile strength was recorded and display to the computer.

3. Result And Analysis

3.1. *Signal-to-Noise Ratio*

The signal-to-noise ratio is an effective tool in measuring sensitivity of the quality being investigated to the uncontrollable factors (error) in the experiment [8]. There are two categories to classify factors that affect system performance which are controllable factors and noise factors. Chamber temperature, humidity, and model temperature are the example of noise factors that cannot be controlled during manufacturing or fabricating process [9]. This indicates signal and the undesirable value (Standard Deviation from Mean) represents noise for the output characteristic [10]. In this experiment, the higher S/N ratio was used because it produced smaller product variance around the target value by means of minimizing the expected loss of function quality. The optimal level of the process parameters can be obtained by the level with the highest S/N ratio which keeps the response on target while making it insensitive to the variation of noise factors. The objective of this experimental plan is to gain the maximum tensile stress of FDM part. Hence, the larger the better quality characteristic was implemented in analysis for better accuracy [11].

$$\eta = -\log\left(\frac{1}{n}\sum_{i=1}^{n}\frac{1}{y_i^2}\right) \qquad (1)$$

Where η is the sample size and y is the tensile stress in MPa.

Table 2. Obtained S/N Ratio Values Of The Tensile Stress Response

Experiment No	Parameters		Responses			
	Layer Thickness (mm)	Fill Angle (°)	Tensile Stress (MPa)		SNRA for Tensile Stress	
			ABS	CFRABS	ABS	CFRABS
1	0.18	30	29.643	15.933	29.4384	24.0460
2	0.18	45	30.478	15.805	29.6797	23.9759
3	0.18	90	32.526	21.619	30.2446	26.6967
4	0.25	30	25.851	16.481	28.2495	24.3397
5	0.25	45	25.815	15.207	28.2374	23.6409
6	0.25	90	29.002	18.795	29.2486	25.4808
7	0.31	30	21.043	16.834	26.4622	24.5237
8	0.31	45	17.090	16.866	24.6548	24.5402
9	0.31	90	22.404	18.441	27.0065	25.3157

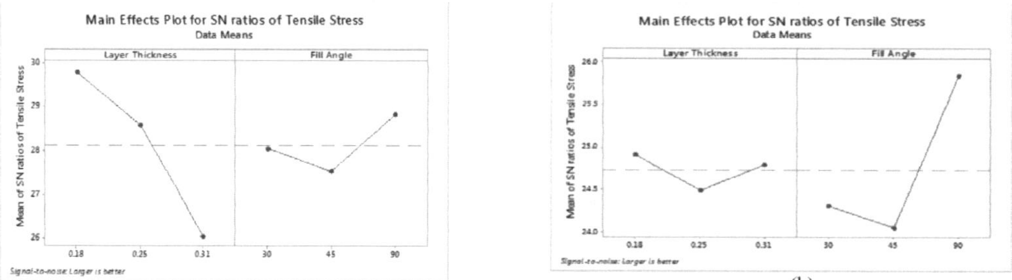

(a) (b)

Figure 1. Graph of S/N Effects of process parameter on Tensile Stress: (a) ABS (b) CRFABS

Table 2 shows that the highest level of S/N ratio for both materials is experiment 3 with 0.18mm layer thickness and 90° fill angle which produced S/N ratio of 30.2446 and 26.6967 respectively prior to their highest values of tensile stress.

As depicted in Fig.1 (a) and (b), the combination of parameters and their levels for ABS and CFRABS, 0.18mm layer thickness and 90° fill angle has produced maximum value of S/N ratio and it yields optimum quality characteristic with maximum variance of targeted value for tensile stress.

Table 3. S/N Ratio Values For Tensile Stress By Factor Level

Level	Control Factors			
	Layer Thickness		Fill Angle	
	ABS	CF-ABS	ABS	CF-ABS
1	29.79	24.91	28.05	24.30
2	28.58	24.49	27.52	24.05
3	26.04	24.79	28.83	25.83
Delta	3.75	0.42	1.31	1.78
Rank	1	2	2	1

Table 4. S/N Ratio Values For Tensile Stress By Factor Level

Source	Df		Adj SS		Adj Ms		F		C%	
Materials	ABS	CF-ABS	ABS	CF-ABS	AB	CF-ABS	ABS	CF-ABS	ABS	CF-ABS
Layer Thickness	2	2	21.936	0.2821	10.9679	0.1410	30.41	0.41	84.42%	3.90%
Fill Angle	2	2	2.604	5.5615	1.3021	2.7808	3.61	8.03	10.02%	76.94%
Error	4	4	1.443	1.3849	0.3607	0.3462			5.55%	19.16%
Total	8	8	25.983	7.2285						

Based on Table 3, tensile stress increases as the layer thickness value increases for ABS material and ranked number 1 compared to layer thickness for CFRABS is ranked as number 2 where its inconsistency of layer thickness performance could be seen. Meanwhile, fill angle ranked as number 1 for CFRABS and number 2 for fill angle of ABS material. The analysis of S/N ratio in Table 4 had shown that the layer thickness is the most significant parameter for ABS which influenced the tensile stress by 84.42% of contribution percentage and fill angle is the most significant for CFRABS which contributed about 76.94%.

3.2. Analysis of Variance (ANOVA)

Analysis of Variance (ANOVA) is often used due to its capability in investigating parameters that significantly affect the output characteristic [8]. The ANOVA computations were tabulated based on procedure in Table 5. F-test was applied to study which parameters have significant effect on the build time response. F ratio is termed as ratio of the mean square error to the residual error [12]. Usually, when $F > 4$, it means that the change of the process parameter has significant effect on the quality characteristic [8].

As depicted in Table 5, results showed that layer thickness gave P-value less than 0.05 which is 0.001 for ABS material and fill angle indicated valid P-value result of 0.046 for CFRABS material. The results were also supported with F-test which both layer thickness and fill angle produced values greater than 4 for ABS. On the other hands, only fill angle for CFRABS that showed valid F-Test result. The contribution percentage of layer thickness marked the highest percentage for ABS which is 86.95% followed by fill angle with lesser contribution of 9.68%. Meanwhile, layer thickness gave least contribution to the output of CFRABS by 4.37% and fill angle showed largest contributions of 75.14%. Based on Table 5, fill angle is found to be negligible to the tensile response of ABS since the P-value is 0.066 and in contrast, layer thickness is found negligible for CFRABS as it produced P-value of 0.679. In this study, the effect of fill angle and interaction between fill angle with layer thickness for ABS were found to be negligible effect and in comparison to CFRABS, the effect of layer thickness and interaction between layer thickness with fill angle were also found to be negligible.

Table 5. Anova Results For Effects Of Process Parameters On Tensile Stress

Source	Df		Seq SS		F		C%		P-Value	
Materials	ABS	CF-ABS	ABS	CF-ABS	ABS	CF-ABS	ABS	CF-ABS	ABS	CF-ABS
Layer Thickness	2	2	175.534	1.387	51.71	0.43	86.95%	4.37%	0.001	0.679
Fill Angle	2	2	19.546	23.852	5.76	7.34	9.68%	75.14%	0.066	0.046
Error	4	4	6.790	6.503			3.36%	20.49%		
Total	8	8	201.870	31.743						
S=	1.30285	1.27508								
R-sq=	96.64%	79.51%								

3.3. *Confirmation Test*

Once the optimal combination and levels of the process parameter had been obtained, the final step was to verify the experimental result against theoretical value. Based on equation obtained from regression analysis, the average percentage error was 0.075% with the reliability of 99.92% for ABS and -0.02% percentage error and reliability of 100% for CFRABS. This verifies that experimental result is strongly correlated with the estimated result.

4. Conclusions

Utilizing Taguchi Approach proved that number of experiments can be simplified and reduced. Based on this method, 9 experiments were sufficient to analyse the results and it was found that layer thickness had a critical effect to the tensile stress of ABS and fill angle had significantly affect CFRABS performances. The effect of fill angle and interaction between fill angle with layer thickness for ABS were found to be negligible effect and in comparison to CFRABS, the effect of layer thickness and interaction between layer thickness with fill angle were also found to be negligible. Referring to ANOVA analysis for ABS, it was found that layer thickness had 86.95% contributions on tensile stress followed by least contribution of fill angle with 9.68%. Meanwhile, ANOVA for CFRABS showed that the effect of layer thickness and interaction between layer thickness with fill angle were also found to be negligible. According to S/N Ratio for both materials, the optimal process parameters and their levels were obtained at level 1 (0.18mm) of layer thickness, level 3 (90°) of fill angle.

Acknowledgment

Authors would like to thank Universiti Teknikal Malaysia Melaka (UTeM) and Ministry of Higher Education Malaysia for supporting this research under Fundamental Research Grant Scheme (FRGS) FRGS/1/2015/TK03/FKM/02/F00268.

References

[1] P. K. Gurrala, and S. P. Regalla, "Friction and Wear Behaviour of ABS Polymer parts made by Fused Deposition Modelling," Technology Letters, Vol.1, pp. 13-17, 2014.
[2] J. Sudeepan, K. Kumar, T. K. Barman and P. Sahoo, "Study of tribological behavior of ABS/ CaCO3 composite using grey relational analysis", Procedia Materials Science, vol. 6, pp.682–691, 2014.
[3] S. Rupinder, S. B. Kamaljit and S. Harwinder, "Wear behaviour of FDM parts fabricated by composite material feed stock filament," Rapid Prototyping Journal, vol. 3, pp. 350–357, 2016.
[4] N. Fuda, C. Weilong, J. Qiu, W. Junhua and W. Shiren, Additive manufacturing of carbon fiber reinforced thermoplastic composites using fused deposition modelling," Composites Part B, Vol. 80, pp. 369-378, 2015.

[5] A.C. Mitra, M. Jawarkar and T.S. Kiranchand, "Implementation of Taguchi Method for Robust Suspension Design," Procedia Engineering, 144, pp.77 – 84, 2016.

[6] S. Kumar, V. N. Kannan and G. Sankaranarayanan, "Parameter optimization of ABS-M30i Parts produced by fused deposition modelling for minimum surface roughness," International Journal of Current Engineering and Technology, vol. 3, pp. 93-97, 2014.

[7] Günay, M., Yücel, E., 2013, Application of Taguchi method for determining optimum surface roughness in turning of high-alloy white cast iron, Measurement, 46(2), 913-919.

[8] B. H. Lee, J. Abdullah and Z. A. Khan, "Optimization of rapid prototyping parameters for production of flexible ABS object, Journal of materials processing technology," Journal of materials processing technology, vol.1, pp. 54-61, 2005.

[9] M. Alhubail, D.Alenezi and B. Aldousiri, "Taguchi-based optimisation of process parameters of fused deposition modelling for improved part quality," International Journal Engineering Research Technology, vol. 2, pp. 2505-2519, 2013.

[10] N. Raghunath and P. M. Pandey, "Improving accuracy through shrinkage modelling by using Taguchi method in selective laser sintering," International journal of machine tools and manufacture, vol.6, pp. 985-995, 2007.

[11] H. Amiruddin, "Application of Taguchi method in optimization of design parameter for turbocharger vaned diffuser," Industrial Lubrication and Tribology, vol. 3, pp. 409-413, 2017.

[12] C. Mendonsa, K.V. Naveen, P. Upadhyaya and V.D. Shenoy, "Influence of FDM Process Parameters on Build Time Using Taguchi and ANOVA Approach," International Journal of Science & Research., vol. 2, pp. 2013-2016, 2013.

IOP Conf. Series: Materials Science and Engineering **409** (2018) 012019 doi:10.1088/1757-899X/409/1/012019

Suppression of Structure-Borne Noise in a Rectangular Enclosure

N.Fathiah Waziralilah[1], Aminudin Abu[2], Lee Kee Quen[4] M Hishamuddin Unuh[5], Lim Meng Hee[*3]

Intelligent Dynamics System, Malaysia Japan International Institute of Technology, Universiti Teknologi Malaysia, Kuala Lumpur, Malaysia
[*]Institute of Vibration and Noise, Universiti Teknologi Malaysia Kuala Lumpur, Malaysia

E-mails: [1]tiah14@gmail.com; [2]aminudin.abu@utm.my; [4]lkquen@utm.my; [5]hishamunuh@gmail.com, [3]mhlim.kl@utm.my

Abstract. To date, there is few study on the relation of the structure mode and noise mode as most of researches are only focusing on the modal analysis of the structure-borne noise. This paper presents the extension of study of structure-borne noise analysis in a developed rectangular enclosure which comprises the reduction of structure-borne noise in the enclosure by mitigation method. The suppression of structure-borne noise is significant due to its agonizing effect of causing annoyance and dizziness thus affecting one's life productivity. The vibrational and sound responses acquired from the modal and sound analysis performed on the enclosure using LMS Spectral Testing. Mitigation is executed by placing the stiffeners made up of aluminum sheet metal perpendicularly to the surface of aluminum sheet-metal of the enclosure. The research purpose is to diagnose the relationship between the structural and noise mode that is obtained from acoustic modal experiment. Resonance between the vibration and noise spectrum needs to be avoided to reduce the structure-borne noise in the enclosure. Before the reduction, the frequency from vibration spectrum is 57.8 Hz while from the noise spectrum is 60.3 Hz. It is found that both of the frequencies are close to each other, thus resulting the occurrence of resonance. After the suppression has been made, the frequency from vibration has declined to 38.8 Hz while the frequency from noise spectrum has increased to 72.8 Hz. The large gap between the frequencies from vibration and noise spectrum indicates that the resonance has been avoided. Therefore, the suppression of the structure-borne noise has successfully attained.

1. Introduction

Vibration and noise are the dual element that persistently present in high-tech millennium by daily basis [1]. Vibration and noise are closely related as noise is initiated from a vibrating body and both vibration and noise have similar mechanism as they are imparted as waves through a medium such as solid and fluid. Intrusion of noise causes vexation that perpetually occurs at home and workplace. Thus, obligation of noise suppression is a matter currently concentrated on by authorities in many countries [2]. Noise can be differentiated into two which are structure-borne noise and air-borne noise. Air-borne noise is transmitted by air and atmosphere while structure-borne noise is transmitted via solid structures. The search for constructible space has impelled the building developer to implement their project much nearer to transportation corridors than previous project. This means the structure-borne noise has become the biggest contributor to the noise from transportation systems transmitted into building [3]. On top of that, as many machineries in a building such as hydraulic equipment can produce a substantial amount of vibration, the structure-borne noise is a never-ending demanding problem in engineering [4]. In our life, we are surrounded by many structure, the noise generated by vibration in the structure could cause annoyance and distraction. In Findeis study case, occupants of an apartment stated that the vibrations near to the walls cause an uneasy feeling in their stomachs which very quickly increased to the point of feeling sick [5]. It is affirmed that structure-borne noise causes dizziness and nausea. Thus, detail study on structure-borne noise in order to reduce the sickness is strongly needed. A deep study and analysis on structure-borne is essential and beneficial for every human that demand a quality and productive life. Studies have shown that the concentration of a person can be incapacitated by certain levels of noise therefore leading to a stress response as a noisy place can increase one's feelings of negative mood [6]. When the structure-borne noise is successfully reduced, the environmental background noise is indirectly reduced thus increasing one's concentration ability which will affect the job and daily activities. Nowadays, the living population of people has concentrated in urban areas and due to insufficiency of land space, development of high building roses rapidly. The use of lightweight structure is the current building construction trend, where it is more economic. Nevertheless, these light-weight wall partitions and floor-ceiling assemblies has substantially less sound insulation. This issue has become one of the significant concerns of the structural noise problem [7]. By improving the structure-borne noise in a building system, for instance, people can have their own acoustical privacy and inner peace by straying away from the noise intrusion. Several studies have been made in suppressing the structure-borne noise. Satish et al. [8] apply structural modification in reducing the structure-borne noise. In Domadiya et al. [9] study, the transmission and emission are minimized by embedment of periodic stiffeners within the wooden panels. Sung [10] presented an analysis on reduction of structure-borne noise using an air-voided elastomer. To date, there is only few studies on the acoustical modal analysis as most of the researches concentrated on modal analysis solely. Therefore, the structure and noise mode and its relationship in the developed enclosure are scrutinized.

2. Theory

2.1. *Theory of Rectangular Enclosure*

In this research, the developed enclosure is in rectangular shape as rectangular volume has a simple analytical solution for its natural frequencies [11].

$$\sqrt[\frac{c}{2}]{(\frac{i}{L_x})^2 + (\frac{j}{L_y})^2 + (\frac{k}{L_z})^2}\ Hz \qquad (1)$$

Where L_x is measured from the front bottom to the centre of the front enclosure, L_y is measured as the average distance between end to end of the enclosure and L_z is the average height of the cavity measured from bottom to top of the enclosure. The speed of sound is calculated at room temperature, c = 343 m/s and the indexes for normal modes of vibration i = 0, 1, 2, j = 0, 1, 2, ... and k = 0, 1, 2, ...

2.2. *Theory of Modal Acoustic*

This section only provide a concise theoretical overview. Numerous references can be consulted for additionally insights regarding the model formulations elucidating the dynamic behaviour of vibro-acoustic system [12,13].

Here boundaries of the enclosure will be modelled as rigid walls, so the dynamic behaviour of a pure acoustic system is described. By considering a three-dimensional closed acoustic system with rigid or finite impedance but non-vibrating boundaries, the governing equation of the system, excited by a point monopole of volume velocity at r_0 can be written as [12]:

$$\nabla^2 p(r,t) - \frac{1}{c^2}\ddot{\mathrm{p}}(\mathrm{r.t}) = -p\dot{\mathrm{q}}\delta(\mathrm{r} - r_0) \qquad (2)$$

where p is the sound pressure, which is a function of space r and time t; c is the speed of sound; $\ddot{p}$ is the density of the medium; and $\dot{q}$ is the volume velocity. The boundary condition over the rigid surface S is

$$\frac{\partial p}{\partial n} = 0 \; over \; s \qquad (3)$$

where n is the internal unit normal to S. The discretization of the continuous acoustic wave equation is based on the finite element formulation. The acoustic domain of volume V is represented by an assemblage of acoustic finite elements. The pressure distribution within an element is interpolated in terms of the nodal pressures by using shape functions.

3. Methodology

The research methodology approach is done experimentally, comprises of two phases of experiment which consist of modal experiment and sound experiment. Validation of the experimental setup has been done using finite element analysis performed in Msc Nastran and Patran [16]. It is prior to validate the experimental setup to ensure great reliability of the data collected. The first phase is the analysis on structure-borne noise in the enclosure and second phase is the re-analysis of enclosure after the suppression of structure-borne noise in the enclosure. Both experiments are conducted simultaneously for data reliability enhancement. LMS Spectral Testing is performed on the enclosure, the acoustical vibration is excited using shaker and the responses are recorded in term of acceleration and sound pressure, respectively. The accelerometer that detect the vibration is attached on the surface of aluminum sheet metal of the enclosure while the microphone that detect the noise is positioned 1 m away from the front surface of the enclosure. The setup of the experiment is illustrated in Fig 1.

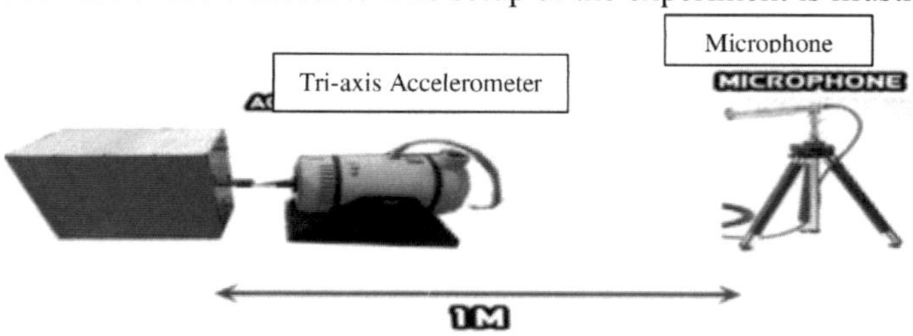

Figure 1. Experimental setup

To suppress the structure-borne noise, a stiffener made by aluminum sheet metal is placed on the inner side of the developed enclosure, which is the back surface of aluminum sheet plates. Predominantly, this mitigation is capable in reducing vibration amplitude, energy of the propagated wave and shifting the frequency content of the instigated vibrations [17]. The developed prototype is modified, three aluminum sheet plates with dimension of 300 mm x 5 mm were placed perpendicularly in vertical arrangement onto the aluminum sheet metal surface of the enclosure. The modified aluminum sheet metal surface of the enclosure is then joined back with the plywood as illustrated in Figure 2.

Figure 2. Modified enclosure

4. Results And Discusion

4.1. Natural Frequencies

The necessity of comparing the natural frequencies is because the experimental work that is previously executed in analysing structure-borne noise has to be validated using simulation work so the data obtained from experimental work is reliable. The result of natural frequencies and the calculated percentage errors are as shown in Table 1 below.

Table 1. Comparison Of Natural Frequencies [16]

Number of mode	Modal Experiment [Hz]	Modal Analysis [Hz]	Percentage Error [%]
1	13.17	13.74	4.33
2	24.74	22.93	7.32
3	30.80	33.50	8.77
4	41.06	40.03	2.51
5	43.34	42.13	2.81
6	59.61	58.32	2.16
7	64.86	66.35	2.30
8	69.11	67.82	2.76
9	75.54	77.69	2.85
10	82.72	84.97	2.72
11	91.05	93.57	2.77

4.2. Mode Shape

The characteristic of structure-borne noise in the enclosure is distinguished by observing the mode shape of the enclosure. The mode shape from modal experiment is then juxtaposed with mode shape from simulation in Msc Patran/Nastran to validate the experimental work. All 11 mode shapes acquired from the experiment and simulation are shown in Figure 3 to Figure 13.

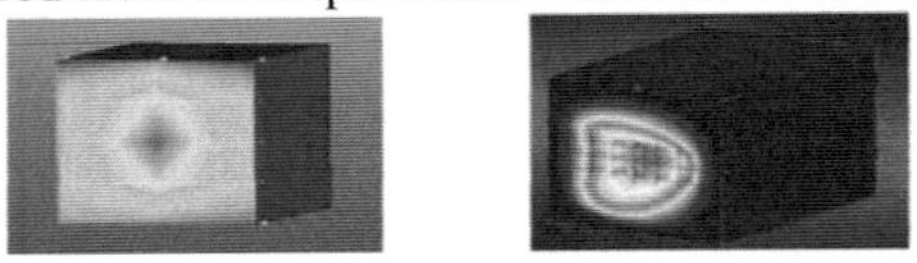

I. Experiment (13.17 Hz) II. Simulation (13.74 Hz)

Figure 3. Mode 1

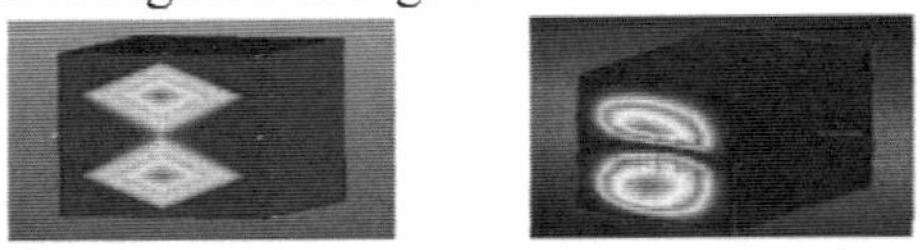

I. Experiment (24.74 Hz) II. Simulation (22.93 Hz)

Figure 4. Mode 2

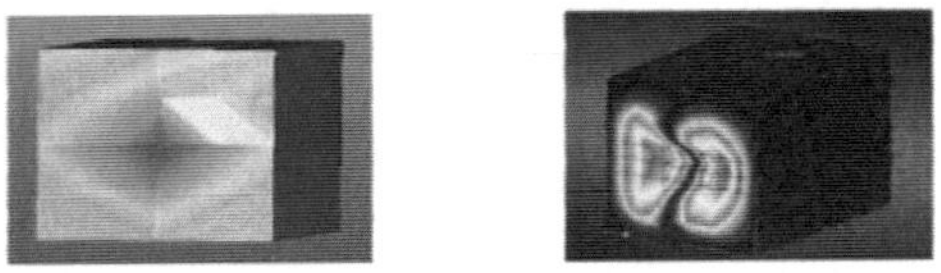

I. Experiment (30.80 Hz) II. Simulation (33.50 Hz)

Figure 5. Mode 3

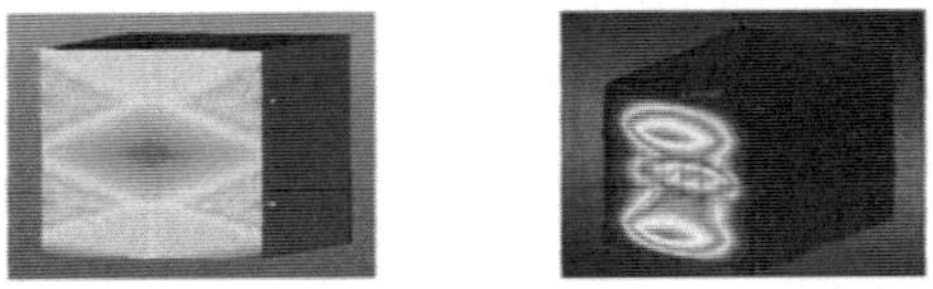

I. Experiment (41.06 Hz) II. Simulation (40.03 Hz)

Figure 6. Mode 4

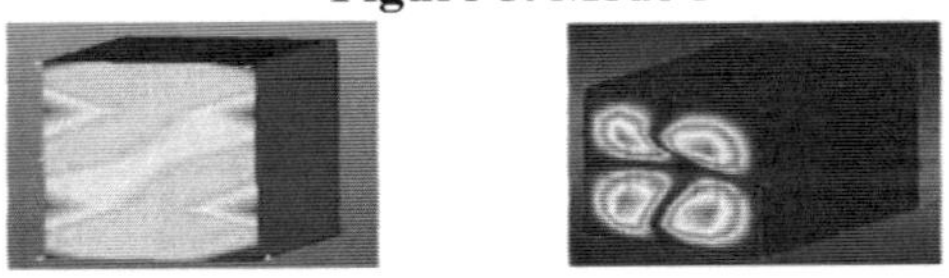

I. Experiment (43.34 Hz) II. Simulation (42.13 Hz)

Figure 7. Mode 5

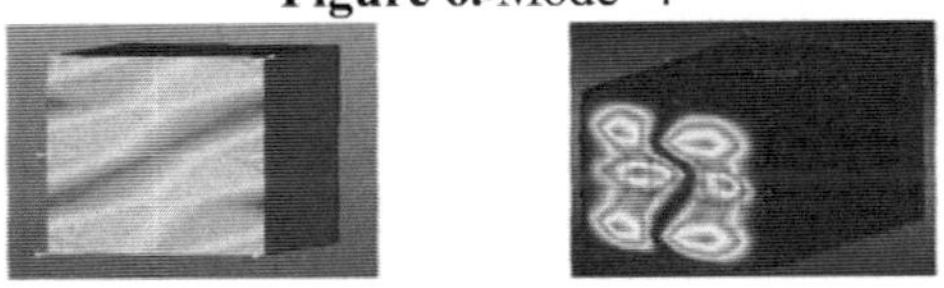

I. Experiment (59.61 Hz) II. Simulation (58.32 Hz)

Figure 8. Mode 6

I. Experiment (64.86 Hz) II. Simulation (66.35 Hz)

Figure 9. Mode 7

I. Experiment (69.11 Hz) II. Simulation (67.82 Hz)

Figure 10. Mode 8

I. Experiment (75.54 Hz) II. Simulation (77.69 Hz)

Figure 11.Mode 9

I. Experiment (82.72 Hz) II. Simulation (84.97 Hz)

Figure 12. Mode 10

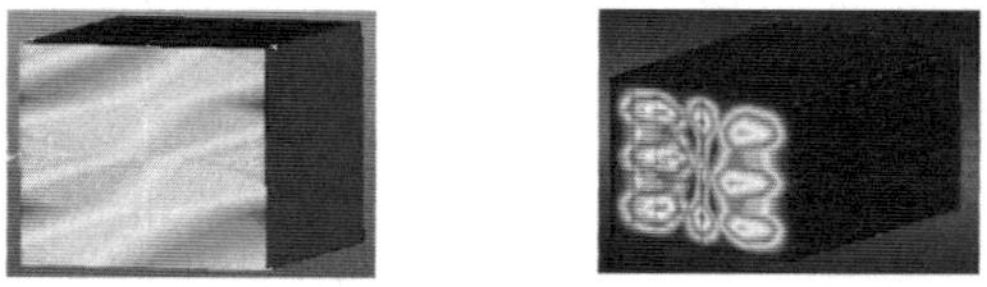

I. Experiment (91.05 Hz) II. Simulation (93.57 Hz)

Figure 13. Mode 11

In this study, the pivotal concern is the analysis of experimental data. All the mode shapes from the experiment results shown that the mode shape is in bending mode shape.

From the mode shape demonstrated in Figure 3 to Figure 13, it can be seen that the mode shape from the modal experiment is similar to the mode shape from the modal analysis performed in the simulation. Nonetheless, in Figure 5, for mode number 3, the mode shape in modal experiment is slightly different from mode shape in simulation. This is because of during experiment, the vibration transmitted through the aluminum plate is possibly be absorbed by the plywood body which hold the aluminum plate. This is due to the stiffness of the plywood which is higher than the aluminum plate. As the stiffness changes, the mode shape will change too [18].

4.3. *Vibration and Noise Spectrum*

After the experimental data is validated, the modal and sound experiments are conducted to obtain the frequency response from vibration and noise. As shown in Fig 14, the frequency in vibration spectrum at the highest peak is closed to the maximum peak of the noise response. This condition is unfavourable as it will cause resonance to the structure. Therefore, mitigation is proposed to avoid the occurrence of resonance. In the present study, the mitigation is implemented and the level of vibration and noise is investigated. The data of comparison of vibration and noise spectrum before and after applying the mitigation are presented in Fig 15 and Fig 16 respectively.

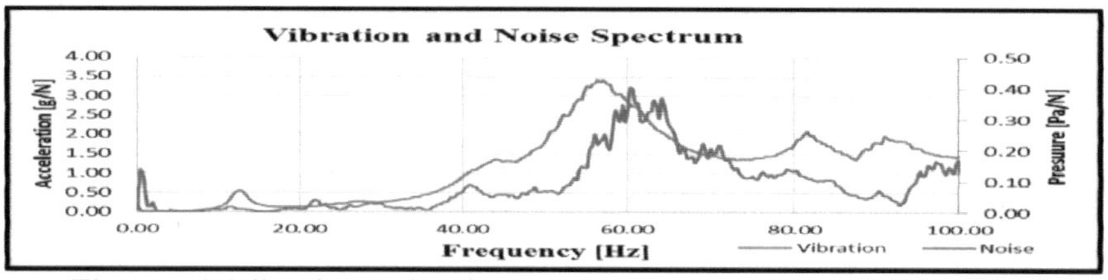

Figure 14. Vibration and Noise Spectrum Comparison Before Mitigation

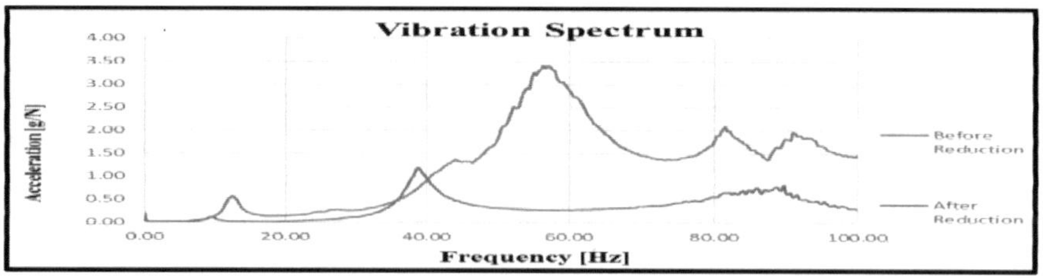

Figure 15. Vibration Spectrum Comparison Before and After Mitigation

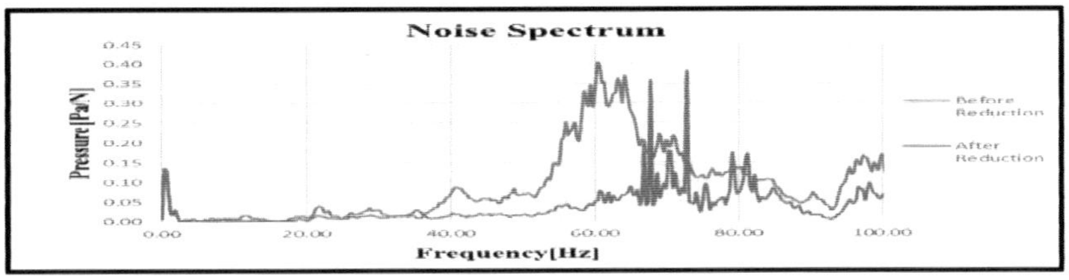

Figure 16. Vibration Spectrum Comparison Before and After Mitigation

From the graphs in Fig 15 and Fig 16, the suppression of vibration and noise spectrum can be seen clearly. For vibration spectrum, significant drop is noticed as the curve of before reduction of structure-borne noise is getting lower in amplitude and has been shifted to the left. In the analysis of before reduction of structure-borne noise, the natural frequency at highest peak is 57.8 Hz while after the reduction the natural frequency drop to 38.8 Hz at the highest peak. The FRF of the vibration is successfully reduced. For noise spectrum, a small decrement is found. All the point in the graph of before the reduction of structure-borne noise has dropped and shifted to the right as shown in after the reduction. In the scope of range of study, for the analysis of after the reduction of structure-borne noise, it is found that the natural frequency at the highest peak is 72.8 Hz. The reduction in FRF of noise is successfully made as the natural frequency at the highest peak has been shifted to a higher frequency range which is from 60.3 Hz to 72.8 Hz. It means it would take a higher frequency for a person to have the possibility to get distracted with the noise that comes from the structure-borne noise. From the data in Fig 15 and Fig 16, the natural frequencies at highest peak of the vibration and noise spectrum after the reduction are 38.8 Hz and 72.8 Hz, respectively. Large difference in range of natural frequency between the vibration and noise spectrum gives a clear view on the absence of the resonance.

5. Conclusion

Suppression of structure-borne noise has been successfully done by placing the stiffeners plates perpendicularly on the aluminum sheet metal of the front face of the enclosure where the excitation force is applied. The vibration and noise spectrum graphs has shown decline in the amplitude of the acceleration and pressure respectively. The natural frequencies at the highest peak from the vibration and noise spectrum has shown a decrement when the graph of before and after reduction of structure-borne noise have been compared. Large difference in range of natural frequency between the vibration and noise spectrum has also given a clear view on the absence of the resonance in the findings. In the future, the study of mitigation may be improved by applying design of experiment to seek the best arrangement and numbers of the stiffeners that should be placed to mitigate the noise. Further investigation via design of experiment will give statistical support to reduce the structure-borne noise. Thus, this will help the researchers in analysing structure-borne noise in the future.

References

[1] HaSPA (Health and Safety Professionals Alliance).(2012). The Core Body of Knowledge for Generalist OHS Professionals. Tullamarine, VIC. Safety Institute of Australia.

[2] Brandt, Noise and Vibration Analysis: Signal Analysis and Experimental Procedures Anders Brandt, John Wiley & Sons, Ltd, 2010.

[3] M. Serra and M. ASSELINEAU, "Solid borne noise in buildings: how does one cope?," in Societe Francaise d Acoustique, Nantes,France, 2012.

[4] M. Vercammen and P. Heringa, "Characterising Structure-Borne Sound From Domestic Appliances," Applied Acoustics, 1989.

[5] H. Findeis and E. Peters, "Disturbing Effects Of Low Frequency Sound Immissions And Vibrations In Residential Building," Noise and Health, vol. 6, no. 23, pp. 29-35, 2004.

[6] H. A, "The open-plan office; A Systematic Study of Employee Reactions to Their Work,"

Environment and Behaviour, vol. 14, pp. 519-542, 1982.
[7] R. D. Berendt and G. E. Winzer, A Guide To Airborne, Impact,andStructure BorneNoise-Control in MultifamilyDwellings, Washington D.C.: U.S. DEPARTMENT OF HOUSINGAND URBAN DEVELOPMENT, 1967.
[8] S. K. Deshmukh and O. S. Madhekar, "Analysis And Experimental Validation Of Structure-Borne Noise From Acoustic Enclosure Of Compressor," in Kirloskar. Pneumatic Co. Ltd, Maharashtra, India, 2014.
[9] Domadiya and P. G, "Mitigation of Structure Borne Noise in the Wooden Panels by Periodic Stiffening," in 13th Internal Danish Center for Applied Mathematics and Mechanics Symposium, Denmark, 2011.
[10] S. H.Ko, "Reduction of Structure-Borne Noise Using An Air-Voided Elastomer," J.Acoust Soc.Am., vol. 101, no. 6, 1997.
[11] R. D. Blevins, “Fluid Systems,” in Formulas for Natural Frequency and Mode Shape, Van Nostrand Reinhold, New York, 1984, Chapter13.
[12] F. Fahy, P. Gardonio, Sound Structural Vibration. Radiation, Transmission and Response, Academic Press, London, 1985.
[13] R. Ohayon, C. Soize, “Structural Acoustics and Vibration. Mechanical Models,” Variational Formulations and Discretizations, Academic Press, London, 1998
[14] B. Kotzen, "Noise Urban Issue, " Architecture, 2010.
[15] Bies, D. A, Hansen and C. H, Fundamentals and basic terminology in Engineering Noise Control, Londin: E & FN Spon, 1988.
[16] N. Fathiah Waziralilah and A. Abu, "Experimental Analysis of Structure-Borne Noise in an enclosure," in ICMAE, Jakarta, 2017.
[17] H. Masoumi, "Mitigation solutions for low frequency," Stockholm, 2012.
[18] B.J.Schwarz and M.H.Richardson,”Experimental Modal Analysis,” CSI Reliability Week, October 1999

IOP Conf. Series: Materials Science and Engineering **409** (2018) 012020 doi:10.1088/1757-899X/409/1/012020

Energy Harvesting, an incredible solution for Structural Health Monitoring Systems of Bridge

Waheed Gul

Department of Mechanical Technology, University of Technology Nowshera, Nowshera, Pakistan

E-mail: waheed@uotnowshera.edu.pk

Abstract. Wireless Sensor network (WSN) has received immense public interest in recent years. Due to ease of installation, optimum size, prolonged life and low cost, Wireless sensors Network (WSN) are widely used for Bridge health monitoring system. The solitary concern of such sorts of systems is the inadequate life of batteries that deliver power to them. In order to overwhelm this problem, energy harvester is an enduring solution. Energy harvesting is essentially seizure free energy from external sources and converts it to useable electrical energy. This paper reviews the supreme topical development of the energy harvesters in the field of structural monitoring system based on vibrations. The PE-VEHs and EM-VEHs are characterized based on their power production capability, resonant frequency, size, power density, base excitation and internal resistance etc. The overall power production range for the developed bridge energy harvesters is from 0.000016 to 31500 μW, conversely, the resonant frequencies of these energy harvesters lying in the range of 1Hz to 13.9 kHz. The power rank produced by the reported PE-VEHs fall in the range of 0.038 to 7700μW and their resonant frequencies are in the varying from 1Hz to 13.9 kHz. The developed EM-VEHs have comparatively low resonant frequencies ranges from (1Hz to 348Hz) and presented the competence of generating power from 0.000016 μW to 31500 μW.

1. Introduction

A WSN as shown in fig.1 comprises numerous components [7], like micro sensors, signal processing unit, power management unit, microcontroller unit, and built-in memory, analog to digital converter (ADC), transmitter, and receiver. Sensor transforms the physical signal, such as pressure, temperature, humidity, or vibration, into a particular electrical signal. The transmitter and receiver are used for transmitting and receiving the information data flanked by WSN and the operator. Memory unit enables on-board data storage. Microcontroller controls the complete performance and process of various components present on board. Power management circuit distributes and manages the power to the WSN components. A battery or a super capacitor is used as the power source in the WSN [8]. Most of the commercial WSNs function on batteries; however, the limited life of batteries restricts the performance and application of WSNs. For a device with 100 μW power consumption, a lithium battery of 1 cm3 volume can be used solitary for one-year process [9 10]

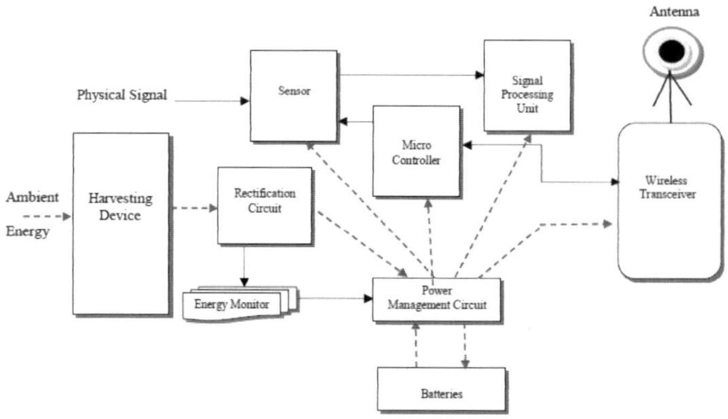

Figure 1. Architecture of an energy harvesting wireless sensor node

2. Bridges Vibrations

Table 1. Tilts The Vibration Data Of Numerous Bridges.

S. No	Bridge	Location	Frequency (Hz)	Acceleration (g)	Reference
1	New Carquinez	California	1–40	0.01–0.102	[11]
2	Komtur	Berlin	2–2.6	0–0.0061	[12]
3	Ypsilanti	Michigan	2–30	0.01–0.035	[13]
4	Golden Gate	San Francisco	0–1.5	0–0.061	[14]
5	RT11 bridge in Potsdam	New York	3.1	0.38	[15]
6	Box girder bridge	Austin	1–15	0.12	[16]
7	3rd Nongro Bridge	South Korea	4.1	0.025	[17]
8	Huanghe Bridge	China	1-2	0.015	[18]
9	Seohae Grand Bridge	South Korea	1	0.0125	[19]
10	IH-35N over Medina River	Texas,USA	3.1	0.15	[20]

3. Results

Figure 2 demonstrates the power produced by the developed bridge vibration energy harvester as a function of base excitation (Acceleration). The EM-VEHs are compared under lower acceleration ranks vary from (0.025 to 8 g), with PE-VEHs, which are exposed to (0.2 to1 g), acceleration levels. Low excitation level (0.02 to 0.29 g) harvesters are stated by [57, 58, 62, 63,65,67,69, 71 and 72]. The developed harvester [61] is characterized under medium acceleration (1 g).Though, the harvesters functioned under high acceleration levels (3 to 8 g) are developed in [70, 73]. The EM-VEH reported in [57] is functioned under the lowermost acceleration level of 0.02 g; still, among the energy harvesters, a PEVEH [70] is exposed to the high acceleration level of 8 g. reasonably, the harvesters [65] and [67] generate high power levels at medium acceleration levels of 0.29 and 0.21 correspondingly.

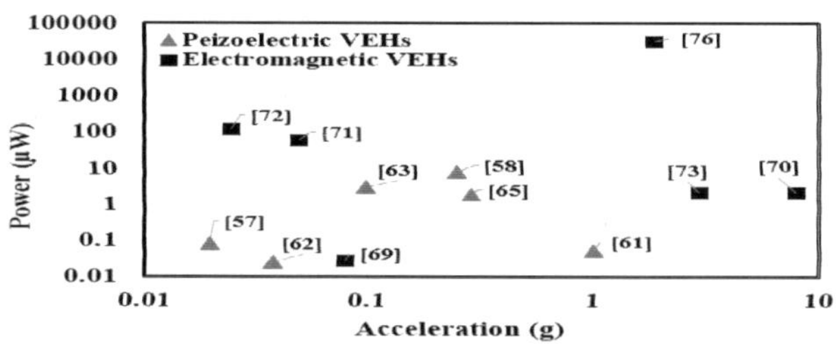

Figure 2. Power vs Acceleration characterization of PVEHs and EMVEHs of Bridges

The power reported for the bridge vibration energy harvesters with respect to the harvester's dimension is presented in Figure 3. In all the reported bridge vibration energy harvesters, the PE-VEH [58] has the smallest size (1 cm^3) and produced a power of 375 μW. However, PE-VEH reported in [66] is of the largest size (12209 cm^3) and it can generate 83.5 μW. By equating the inclusive size, the EM-VEHs are comparatively smaller in size than PE-VEHs except reported in [58].

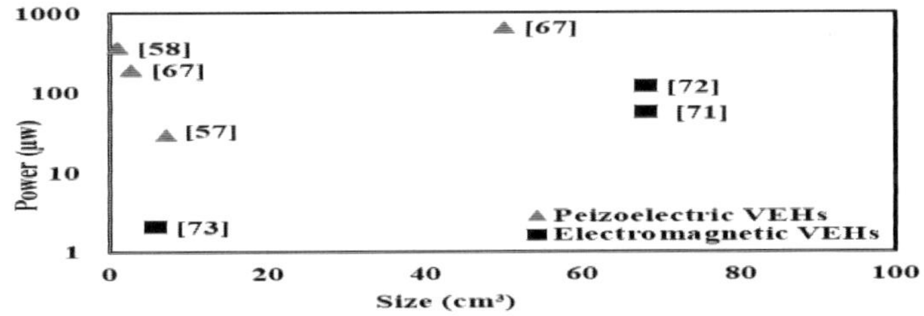

Figure 3. Power vs Size characterization of PVEHs and EMVEHs of Bridges

For the developed bridge vibration energy harvesters, the power density schemed against the harvester's resistance is presented in Figure 4.The resistance of the EM-VEHs is in the range of (3.6 to 457Ω), conversely, for the PE-VEHs the resistance is comparatively high and it ranges from 9.7 to 5200 kΩ.

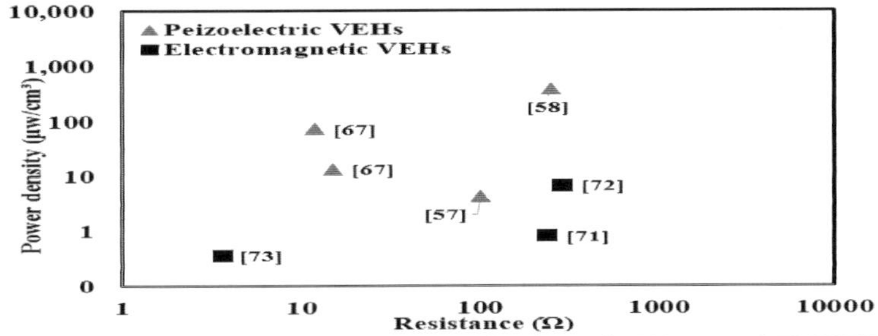

Figure 4. Power density vs Resistance characterization of PVEHs and EMVEHs of Bridges

All of the reported bridge vibration energy harvesters would produce peak power at resonance state. Though, at off-resonance action the power generation by the resonant energy harvesters is usually on the lower side. Furthermore, these energy harvesters have a slender bandwidth, which is also a subject of matter. Power plotted resonant frequency is shown in Fig. 5.

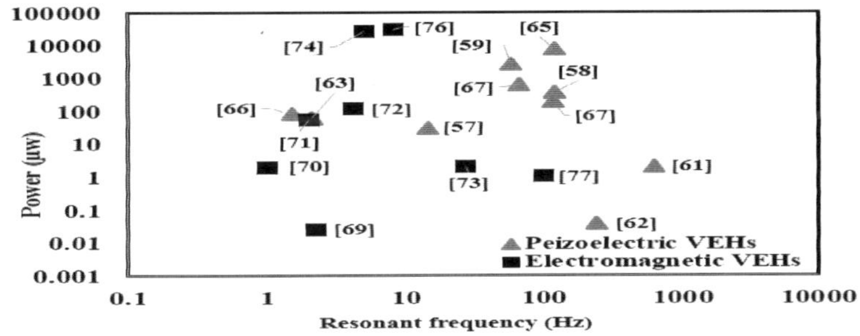

Figure 5. Power vs Resonant frequency characterization of PVEHs and EMVEHs of Bridges

4. Conclusions

The power producing capability of the developed bridge vibration energy harvesters is pretty enough to function maximum of the commercially accessible wireless sensor nodes conferred in the introduction. Furthermore, the resonant frequencies of most of the developed bridge energy harvesters placed in the narrowband which is quite suitable for vibrations available at bridge assemblies.

References

[1] Charles R. Farrar and Keith warden," An introduction to structural health monitoring", The Royal Society Publishing, 2016.
[2] Guang-Dong Zhou1 and Ting-Hua Yi," Recent Developments on Wireless Sensor Networks Technology for Bridge Health Monitoring", Hindawi Publishing Corporation Mathematical Problems in Engineering Volume 2013, Article ID 947867, 33 pages http://dx.doi.org/10.1155/2013/947867.
[3] CheeKian Teng," Structural Health Monitoring of a Bridge Structure Using Wireless Sensor Network (2012).Master's Theses. Paper 79.
[4] K. Wardhana and F. C. Hadipriono, "Analysis of recent bridge failures in the United States," Journal of Performance of Constructed Facilities, vol. 17, no. 3, pp. 144–150, 2003.
[5] B. F. Spencer Jr. and S. Cho, "Wireless smart sensor technology for monitoring civil infrastructure: technological developments and full-scale applications," in Proceedings of the World Congress on Advances in Structural Engineering and Mechanics (ASEM '11), Seoul, Republic of Korea, 2011.
[6] M. Bocca, E. I. Cosar, J. Salminen, and L. M. Eriksson, "A reconfigurable wireless sensor network for structural health monitoring," in Proceedings of International Society for Structural Health Monitoring of Intelligent Infrastructures (ISHMII'09), Zurich, Switzerland, 2009.
[7] K. Sohrabi, J. Gao, V. Ailawadhi, and. J. Pottie, "Protocols for self-organization of a wireless sensor network," IEEE Personal Communications, vol. 7, no. 5, pp. 16–27, 2000.
[8] M. V. Gangone, M. J. Whelan, and K. D. Janoyan, "Wireless monitoring of a multispan bridge superstructure for diagnostic load testing and system identification," Computer-Aided Civil and Infrastructure Engineering, vol. 26, no. 7, pp. 560–579, 2011.
[9] N. Kaur, V.Walia, and R.Malhotra, "Activation matrix oriented base station implementation for energy optimization in wireless sensor networks," International Journal of Computing and Corporate Research, vol. 4, no. 4, 2014.
[10] M. Enckell, Lessons learned in structural health monitoring of bridges using advanced sensor technology [Ph.D. thesis],Division of Structural Engineering and Bridges, Department of Civil and Architectural Engineering, Royal Institute of Technology(KTH), Stockholm, Sweden, 2011.
[11] Susan M. Lashomb, Jine-Wen Kou, Edward V. Gant, John T. Dewolf, Oct., 1985,"Study of Bridge vibrations for Connecticut".
[12] J. Kala, V. Salajka, and P. Hradil, "Footbridge response on single pedestrian induced vibration analysis," World Academy of Science, Engineering and Technology, vol. 3, no. 2, pp. 548–559, 2009.
[13] F. Neitzel, B. Resnik, S.Weisbrich, and A. Friedrich, "Vibration monitoring of bridges," Reports on Geodesy, vol. 1, no. 90, pp.331–340, 2011.
[14] T. Galchev, J. McCullough, R. L. Peterson, and K. Najafi, "A vibration harvesting system for bridge health monitoring applications, "in Proceedings of the 10th International Workshop on Micro and Nanotechnology for Power Generation and Energy Conversion Applications (Power MEMS '10), pp. 179–182, Belgium, Leuven, Belgium, November-December 2010.
[15] A. M. Abdel-Ghaffar and R. H. Scanlan, "Ambient vibration studies of golden gate bridge: I. Suspended structure," Journal of Engineering Mechanics, vol. 111, no. 4, pp. 463–482, 1985.
[16] E. Sazonov, H. Li, D. Curry, and P. Pillay, "Self-powered sensors for monitoring of highway bridges," IEEE Sensors Journal, vol. 9, no. 11, pp. 1422–1429, 2009.
[17] T. McEvoy, E. Dierks, J. Weaver et al., "Developing innovative energy harvesting approaches for infrastructure health monitoring systems," in Proceedings of the 37th Design Automation

Conference, Parts A and B, pp. 325–339,Washington, DC, USA,August 2011.

[18] S.-D. Kwon, J. Park, and K. Law, "Electromagnetic energy harvester with repulsively stacked multilayer magnets for low frequency vibrations," Smart Materials and Structures, vol. 22 no. 5, Article ID 055007, 2013.

[19] L. Zhang, X. Yan, and X. Yang, "Vehicle-bridge coupled vibration response study of Huanghe cable-stayed bridge due to multiple vehicles parallel with high speed," in Proceedings of the International Conference on Mechanic Automation and Control Engineering (MACE '10), pp. 1134–1137, Wuhan, China, June,2010.

[20] B. Jo, Y. Lee, G. Yun, C. Park, and J. Kim, "Vibration-induced energy harvesting for green technology," in Proceedings of the International Conference on Chemical, Environment and Civil Engineering (ICCECE '12), pp. 167–170, Manila, Philippines, November 2012.

[21] N.G.Elvin, and A. E. N.Lajnef, "Feasibility of structural monitoring with vibration powered sensors," Smart Mater. Struct, vol. 15, pp. 977–986, 2006.

[22] S. Priya, "Modeling of electric energy harvesting using piezoelectric windmill," Appl. Phys. Lett, vol. 87, pp. 184101–184103, 2005.

[23] Farid Ullah Khan and Iftikhar Ahmad," Review of Energy Harvesters Utilizing Bridge Vibrations", Hindawi Publishing Corporation Shock and Vibration Article ID 592346, 2015.

[24] M. Peigney and D. Siegert, "Piezoelectric energy harvesting from traffic-induced bridge vibrations," Smart Materials and Structures, vol. 22, no. 9, Article ID095019, 2013.

[25] S Roundy and P K Wright," A piezoelectric vibration based generator for wireless electronics", Smart Mater. Struct. 1131–1142, 2004.

[26] Gyuhae Park, Tajana Rosing, Michael D. Todd, Charles R. Farrar, William Hodgkiss," Energy Harvesting for Structural Health Monitoring Sensor Networks", LA-UR-07-0365, ASCE Journal of Infrastructure Systems, Vol. 14 (1), pp. 64-79, 2008.

[27] Y.B. Jeon, R. Sood, J.-h. Jeong, S.-G. Kimd," MEMS power generator with transverse mode thin film PZT", Sensors and Actuators A xxx (2005) xxx–xxx, 2005.

[28] Hua-Bin Fanga,, Jing-Quan Liua,, Zheng-Yi Xub, Lu Donga, Li Wangb, Di Chena, Bing-Chu Caia, Yue Liub," Fabrication and performance of MEMS-based piezoelectric power generator for vibration energy harvesting", Microelectronics Journal 37 1280–1284,2006.

[29] Marcin Marzencki, Skandar Basrour et.al,"A MEMS Peizoelectric vibration energy harvesting device", the fifth international workshop on micro and Nano technology for power generation and energy conversion applications,Nov,28-30,2005.

[30] T. McEvoy, E. Dierks, J. Weaver et al., "Developing innovative energy harvesting approaches for infrastructure health monitoring systems," in Proceedings of the 37th Design Automation Conference, Parts A and B, pp. 325–339,Washington, DC, USA, August 2011.

[31] M. Rhimi and N. Lajnef, "Tunable energy harvesting from ambient vibrations in civil structures," Journal of Energy Engineering, vol. 138, no. 4, pp. 185–193, 2012.

[32] Y. Zhang, C. S. Cai, and W. Zhang, "Experimental study of a multi-impact energy harvester under low frequency excitations, "Smart Materials and Structures, vol. 23, no. 5, Article ID 055002, 2014.

[33] J. D. Baldwin, S. Roswurm, J. Nolan, and L. Holliday, "Energy harvesting on highway bridges," Final Report FHWA-OK-11-01,2011,http://www.okladot.state.ok.us/hqdiv/p-r-div/spr-rip/library/reports/radspr2-i2224-fy2010-rpt-finalbaldwin.pdf.

[34] J. Li, Structural health monitoring of an in-service highway bridge with uncertainties [Ph.D. thesis], University of Connecticut, Mansfield, Conn, USA, 2014.

[35] C. E. Dierks, Design of an electromagnetic vibration energy harvester for structural health monitoring of bridges employing wireless sensor networks [M.S. thesis], Department of Mechanical Engineering, University of Texas at Austin, Austin, Tex,USA, 2011.

[36] T. McEvoy, E. Dierks, J. Weaver et al., "Developing innovative energy harvesting approaches for infrastructure health monitoring systems," in Proceedings of the 37th Design Automation Conference, Parts A and B, pp. 325–339,Washington, DC, USA,August 2011.

[37] B. Jo, Y. Lee, G. Yun, C. Park, and J. Kim, "Vibration-induced energy harvesting for green technology," in Proceedings of the International Conference on Chemical, Environment and Civil Engineering (ICCECE '12), pp. 167–170, Manila, Philippines,November 2012.

[38] T. Galchev, H. Kim, and K. Najafi, "Micro power generator for harvesting low-frequency and non-periodic vibrations," Journal of Microelectromechanical Systems, vol. 20, no. 4, pp. 852–866, 2011.

[39] S.D. Kwon, J. Park, and K. Law, "Electromagnetic energy harvester with repulsively stacked multilayer magnets for low frequency vibrations," Smart Materials and Structures, vol. 22, no. 5, Article ID 055007, 2013.

[40] F. U. Khan and I. Ahmad, "Vibration-based electromagnetic type energy harvester for bridge monitoring sensor application," in Proceedings of the 10th International Conference on Emerging Technologies (ICET '14), pp. 125–129, Islamabad, Pakistan,December 2014.

[41] Suyoung Yang, Kiyoung Kim, Peipei Liu and Hoon Sohn," two degrees-of-freedom electromagnetic system for harvesting energy from low-frequency bridge vibration",The 22nd International Congress on Sounds and Vibrations,Florence Italy,12-16 July,2015.

[42] Huicong Liu, Tao Chen , Lining Sun and Chengkuo Lee," An Electromagnetic MEMS Energy Harvester Array with Multiple Vibration Modes", Micromachines 2015, 6, 984-992; doi:10.3390/mi6080984.

[43] Wenai Shen,Wenai Shen, Songye Zhu and Hongping Zhu," Experimental study on using electromagnetic devices on bridge stay cables for simultaneous energy harvesting and vibration damping", Smart Mater. Struct. 00 (2016) 000000 (17pp).

[44] Ankita Kumar ,S.S.Balpande and S.C. Anjankar," Electromagnetic Energy Harvester for Low Frequency Vibrations using MEMS, 7th International Conference on Communication, Computing and Virtualization 2016 held in India, Procedia Computer Science 79 (2016). 785 – 792.

IOP Conf. Series: Materials Science and Engineering **409** (2018) 012021 doi:10.1088/1757-899X/409/1/012021

Properties and Common Industrial Applications of Polyvinyl fluoride (PVF) and Polyvinylidene fluoride (PVDF)

M. H. Alaaeddin[1], S. M. Sapuan[2], M.Y.M Zuhri[2], E.S Zainudin[2] and Faris M. AL- Oqla2

[1]Department of Mechanical and Manufacturing Engineering,UPM, Serdang, Malaysia
[2]Department of Mechanical Engineering, The Hashemite University, Zarqa, Jordan

E-mails: m.h.alaaeddin@gmail.com; sapuan@upm.edu.my; zuhri@upm.edu.my; edisyam@upm.edu.my; farisv9@yahoo.com

Abstract. This work features the properties and the common industrial applications of two thermoplastic polymers which are the Polyvinyl fluoride (PVF) and the Polyvinylidene fluoride (PVDF), the polymerization and the manufacturing processes of their monomers have been emphasized in this work. It is believed that these two materials are having a number of shared properties and applications. Their distinguishing characteristics allow them to be utilized in various applications with immense interest to the industrial world. They excel, converge, and slightly diverge in most of their properties with diverse transition phases as well as piezoelectric and pyroelectric effects. Their eminent properties qualify them to be used in a number of industrial and outdoor applications e.g. insulations, sensing materials, laminations, encapsulations, coatings, membranes, biomaterials, aircraft interiors and photovoltaic applications. In conclusion, this work recommends further in-depth analysis to investigate the correlations between these two polymers and to provide oriented numerical information on their performance.

1. Introduction

Polyvinyl fluoride (PVF) and polyvinylidene fluoride (PVDF) are considered to be among the closest thermoplastic polymers in terms of their chemical, mechanical and physical properties [1–3]. Both fall under the tree of fluoropolymers which consists of fluorine atoms in the structural formulas [2,4]. The PVF and the PVDF were used competitively in the last few decades due to their distinguished properties and high performance; they have been used in different and similar applications [5,6]. Moreover, the two polymers are believed to share relevant optical, piezoelectric and pyroelectric properties as well as isomorphic characteristics [7]. This convergence in their properties promotes the utilization of these two polymers in certain applications e.g. PVF/PVDF films, hybrid membranes, coatings, encapsulation materials and outdoor applications [8,9]. In addition, possessing piezoelectric and pyroelectric properties widens the use of these polymers to include other sensitive and advanced applications [7,10].

2. PVF And PVDF Structure

PVF is a thermoplastic polymer which contains on a repeating vinyl fluoride (VF) unit, it is comparable in its structure to the PVDF [2]. The chemical formula of PVF is (C2H3F)n with one fluorine atom for two carbon atoms, whereas the PVDF chemical formula is -(C2H2F2)n- with two fluorine atoms and two carbon atoms [11,12]. However, there has been considerable progress in the development of copolymers and terpolymers in vinyl fluoride with extra fluorinated monomers e.g. tetrafluoroethylene, hexafluoropropylene, perfluoro-butylethylene [12]. Furthermore, fluoropolymers can be considered as homopolymers and copolymers, the fluorinated fluoropolymers usually encompass the atoms of carbon (C), hydrogen (H) and fluorine (F) or chlorine (CI). Figure 1 provides ball-and-stick model of PVF and PVDF structure.

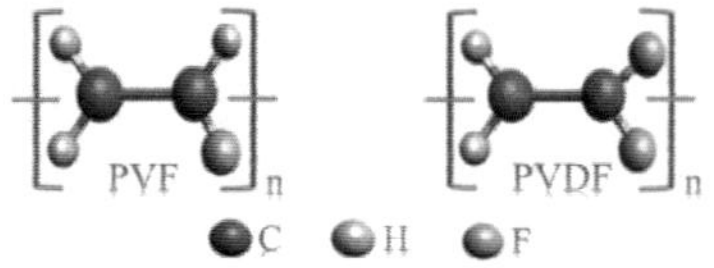

Figure 1. Ball-and-stick model for the structure of the PVF and the PVDF [13]

The PVF and the PVDF are manufactured according to a specific use with varied controllable specifications. These specifications vary according to the functionality and the nature of product [2,11,13]. The two polymers have exhibited high attributes and good mechanical properties, both polymers have been confirmed to have a good tensile strength, although there is a slight difference in favor of the PVF, the PVDF also showed eminent collective properties with strong reliability [1,10,13]. On the other hand, the polyblends between these two polymers have showed different mechanical properties which can be investigated separately [1]. The properties of the PVF and the PVDF can be slightly altered by certain chemical treatments and able to adapt in a homogeneous manner, the modifications can be applied on the mechanical strength, mass transfer, thermal stability, chemical stability, and the degree of crystallinity [3,14]. Table I. provides information on the most common and shared applications of the PVF and the PVDF.

Table 1. A Comparison Of The Most Common Applications Of PVF And PVDF

APPLICATION	PVF	PVDF
Encapsulations	Photovoltaic applications, encapsulation materials and commercial products [4].	Photovoltaic applications, sensing materials and Melt-Bonding Laminations [15,16,17].
Release films & adherent materials	Carbon fiber reinforced polymer composite (CFRP), Fiber reinforced polymer (FRP), peel ply and many relevant applications [18].	Contact surfaces, Piezoelectric materials, peel ply, sealant-adhesives and Products with external protection layers [19,20].
Insulations & interiors	Aircraft parts, Wiring and cabling, sidewalls and stow pin doors, ceiling panels and products with insulation covers [4,10,21].	Premium wire, products with protective barriers, insulation coatings and aircraft interiors [14,22].
Membranes	Hybrid membranes, roofing membranes, antifouling membranes and translucent design elements [8,23].	Nanofibrous membranes, Hybrid membranes, composite membranes, antifouling membranes and super-hydrophobic membranes, micro and ultra-filtration membranes [24-27].

3. Polymerization And Manufacturing Of Vinyl Fluoride and Vinylidene Fluoride

The polymerization process of vinyl fluoride (VF) relies on a number of different techniques, adjutants and initiators, the use of each one of them should be according to the polymerization procedure. The suspension polymerization method uses the technique where the vinyl fluoride is suspended in water, the recommended adjutant for this method is the dispersion stabilizer and the organic peroxide as an initiator. The graft polymerization applies the radiation technique considering the grafting with monomers and the γ-rays. While the radiation-induced method decomposes into free radical. The bulk polymerization technique uses the irradiation with UV method and the Di-tert-butyl peroxide to support the interaction, whereas the emulsion polymerization functions through the procedure of reduced pressure emulsifiers at lower temperatures; the reaction occurs with radical initiators in aqueous emulsion as preferred [5,10,12,13]. On the other hand, the vinylidene fluoride (VDF) can be polymerized in an aqueous medium in either suspension or emulsion procedures. The polymerization of VDF in nonaqueous mediums and with copolymers in a collection of fluorinated and nonfluroinated monomers have been in a continuous development and research area. However, the emulsion and suspension polymerizations are the most two recommended and commercially practiced procedures. This is due to the sufficient heat sink provided by water for the large heat release. In the VDF emulsion procedure, the water-soluble perfluorinated-emulsifying agent can be applied to prevent radical scavenging reactions in the polymerization process and in order to produce stabilized latex with good solid substance. There are different components and methods to be used in the initiating process of the VDF polymerization e.g. the organic peroxides, the inorganic peroxy compounds, fluoroaliphatic sulfinate and bromates combinations as well as redox methods. Numerous chain-transfer agents can be applied in the polymerization process i.e. the mercaptans, esters, chlorofluorohydrocarbons and fluorohydrocarbons. Nevertheless, the temperature, pressure, polymerization procedures, process structure, monomer feeding strategy and the post-polymerization methods are considered as the variables that directly influence the properties and the final form of the PVDF. The recognized degree of polymerization of PVDF ranges from 1000 to 2500 VDF units in the commercial category. The suspension polymerization procedure can be applied using radical photoinitiators with visible light (UV), the photogeneration reduces the polymerization temperature. Throughout the radical-initiated polymerization process of the VDF, the head-to-tail addition preponderates the structure in the monomer units, the Head is CF2 and the tail is CH2, the reversed monomeric addition leads to head-to-head and tail-to-tail structure, which designates the CF2 CF2 and CH2 CH2 respectively, and this is the point where the defect structure occurs. The rate increases when the polymerization temperature increased which has a significant effect on the degree of crystallinity and the ultimate morphology. Lowering the temperature of the polymerization process reduces the number of defect structures in the PVDF. However, the microemulsion polymerization can be held for the VDF and some specific comonomers in the case where the fluorinated surfactant and the fluorinated oil-in-water microemulsion were introduced in the process [22].

4. PVF And PVDF As Competing Polymers

There are specific polymers that differ partially from PVF in their chemical structure i.e. PVDF, PTFE, PVC, and other relevant fluoropolymers with a semicrystalline component [18,28]. These polymers share distinctive features with each other in some mechanical, chemical, physical, thermal properties as well as piezoelectric and pyroelectric properties [29,30]. However, PVDF has reliable Piezo-pyroelectric coefficients [20,29] and therefore contributes to the heat-generated electricity concept with a wide range of applications [29]. The isomorphic features of the PVF and the PVDF make them more homogeneous as vinyl floropolymers [1], these features prioritize them to be used in the applications where their distinctive properties perform better. They are suitable candidates for a good number of applications e.g. encapsulations, release films, coatings, outdoor applications, insulations, photovoltaic applications, aircraft interiors, laminations and membranes [8,14,30]. However, the solubility of PVDF is higher, the PVF is harder to solubilize due to its high polarity and melting temperature which interprets the strength of its monomer-monomer interactions [2]. The two polymers exhibit strong correlation with their morphological structures, the isomorphic blends of the PVF and the PVDF can be maintained with ideal weight proportion and outstanding mechanical properties [1].

5. Properties And Distinguished Features

The PVF and the PVDF are known as semicrystalline polymers with a significant variation that ranges between 20-60% and 35-70% respectively. The transition of PVF is below the melting temperature which is 190 °C, and there are several transition phases at Lower T_g which occurs at –15 to –20 °C and the Upper T_g which can be maintained in the range of 40–50 °C, and another two phases were reported between –80 and 150°C. While the PVDF has four named phases as its density differs from each polymorph phase, these phases are: α, β, γ, and δ. However, α-PVDF phase is the most common state that displays transitions as: (γ) at –70 °C, (β) at –38 °C, (α") at –50 °C, (α') at –100 °C [31]. Whereas the polarized β-PVDF is mostly known as a potential polymer with good piezoelectric properties [16].

The PVF exhibits strong mechanical, chemical and general properties e.g. tensile strength, Young's modulus, impact strength, elongation, flexural strength, color stability, weathering performance, melting point, heat deflection temperature and thermal transition temperature [7,8,32]. Whereas the PVDF exhibits high impact strength, hardness, thermal expansion, altered density, higher degree of crystallinity, solubility and high limiting oxygen index. In addition, the PVDF exposes better piezoelectric and pyroelectric coefficients [24,26]. Table II shows the fluorine to carbon ratios, the coefficients of friction and the critical surface tension of the commercial PVF and PVDF.

Table 2. Fluorine To Carbon Ratio, Coefficients Of Friction and Surface Energy Of the Commercial Pvf And Pvdf [32,33]

Fluoropolymers	Monomer	Fluorine to Carbon Ratio	Coefficient of Friction (dynamic)	Critical Surface Tension, dyne/cm
Polyvinyl fluoride (PVF)	CH_2=CHF	0.5	0.3	28
Polyvinylidene fluoride (PVDF)	CH_2=CF_2	1	0.3	25

6. piezoelectric and pyroelectric properties

The piezoelectricity effect in materials has been defined as the electric polarization that generated in specific crystalline materials at certain mechanical deformation, and the polarity is subjected to a change according to the changes that occurred in the deformation procedure. However, the piezoelectric materials also retain pyroelectric effect at temperature changes [14]. The PVF is relatively lower than the PVDF in terms of the Piezo/Pyroelectric effect. The dipole moment per-repeat unit in the PVDF is greater than the one in the PVF, and the net moment of trans in the PVF is nearly 20% of that found in the PVDF, whereas the Pyroelectric coefficient of the PVF is about 40% of that found in the PVDF [7]. The PVDF demonstrates unusual piezoelectric and pyroelectric properties, which makes it as one of the most attractive and promising polymers [14,33]. The piezoelectric and pyroelectric properties open new potentials and research prospects in these polymers particularly, and can be employed in advanced and piezoelectric applications, such as nanogenerators, sensors, photonic devices, hydrophones, actuators and transducers [22,34,35]. In addition, numerous observations have categorized the PVDF as a ferroelectric polymer with undetected Curie point. The *Curie point* is the temperature point where the transition phase appears in ferroelectric materials which transform into paraelectric state after surpassing this particular temperature [14]. However, table III provides information about the piezoelectric & the pyroelectric constants of the PVF and the PVDF β-phase as well as δ-phase respectively.

Table 3. The Piezoelectric and the Pyroelectric Constants (D_{31}) Of PVF And PVDF [10]

Polymers	Structure	Piezoelectric Coefficient, pC/N	Piezoelectric Coefficient, μC/K.m^2
PVF	–CH_2–CFH–CH_2–CFH–	1	10
PVDF (β-phase)	–CF_2–CH_2–CF_2–CFH–	20-30	30-40
PVDF (δ-phase)	-	10-17	10-15

However, the stability of these two polymers are noticeably high and can keep performing in a good manner, specially in the most extreme chemical environments, the limiting oxygen index (LOI) in both polymers is observed higher in the PVDF and lower in the PVF due to the large proportion of hydrogen atoms in its chain [36].

7.Conclusions

The two thermoplastic polymers are foreseen to have a good number of common properties which qualify them to be utilized in a wide range of industrial and outdoor applications e.g. commercial products, insulations, encapsulations, sensing materials, coatings, laminations, interiors, outdoor applications, aircraft parts and photovoltaic applications. Substantially, some discrepancy might take place in the categorization of PVF and PVDF properties due to the variances occurred in the applied methods. The PVF have shown a wide range of good and reliable properties e.g. the flexural strength, tensile strength, melting point, heat deflection temperature and thermal transition temperature, whereas the PVDF exhibits better piezoelectric and pyroelectric properties, higher impact strength, hardness, thermal expansion, altered density, higher degree of crystallinity, solubility and higher limiting oxygen index. However, further in-depth researches and oriented numerical information are needed to investigate the correlations and contributions of these two distinctive polymers.

Acknowledgment

The authors wish to acknowledge the Ministry of Higher Education-Malaysia for the research grant HiCoE project vote number 6369107, and Universiti Putra Malaysia for facilitating the necessary means to accomplish this research work, as well as providing financial support through the IGSS scholarship.

Reference

[1] A. K. Gupta, R. Bajpai, and J. M. Keller, "Morphological, crystallization and plasticization studies on isomorphic blends of poly (vinyl fluoride) (PVF) and poly (vinylidene difluoride) (PVDF): Using microhardness, XRD and SEM techniques," J. Mater. Sci., vol. 41, Jun. 2006, pp. 5857–5861, doi: 10.1007/s10853-006-0359-2.

[2] M. Lora, J. S. Lim, and M. A. Mchugh, "Comparison of the Solubility of PVF and PVDF in Supercritical CH2F2 and CO2 and in CO2 with Acetone, Dimethyl Ether, and Ethanol," J. Phys. Chem., vol. 103, Mar. 1999, pp. 2818–2822, doi: 10.1021/jp9844462.

[3] D. M. Brewis, I. Mathieson, I. Sutherland, R. A. Cayless, and R. H. Dahm, "Pretreatment of poly(vinyl fluoride) and poly(vinylidene fluoride) with potassium hydroxide," Int. J. Adhes. Adhes., vol. 16, 1996, pp. 87–95, doi: 10.1016/0143-7496(95)00053-4.

[4] M. DeBergalis, "Fluoropolymer films in the photovoltaic industry," Journal of Fluorine Chemistry, Vol.125, Aug. 2004, PP.1255-1257, doi: doi.org/10.1016/j.jfluchem.2004.05.013.

[5] S. Ebnesajjad, Manufacturing of Oriented Polyvinyl Fluoride Film, 3rd ed., Polyvinyl Fluoride Technol. Appl. PVF, William Andrew, 2012, pp. 117–139.

[6] G. dong Kang and Y. ming Cao, "Application and modification of poly(vinylidene fluoride) (PVDF) membranes - A review," J. Memb. Sci., vol. 463, Mar. 2014., pp. 145–165, doi: 10.1016/j.memsci.2014.03.055.

[7] A. S. de Reggi, M. G. Broadhurst, and G. T. Davis, "Effects of poling field and time on pyroelectric coefficient and polarization uniformity in polyvinyl fluoride," Ferroelectrics, vol. 33, 1981, pp. 119–125, doi: 10.1080/00150198108008077.

[8] I. Kaur, V. Kumari, B. Sharma, and N. Gupta, "Characterization and applications of PVF film grafted with binary mixture of methacrylic acid and 4-vinyl pyridine by gamma radiations: Effect of swift heavy ions," Appl. Radiat. Isot., vol. 79, May 2013, pp. 118–130, doi: 10.1016/j.apradiso.2013.05.004.

[9] Z. Liu, H. Wang, X. Zhang, C. Lv, Z. Zhao, and C. Wang, "Durable and self-healing superhydrophobic polyvinylidene fluoride (PVDF) composite coating with in-situ gas compensation function," Surf. Coatings Technol., vol. 327, Aug. 2017, pp. 18–24, doi: 10.1016/j.surfcoat.2017.08.003.

[10] S. Ebnesajjad, Fluoroplastics: Melt Processible Fluoropolymers-The Definitive User's Guide and Data Book, Vol. 2. William Andrew, 2015.

[11] S. Huber, M. Moe, N. Schmidbauer, G. Hansen, and D. Herzke, "Emissions from incineration

of fluoropolymer materials–A literature survey," Norwegian Institute for Air Research: Kjeller, Norway, 2009.
[12] S. Ebnesajjad, P. R. Khaladkar, Fluoropolymer Applications in the Chemical Processing Industries -The Definitive User's Guide and Data Book, 2nd ed. Vol. 2. William Andrew, 2017.
[13] S. Ebnesajjad, Introduction to fluoropolymers, In Applied Plastics Engineering Handbook, 2nd ed. 2017, pp. 55-71.
[14] M. Chanda, and S. K. Roy, Industrial polymers, specialty polymers, and their applications, Vol. 74. CRC press, 2008.
[15] N. An, H. Liu, Y. Ding, M. Zhang, and Y. Tang, "Preparation and electroactive properties of a PVDF/nano-TiO2 composite film," Appl. Surf. Sci., vol. 257, Feb. 2011, pp. 3831–3835, doi: 10.1016/j.apsusc.2010.12.076.
[16] G. Melilli, D. Lairez, D. Gorse, E. Garcia-Caurel, A. Peinado, O. Cavani, B. Boizot, and M.C. Clochard "Conservation of the piezoelectric response of PVDF films under irradiation," Radiat. Phys. Chem., Vol. 142, Mar. 2017, pp. 1–6, doi: 10.1016/j.radphyschem.2017.03.035.
[17] K. Kim, M. Yoo, K. Ahn, and J. Kim, "Thermal and mechanical properties of AlN / BN- fi lled PVDF composite for solar cell backsheet application," Ceram. Int., vol. 41, Jan. 2015, pp. 179–187, doi: 10.1016/j.ceramint.2014.08.056.
[18] F. Mazille, T. Schoettl, and C. Pulgarin, "Synergistic effect of TiO2 and iron oxide supported on fluorocarbon films. Part 1: Effect of preparation parameters on photocatalytic degradation of organic pollutant at neutral pH," Appl. Catal. B Environ., vol. 89, Feb. 2009, pp. 635–644, doi: 10.1016/j.apcatb.2009.01.027.
[19] E. D. Burnham-Fay, T. Le, J. A. Tarbutton, and J. D. Ellis, "Strain characteristics of additive manufactured polyvinylidene fluoride (PVDF) actuators," Sensors Actuators A Phys., Oct. 2017, pp.85-92, doi: 10.1016/j.sna.2017.08.053.
[20] T. Tao, Y. Deng, Z. Liang, B. Liang, Y. Yao, P. Li, S. Lu, and Y. Chen, "Synthesis of porous polyvinylidene fluoride (PVDF) microspheres and their application in lithium sulfur batteries," Mater. Lett., vol. 188, Oct. 2017, pp. 180–183, doi: 10.1016/j.matlet.2016.11.023
[21] S. Ebnesajjad, and A. H. Landrock, Adhesives technology handbook, 3rd ed., William Andrew, 2012.
[22] J. S. Humphrey, R. Amin-Sanayei, Vinylidene fluoride polymers, Encyclopedia of polymer science and technology, 2004.
[23] M. Muti, F. Kuralay, and A. Erdem, "Single-walled carbon nanotubes-polymer modified graphite electrodes for DNA hybridization," Colloids and Surfaces B: Biointerfaces, vol. 91, Mar. 2012, pp. 77–83, doi: 10.1016/j.colsurfb.2011.10.042.
[24] J. Ji, F. Liu, N. A. Hashim, M. R. M. Abed, and K. Li, "Poly(vinylidene fluoride) (PVDF) membranes for fluid separation," React. Funct. Polym., vol. 86, Jan. 2015, pp. 134–153, doi: 10.1016/j.reactfunctpolym.2014.09.023.
[25] C. N. B. Elizalde, S. Al-Gharabli, J. Kujawa, M. Mavukkandy, S. W. Hasan, and H. A. Arafat, "Fabrication of blend polyvinylidene fluoride/chitosan membranes for enhanced flux and fouling resistance," Sep. Purif. Technol., vol.190, Jan. 2018, pp.68-76, doi:10.1016/j.seppur.2017.08.053.
[26] F. Liu, N. A. Hashim, Y. Liu, M. R. M. Abed, and K. Li, "Progress in the production and modification of PVDF membranes," J. Memb. Sci., vol. 375, Jun. 2011, pp. 1–27, doi: 10.1016/j.memsci.2011.03.014.
[27] D. Ghazanfari, D. Bastani, and S. A. Mousavi, "Preparation and characterization of poly (vinyl chloride) (PVC) based membrane for wastewater treatment," J. Water Process Eng., vol. 16, Apr. 2017, pp. 98–107, doi: 10.1016/j.jwpe.2016.12.001.
[28] M. Biron, Plastics Solutions for Practical Problems, 2nd ed. Thermoplastics and Thermoplastic Composites, William Andrew Publishing, 2013, pp831-984.
[29] K. Al Abdullah, M. A. Batal, R. Hamdan, T. Khalil, J. Zaraket, M. Aillerie, and C. Salame,

"The Enhancement of PVDF Pyroelectricity (Pyroelectric Coefficient and Dipole Moment) by Inclusions," Energy Procedia, vol. 119, Jul. 2017, pp. 545–555, doi: 10.1016/j.egypro.2017.07.074.
[30] K. J. Geretschläger, G. M. Wallner, and J. Fischer, "Solar Energy Materials & Solar Cells Structure and basic properties of photovoltaic module backsheet films," Sol. Energy Mater. Sol. Cells, vol. 144, Jan. 2016, pp. 451–456, doi: 10.1016/j.solmat.2015.09.060.
[31] M. Kutz, ed., Applied plastics engineering handbook: processing and materials, William Andrew, 2011.
[32] S. Ebnesajjad, "Manufacturing of Unoriented Polyvinyl Fluoride Films and Coatings," Polyvinyl Fluoride Technol. Appl. PVF, 2012, pp. 141–150, doi: 10.1016/B978-1-4557-7885-0.00006-5.
[33] S. Ebnesajjad, Introduction to fluoropolymers: materials, technology and applications, William Andrew, 2013.
[34] A. Baji, Y. W. Mai, Q. Li, and Y. Liu, "Nanoscale investigation of ferroelectric properties in electrospun barium titanate/polyvinylidene fluoride composite fibers using piezoresponse force microscopy," Compos. Sci. Technol., vol. 71, Jul. 2011, pp. 1435–1440, doi: 10.1016/j.compscitech.2011.05.017.
[35] S. Biswas and S. Bhattacharya, "Effect of Ag nanoparticle embedment on electroactive phase nucleation of PVDF and PVDF-HFP under non-isothermal condition: A comparative study," Thermochim. Acta, vol. 649, Mar. 2017, pp. 69–81, doi: 10.1016/j.tca.2017.01.009.
[36] R. R. Mather, and R. H. Wardman, The chemistry of textile fibres, 2nd ed., R. Soc. Chem: Cambridge, 2015.

IOP Conf. Series: Materials Science and Engineering **409** (2018) 012022 doi:10.1088/1757-899X/409/1/012022

A study on the Relative Slip Chracteristics in Continuous Rolling Process of Rail with Forward Pull

Jian-feng Song[a],Hui Yu[b], Heng Zhu[c] and Yong-gang Dong[d*]

College of mechanical engineering, Yanshan University, Qinhuangdao, China, 066004

E-mails: [a]jfsong2003@163.com; [b]yuhui@ysu.edu.cn; [c]kixhery@163.com; [d*]d_peter@163.com

Abstract. In continuous rail rolling by universal mill, the forward-slip coefficient is one of the most important parameters to control the rollling process. Firstly the forward-slip coefficient of the web in a single stand without tension has been proposed by considering the influencing factors as many as possible. Then the tension between two adjacent stand has been represented and the forward-slip coefficient with forward pull has been derived. Moreover, the experiments of two-stand continuous rolling of rail by universal mill has been accompolished to get the forward slip coefficient of web of rail with forward pull. The theoretical results is in agreement with the experimental data basically and approaches that well.

Notation

α_h, α_{vb}, α_{vt} Bite angle of the horizontal roll,the vertical roll for rolling the base of rail and the top of rail respectively

H_{w0}, H_{w1} The height of the web of rail at the entrance sectionThe height of the web of rail at the exit section respectively

H_{b0}, H_{b1} The height of the web of rail at the entrance section The height of the base of rail at the exit section

W_{b0}, W_{b1} The width of the base of rail at the entrance sectionThe width of the base of rail at the exit section

W_{t0}, W_{t1} The width of the head of rail at the entrance section The width of the head of rail at the exit section

R_h The radius of horizontal roll R_{vb} The radius of vertical roll for rolling the base of rail

$\overline{G_{vt}}$, G_{vt} The equivalent gap of the vertical roll rolling the top of railThe gap of the vertical roll rolling the top of rail

$\overline{D_{vt}}$ The equivalent diameter of vertical roll for rolling the base of rail

$\overline{p_h}$ Mean unit pressure acting on the web from the horizontal roll

$\overline{p_{vt}}$, $\overline{p_{vb}}$ Mean unit pressure acting on the top of rail and the base of rail from the vertical roll

$\overline{p_{ht}}$, p_{hb} Mean unit pressure acting on the top of rail from the flank of horizontal roll acting on the rail bas

μ_{hw} Friction coefficient between the web and horizontal roll

μ_{ht}, μ_{hb} Friction coefficient between the top of rail ,the rail base and the flank of horizontal roll

μ_{vt} , μ_{vb} Friction coefficient between the top of rail, the base of rail and vertictal roll

X_w , X_v the horizontal resultant force acting on the web of rail and the base of rail

X_{ft}, X_{fb} the horizontal resultant force acting on the top of rail and the base of rail from the flank of horizontal roll

ρ the proportion factor about horizontal resultant force X_{ft} and X_{fb}

L_t the length of deformation zone in rolling the base and top of rail

1. Introduction

Since 1970s, the universal rolling methods has been applied in H-beam rolling widely and the theoretical research about that has been carried out, and the numerical simulation by FEM has been used in universal rolling prosess[1-4].Then the theory of universal rolling methods has been developed and improved. Although the universal rolling methods has been also applied in rail rolling for 30 years, there are few theoretical research about the rail rolling by universal mills.W.Zhang and Q.Zhou[5,6] studied the deformation mechanism and force-energy parameters by energy principle and variational methods during H-Beam universal rolling in single stand. X.Jin [7]built the theoretical model of velocity field, plastic flow and relative slip in the Continuous Rolling Process of H-beam, and accomplished the H-beam universal rolling experiments. In previous work of author, Y.Dong and J.Song proposed the theoretical model for calculating the rolling force[8], plastic flow law[9-11], cross-section temperature[12] and forward slip coefficient[13,14] during the rail universal rolling in single stand by upper-bound principle and energy methods. At present, the application of universal mill becomes more and more prevalent in producing the high-precision rail, and it has been taking the place of conventional producing methods gradually. Continuous rolling method is a high efficiency rolling method and it has been applying in strip rolling and rod rolling generally.However, the continuous rolling method has not been applied in rail universal rolling widely since the relative slip characteristics of rail universal rolling is hard to be predicted precisely and then the rolling speed of two adjacent stand is hard to set. So the .For the certain similarity of Heavy Rail rolling and H-beam rolling, the result of theoretical research about H-beam rolling can be applied as an available reference in Heavy Rail rolling.

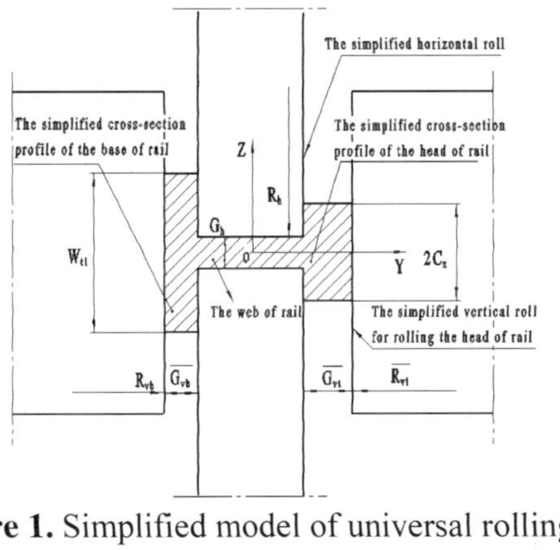

Figure 1. Simplified model of universal rolling of Heavy Rail

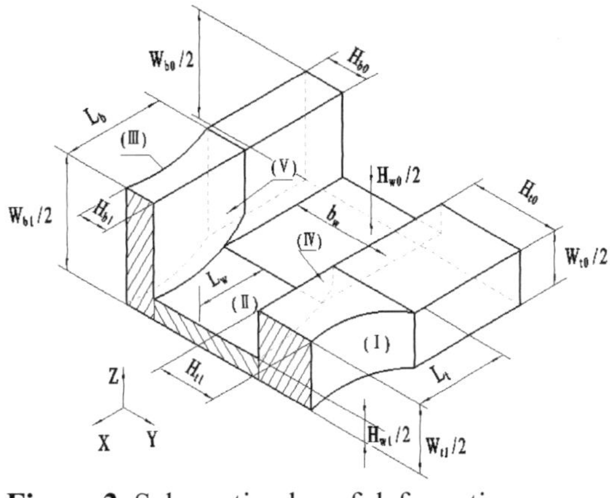

Figure 2. Schematic plan of deformation zone

The universal mill consists of four rolls, two horizontal drive rolls and two vertical driven rolls (**Fig.1**). The size of two horizontal rolls is identical, but the vertical roll of rolling the top of rail is different from that of rolling the base of the rail. The vertical roll of rolling the top of rail is flat, but the box pass has been grooved on the vertical roll of rolling the top of the rail. In Heavy Rail rolling, there are 5 deformation zone existing in the deformed workpiece (**Fig.2**): (Ⅰ)the deformation zone between the vertical roll and the top of rail; (Ⅱ) the deformation zone between the horizontal roll and the web of rail ;(Ⅲ) the deformation zone between the vertical roll and the base of rail; (Ⅳ) the deformation zone between the side of horizontal roll and the top of rail; (Ⅴ) the deformation zone between the side of horizontal roll and the top of rail. The top view of rail universal rolling was shown as **Fig.3**, the two rolls were vertical roll for rollling the base of rail and head of roll.

2. The Simplified Model Of Vertical Roll Of Rolling The Top Of Rail

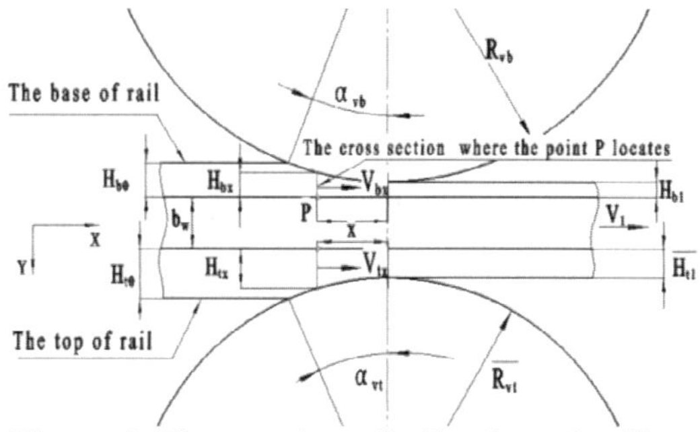

Figure 3. The top view of rail universal rolling

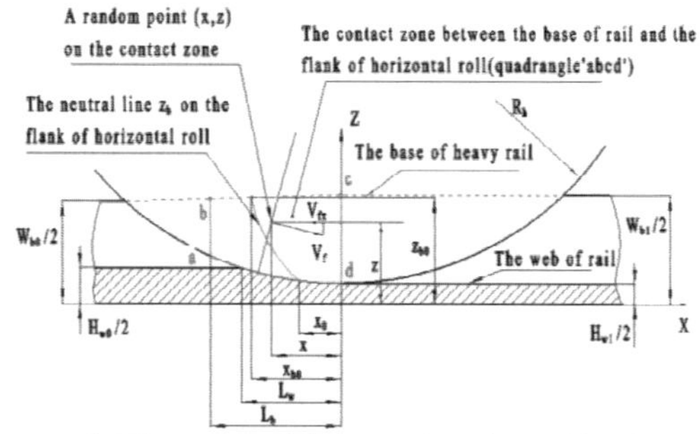

Figure 4. The neutral line on the side of horizontal roll

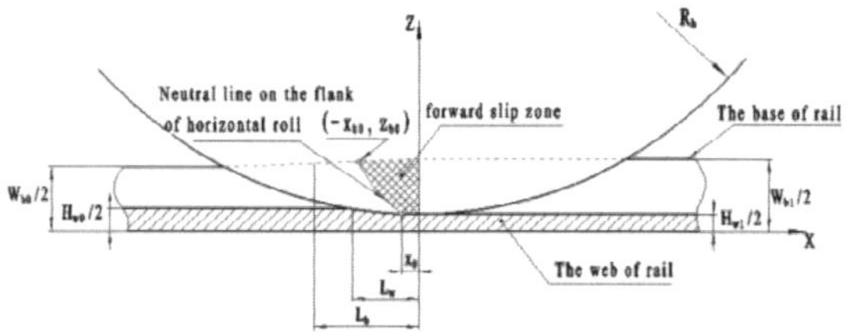

Figure 5. forward slip zone on the side of horizontal roll

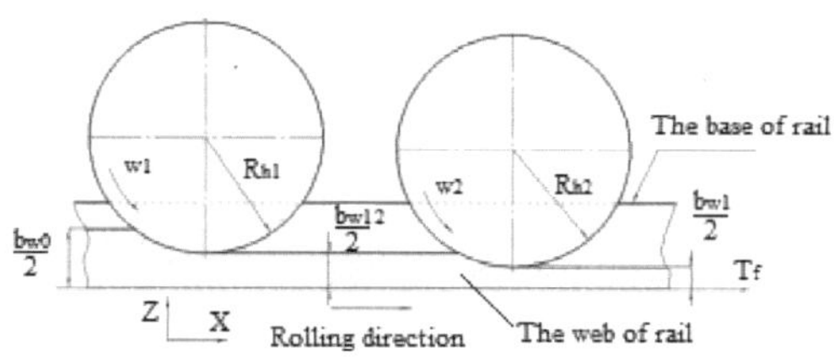

Figure 6. Continuous rolling of rail in two stand

In earlier work by Yonggang Dong and Wenzhi Zhang[10-12], the influence of base of rail and top of rail on the forward coefficient of the web of rail was studied and the neutral line on the deformation (Ⅳ) and deformation (Ⅳ) were determined to distinguish the forward slip zone and backward slip zone on this two deformation zones. The deformation zones was shown in **Fig.2** and the derivation of neutral line on the side of horizontal roll were shown as **Fig.4** and **Fig.5**.

3. Forward-Slip Coefficient Of The Web In Single Stand Without Forward Pull

Forward-slip coefficient is one of the most important parameters in the automatic tension control of the continuous rail rolling process by universal mill. For the great differences in the conditions of roll and workpiece, there are two different metal flow existing in the deformation zone. When the forward-slip coefficient of the web δ_w is greater than unity, forward-slip and backward-slip occur synchronously. However, when δ_w is less than unity, the complete backward-slip occurs.

The forward-slip coefficient of the web is always less than unity on the condition that a thicker web, wider base or top of rail, large reduction ratio and an backward pull are present. On the contrary, the coefficient is greater than unity in all other cases. However, whatever are the rolling conditions, the forward coefficient between the vertical roll and base is always greater than unity.

In reference[13,14], if the forward slip coefficient between the horizontal roll and the web of rail in a single stand without tension is not less than1, it can be obtained by Eq.1 and Eq.2

If $x_{b0} < L_b$

$$\delta_{w0}=1+\frac{1}{b}\left[\frac{\overline{p_h}b_w L_w\mu_{hw}}{p_{hb}\mu_{hb}}\left(1-\frac{\alpha_h}{2\mu_{hw}}\right)+(1+\rho)L_f\left(W_{b0}-H_{w1}\right)\left(1-\frac{2}{L_f}\sqrt{\frac{H_{b1}\cdot 2R_{vb}\left(W_{b0}-H_{w1}\right)}{2R_h-W_{b0}+H_{w1}}}\right)+\frac{X_v}{p_{hb}\mu_{hb}}\right]$$

where b is expressed as

$$b=(1+\rho)\left(50\sqrt{\frac{\left(0.01R_h+W_{b0}-H_{w1}\right)^3 H_{b1}\cdot 2R_{vb}}{2R_h-W_{b0}+H_{w1}}}-0.28R_h\sqrt{R_h H_{w1}}-50\left(W_{b0}-H_{w1}\right)\sqrt{\frac{H_{b1}\cdot 2R_{vb}\left(W_{b0}-H_{w1}\right)}{2R_h-W_{b0}+H_{w1}}}\right)+\frac{14\overline{p_h}b_w\mu_{hw}}{p_{hb}\mu_{hb}}\sqrt{H_{w1}R_h}$$

If $x_{b0} \geq L_b$

$$\delta_{w0}=1+\frac{\frac{\overline{p_h}b_w L_w\mu_{hw}}{p_{hb}\mu_{hb}}\left(1-\frac{\alpha_h}{2\mu_{hw}}\right)+(1+\rho)L_f\left(W_{b0}-H_{w1}\right)\left(1-\frac{2}{L_f}\sqrt{\frac{H_{b1}\cdot 2R_{vb}\left(W_{b0}-H_{w1}\right)}{2R_h-W_{b0}+H_{w1}}}\right)+\frac{X_v}{p_{ht}\mu_{ht}}}{2(1+\rho)\left(R_h L_f-0.14R_h\sqrt{H_{w1}R_h}\right)+\frac{14\overline{p_h}b_w\mu_{hw}}{p_{hb}\mu_{hb}}\sqrt{H_{w1}R_h}}$$

Forward Slip Of The Web Between Two Adjacent Stand With Tension

The tension force during H-beam continuous rolling had been proposed by X.Jin[6]. By the similar method, the tensile force of continuous rail rolling between two adjacent stand can be represented **(Fig.6)**

$$\sigma_{12}=\frac{\delta_{w0.2}R_{h2}\left(W_2/W_1\cdot S_2/S_1\right)-\delta_{w0.1}R_{h1}}{G_2R_{h2}\left(W_2/W_1\right)+G_1R_{h1}}$$

Where $\delta_{w0.1}$, $\delta_{w0.2}$ is the forward-slip coefficient of the web in stand 1 and stand 2 without tension, respectively, and can be obtained by Eq.(1), Eq.(2) or Eq.(3) and Eq.(4) according to the position of neutral line. $W_2, W_1, S_2, S_1, R_{h1}, R_{h2}, G_2, G_1$ are the angle velocity of horizontal roll in stand 1 and stand 2 , the cross-section area of deformed workpiece in stand 1 and stand 2, the radius of horizontal roll in stand 1 and stand 2, the influence coefficient of vertical roll in stand 1 and stand 2, respectively. G_1, G_2 can be shown as

$$G_1 = S_1 / \left[2(1+\rho)\overline{p_{ht}}L_{b1}f_{ht}\right]$$

$$G_2 = S_2 / \left[2(1+\rho)\overline{p_{ht}}L_{b2}f_{ht}\right]$$

Then, the forward-slip coefficient of the web with tension can be shown as

$$\delta_{w1} = \delta_{w0.1} + \frac{\sigma_{12}\cdot S_1}{2(1+\rho)\overline{p_{ht}}L_{b1}f_{ht}}$$

$$\delta_{w2} = \delta_{w0.2} + \frac{T_f-\sigma_{12}\cdot S_2}{2(1+\rho)\overline{p_{ht}}L_{b1}f_{ht}}$$

4. Results and discussions

According to Eq.(1), Eq.(3), the forward slip coefficient during single stand universal rolling without tension can be obtained, then the forward slip coefficient during two adjacent stand universal rolling with tension can be calculated by Eq.(7) and Eq.(8).The size of rolling pieces was shown as Fig.7, and the corresponding parameters was shown as Table1.

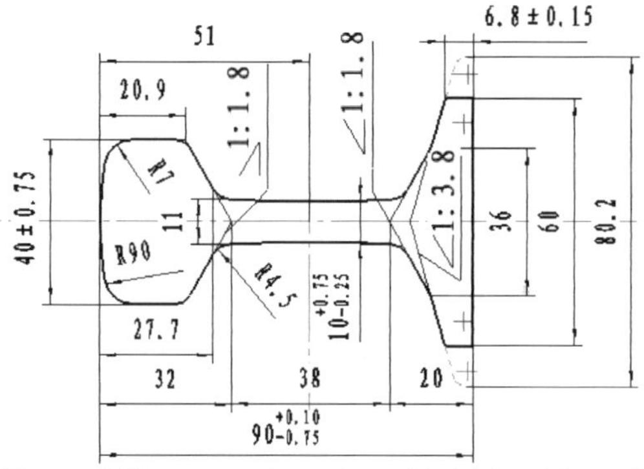

Figure 7. Cross-section size of initial rolling pieces

Figure 8. The three-stand universal experimental mills

Figure 9. Figure 10. Measurment of hump marks on the web of rail

Table 1. Calculation Result From The Theoretical Model

Parameters / Stand No.	Parameter of simplifying the vertical roll(mm)				Neutral angle(rad)			Neutralline (mm)		Forward slip coefficient
	R_{vb}	R_h	$\overline{D_{vt}}$	$\overline{H_{vt}}$	γ_h	γ_{vb}	γ_{vt}	x_{t0}	x_{b0}	δ_{w0}
Stand1	210	140	200	40	0.0328	0.0482	0.0522	10.2	8.8	0.972
Stand2	190	120	180	36	0.0333	0.0490	0.0531	9.3	7.4	0.983

Table 2. The Rolling Schedule For 18kg/M Light Rail Universal Rolling Experiments

Workpiece No.	Reduction (mm)			Main motor speed (rpm)	Horizontal roll rotating velocity (rad/s)	Workpiece's temperature(℃)	
	The web of rail	The head of rail	The base of rail			Heating temperature	Initial rolling temperature
Stand1	1.0	2.5	1.5	1200	3.9873	1050	1000
Stand2	0.8	2.0	1.2	1000	3.3228	1040	995

Table 3. Calculation Result Of Tension In Continuous Rolling

Parameter	G_1 (mm^3/ N)	G_2 (mm^3/ N)	L_{f1} (mm)	L_{f2} (mm)	S_1 (mm^2)	S_2 (mm^2)	f_{ht}	$\overline{p_{ht}}$ (N/mm^2)	σ_{12} (N/mm^2)
Value	0.79	1.15	8.9	8.1	2100	1920	0.15	208	0.04

For verifying results from the theoretical model, a piece of 18kg/m light rail(Fig.8)was heated and rolled in the three stand universal mill of Yanshan University, the rolling equipments were shown as Fig.8(a) and Fig.8(b). The rolling schedule was shown as Table.2

The forward pull can be imposed on the workpiece and adjusted by the tension control system of three-stand universal rolling mills. Two little dents with constant distance was chisled on the surface of horizontal rolls, and the forward slip coefficient can be obtained by mearsuring the distance between two hump marks on the web of rail after rolled(Fig.9) and and calculated the diatance ratio between the two dents on the roll and two hump marks on the workpiece.

The experimental data and theoretical results were shown in **Fig.11.**

In continuous rolling process, the tension between two adjacent stand has been calculated and results is shown in **Table 3.** And then the forward-slip coefficient of the web with backward pull and forward pull is expressed in **Fig.11(a) and Fig.11(b).** As can be seen in **Fig.11,** there exists **a little** discrepancy between the theoretical results and experimental data, and either theoretical results or experimental data, the forward-slip coefficient of the web becomes greater with the increment of forward pull.

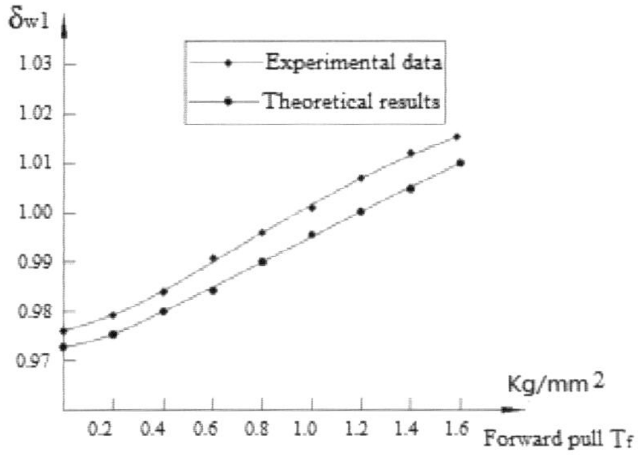

(a)First Stand

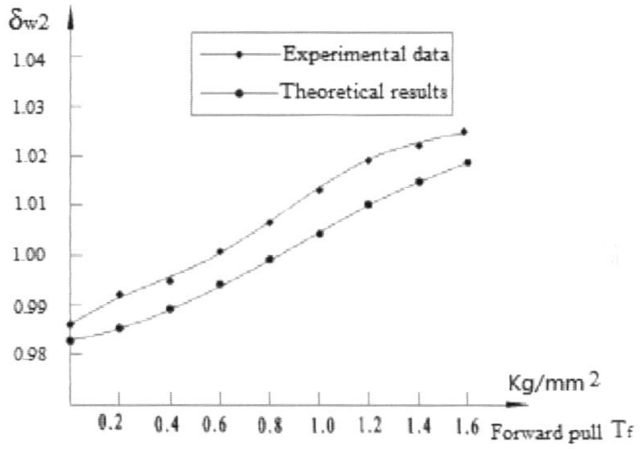

(b)Second Stand

Figure 10. The forward-slip coefficient of the web with forward pull:

5. Conclusion

In this paper, the method of force equilibrium has been employed to derive a series of equations for forward-slip coefficient, which is applicable to different rolling conditions, for example the size of workpiece and pass schedule.

(1) The forward pull influences the forward-slip coefficient of the web greatly. The forward-slip coefficient of the web becomes greater with the increment of forward pull;

(2) With the incresement of forward full, the relative slip status of the web of rail changes from the backward slip to forwad slip against the horizontal roll;

(3) The theoretical results is in agreement with the experimental data basically and approaches that well. So, thistheoretical model can be applied in universal continuous rolling of rail.

IOP Conf. Series: Materials Science and Engineering **409** (2018) 012022 doi:10.1088/1757-899X/409/1/012022

References

[1] T.Iguchi, H.Hayashi, I.Yarita:Proc. 1st Int. Conf. On modeling of metal Rolling Processes, The Institute of Materials ,London,(1993),p707-p719.

[2] Kiuchi, J.Yanagimoto:*ISIJ Int.*,**30**(1990),p142-p149.

[3] Kazutake Komori, Katsushi Koumura: *J.Mater.Proc.Tech.*,105(2000),p24-31.

[4] Kazushige,IKUTA et.al: *ISIJ Int.*,**35**(1995),p1089-p1093.

[5] W.Zhang ,Q.Zhou,C.Zhu et.al:*J.Mater.Proc.Tech.*,**94**(1999),p123-p127.

[6] W.Zhang , Q.Zhou,Y.Li et.al:*J.Mater.Proc.Tech.*,**101**(2000),115-123.

[7] X.Jin: Experimental and Theoretical Study on the Continuous Rolling Process of H-beam, Dissertation of the Master Degree in Engineering, Yanshan University,1990.3:20-35.(in Chinese)

[8] Y.Dong, W.Zhang, H.Cao:Journal of Iron and Steel Research, International. 15(2008),p32-p38.

[9] Y.Dong, G.Luo and Z.Ren. Research on the Metal's Three-dimensional Plastic Flow Mechanism in Rail Universal Rolling. 2nd Annual International Conference on Advanced Material Engineering, Part1:444-449，Wuhan，China，April 15-17，2016.

[10] Y.Dong, J.Zhang, L.Zhang and J.Song.Study on the lateral plastic flow mechanism during rail universal rolling,Material Science and Technology,22(2014),p96-p100.(In Chinese)

[11] Y.Dong, W.Zhang and J.Song.Theoretical and experimental research on the elongation law during rail universal rolling,Chinese Journal of Mechanical Engineering,46(2010),p87-p92. (In Chinese)

[12] Y.Dong, W.Zhang, J.Song:CHINESE JOURNAL OF MECHANICAL ENGINEERING. 2009, 22(3):376-383.

[13] Y.Dong, W.Zhang, J.Song:ISIJ INTERNATIONAL. 49(2009),p385-p394.

[14] Y.Dong, W.Zhang, J.Song:Journal of Iron and Steel Research, International, 17(2010),p27-p32.

The performance of a piezoelectric micro-pump for 3D additive manufacturing

Kai Li, Jun-kao Liu, Wei-shan Chen, Lu Zhang

State Key Laboratory of Robotics and System, Harbin Institute of Technology, Harbin, China
Harbin Dongan Engine Corporation LTD, Aero Engine Corporation of China, Harbin, China

E-mails: sdcxlikai@126.com (K.L), jkliu@hit.edu.cn (J.L), cws@hit.edu.cn (W.C), Zhanglu916hit@163.com

Abstract. The piezoelectric micro-pump is one kind of electronic device which can achieve supplying fluid on demand, and the total volume of fluid per time is at μL level. For fluid, such as fluid base on Cyanoacrylate, with low dynamic viscosity and can be solidified easily, the micro-pump is effectively to be used for 3D additive manufacturing. In this paper, we show a valve less piezoelectric micro-pump with the advantages of simple structure and stable performance. And we analyze the performance of the piezoelectric micro-pump which can be used for 3D additive manufacturing. A fluid solid coupling method is proposed for simulating the working progress of the piezoelectric micro-pump. The fluid transmission performance of the piezoelectric is mainly analyzed, such as the mass flow, flow rate of the outlet and the internal pressure distribution. The working principle of the piezoelectric micro-pump is discussed as well.

1. Introduction

The piezoelectric micro-pump is mainly used for trace fluid transmission, and is wildly used in varies kind of industrial field thank to its advantages of simple structure, easily control and good stability [1-4]. By the structure of the piezoelectric micro-pump, it can be divided into two categories, one is valve piezoelectric micro-pump and the other one is valve less piezoelectric micro-pump [5-13]. For 3D additive manufacturing, we need to control the supply solidable fluid material, in traditional, we push the fluid with a high pressure pump and control the accuracy by the control of the opening and closing of the nozzle. Fortunately, the piezoelectric micro-pump is a kind of micro electronic device which can achieve suppling a bit of fluid to the target surface. Though, there exist shortages when a high viscosity fluid is selected, for ejecting the solidable fluid which is based on Cyanoacrylate, the piezoelectric micro-pump show its advantages. The fluid material based on Cyanoacrylate is a kind of material using for the medical and biological sciences, thanks to its fast solidifying and biodegradation [14-16].

Thus, in this paper we proposed using the piezoelectric micro-pump for 3D additive manufacturing, and we analyze the performance of the micro-pump to verify the feasibility. For easy assembly and long working life, we select the valve less piezoelectric micro-pump as the device for 3D additive manufacturing, as the valve of the micro-pump is often damaged. In this paper, we taking a fluid solid coupling method to simulate the piezoelectric micro-pump. The deformations of the piezoelectric vibrator with the coupling effect at different time are given, the mass flow and fluid velocity of the outlet of the piezoelectric micro-pump are shown as well. The pressure distribution statuses at different time when the piezoelectric is working are given. The analysis results show the piezoelectric micro-pump can be used as the device for 3D additive manufacturing.

2. Structure And Working Principle

The piezoelectric micro-pump is composed of the piezoelectric vibrator and its chamber. The piezoelectric vibrator is made by gluing the piezoelectric ceramic on a metal substrate. When pulse voltage is applied on the piezoelectric vibrator, the vibrator deforms and creates volume change of the chamber. As shown in Fig.1, due to the deformation of the piezoelectric vibrator, the chamber of the micro-pump produces two forms as shown in Fig.1 (a) and Fig.1 (b). In Fig.1, the Φ1 is the mass flow of the inlet and Φ2 is the mass flow of the outlet. Under the excitations, the piezoelectric vibrator vibrates, and the volume of the chamber reciprocating changes. When the chamber volume increases, the mass flow of the inlet is large than that of outlet, and when the chamber volume decreases, the mass flow of the inlet is less than that of outlet. Thus, due to the mass flow difference, the fluid keeps moving from the inlet to the outlet.

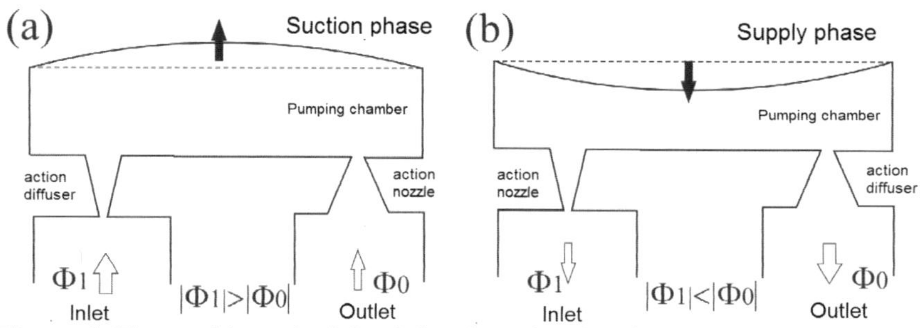

Figure 1. The working principle of the piezoelectric micro-pump [17].

The structure of the piezoelectric micro-pump used in this paper is shown in Fig.2. The piezoelectric ceramic used here is PZT-5H, the material of the ceramic is PZT-5H, and the physical properties (elastic stiffness constant matrix $[c^E]$, piezoelectric stress constant matrix $[e]$ and dielectric constant matrix $[\varepsilon^T]$) of the PZT in the simulation are set as:

$$[c^E]=\begin{bmatrix} 13.2 & 7.3 & 7.1 & 0 & 0 & 0 \\ 7.3 & 13.2 & 7.1 & 0 & 0 & 0 \\ 7.1 & 7.1 & 11.5 & 0 & 0 & 0 \\ 0 & 0 & 0 & 2.6 & 0 & 0 \\ 0 & 0 & 0 & 0 & 2.6 & 0 \\ 0 & 0 & 0 & 0 & 0 & 3 \end{bmatrix}\times 10^{10}(\mathrm{N/m^2})$$

$$[e]=\begin{bmatrix} 0 & 0 & -2.4 \\ 0 & 0 & -2.4 \\ 0 & 0 & 17.3 \\ 0 & 0 & 0 \\ 0 & 12.95 & 0 \\ 12.95 & 0 & 0 \end{bmatrix}(\mathrm{C/m^2})$$

$$[\varepsilon^T]=\begin{bmatrix} 804.6 & 0 & 0 \\ 0 & 804.6 & 0 \\ 0 & 0 & 659.7 \end{bmatrix}\times 10^{-11}(\mathrm{F/m})$$

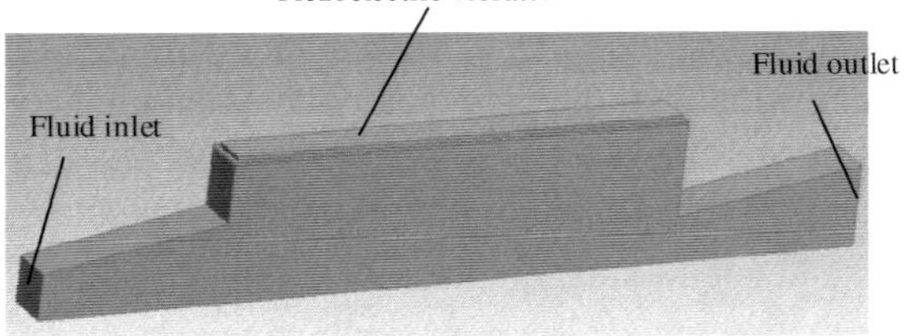

Figure 2. The structure of the piezoelectric micro-pump

As the piezoelectric micro-pump is used for ejecting the solidable fluid which is based on Cyanoacrylate, the fluid properties in the simulations are set depend on its real parameters. The piezoelectric vibrator used in this paper is made by gluing the piezoelectric ceramic on a copper substrate. The outer of the copper substrate is set as fixed condition, the outer the fluid domain is set as wall boundary condition, the Fluid inlet is set as pressure inlet, and the Fluid outlet is set as pressure outlet. The upper and bottom of the piezoelectric ceramic are applied on voltage boundaries as 0 V and 5 V respectively, the applied pulse voltages are shown in Fig.3.

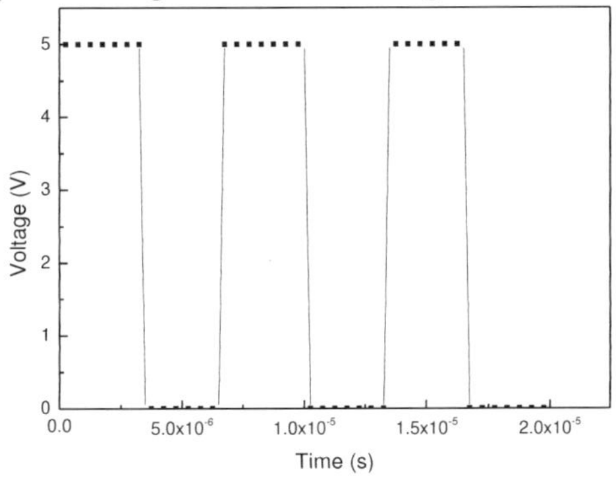

Figure 3. The applied pulse voltages.

3. Results And Discussions

3.1. Coupling results of the piezoelectric vibrator

In the simulations, we carry two way coupling analysis, when the piezoelectric vibrator is applied on pulse voltage, it deforms and create pressure on the fluid solid interface. The fluid is pushed by the piezoelectric vibrator, and the vibrator obtains the counterforce of the fluid as well. The deformation of the piezoelectric vibrator is shown in Fig.4. As we can see that, due to the fixed boundary of the outer ring of the copper substrate, the maximum deformation is created at the center part of the vibrator. The deformation of the piezoelectric vibrator causes the volume change of the chamber.

Figure 4. The deformation of the piezoelectric vibrator after coupling

The deformation values of the piezoelectric vibrator varies with time is shown in Fig.5. Combined with Fig.3, we can see that, the deformation is created along with the applying of the pulse voltages. And when the voltage value reaches peak value, the deformation of the vibrator obtains the maximum deformation. Although the voltage is kept constant for a period of time, the deformation of the piezoelectric vibrator does not remain stable. The reason is that, due to the action of the inertia force, the piezo vibrator continues to deform when the voltage is constant.

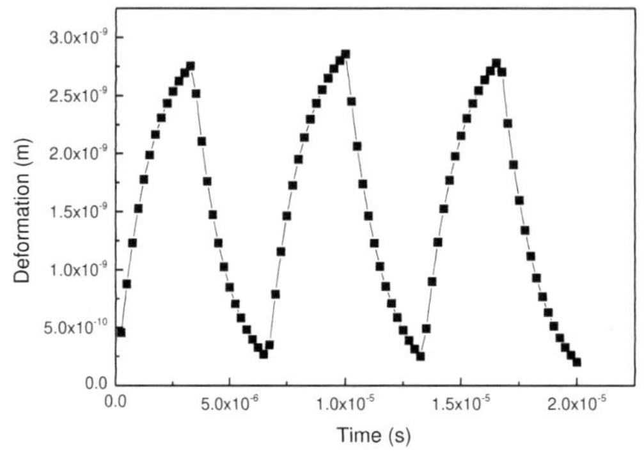

Figure 5. The deformation values of the piezoelectric vibrator varies with time

3.2. Coupling results of the fluid

In order to better understand the working principle of the piezoelectric micro-pump, we obtain the pressure distribution statuses at different time inside the chamber of the micro-pump with two-way fluid solid coupling effect, as shown in Fig.6. We can see that, the coupling pressure, which is created by the coupling effect between piezoelectric vibrator and fluid, first occurs on the fluid solid coupling interface. Then, the pressure spreads in the fluid domain, and then spreads to the inlet and outlet of the piezoelectric micro-pump. Because the structure of the outlet part and inlet part is different, the pressure distribution inside the chamber is asymmetric, which causes the directional flow of the fluid.

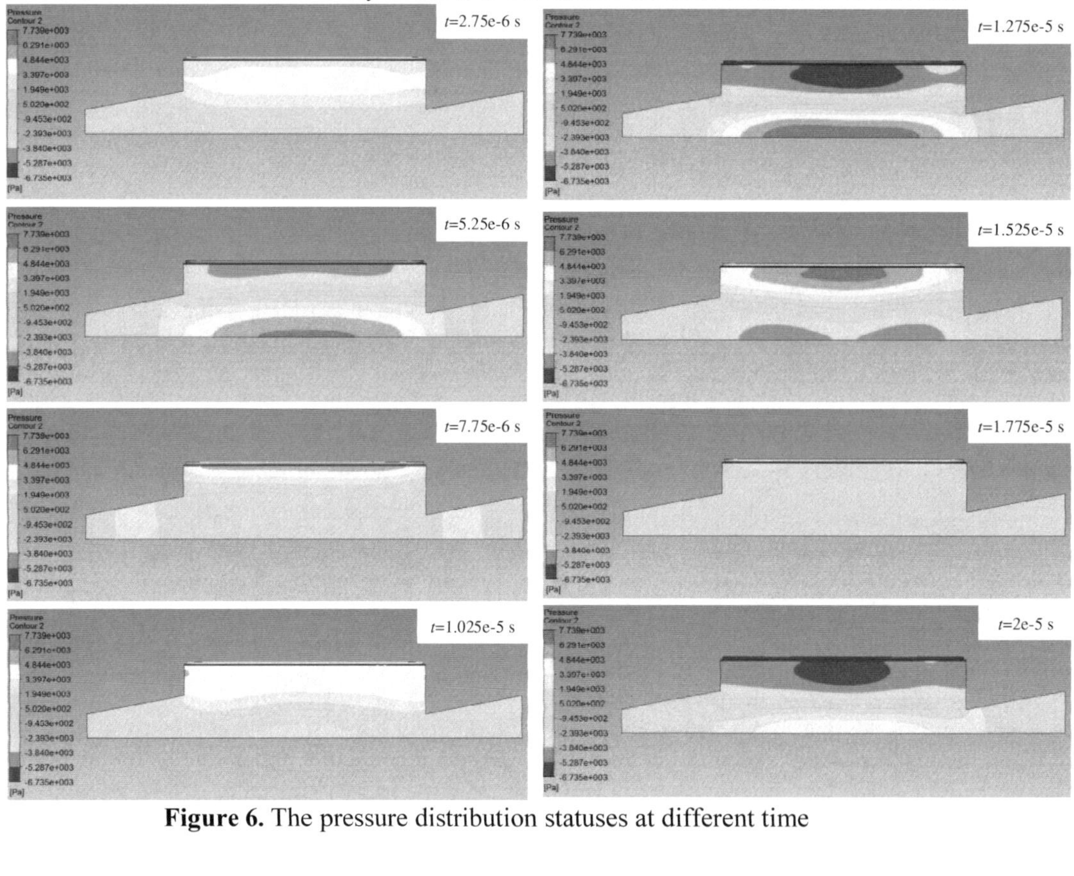

Figure 6. The pressure distribution statuses at different time

What we care about is the mass flow at the outlet of the piezoelectric micro-pump, as the mass flow reflect the fluid transform performance of the micro-pump. The coupling results of the mass flow of the outlet is shown in Fig.7. As we can see that, under three pulse voltages the mass flow of the outlet only create two peaks. It demonstrate the delay of the pressure propagation, as the fluid be regard as an elastomer. When pulse voltages are applied on the vibrator, the vibrator vibrates, and pressure creates in the fluid, then the pressure propagation in the chamber of the micro-pump, when it propagate to the outlet of the micro-pump, changes are created in the outlet. That also shows why the outlet mass flow has not changed at the beginning. Besides, we can see that the mass flow at outlet creates positive and negative values, which is consistent with the previous. At the same time, it can be seen that the total mass flow of fluid which out of the outlet is large than the fluid which is back flow into the outlet. Thus, the piezoelectric micro-pump can realize the quantitative output of fluid.

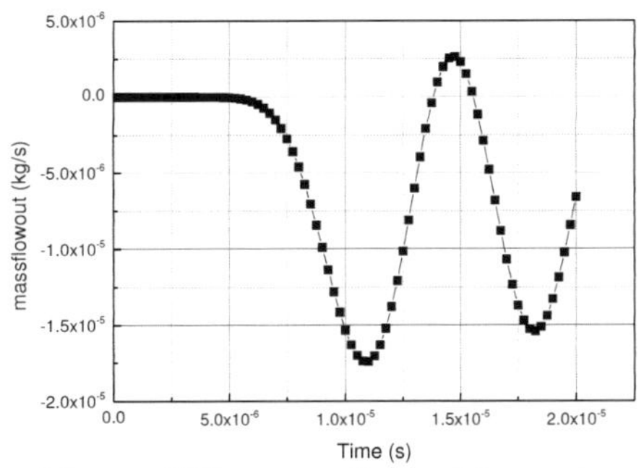

Figure 7. The mass flow of the outlet at different time

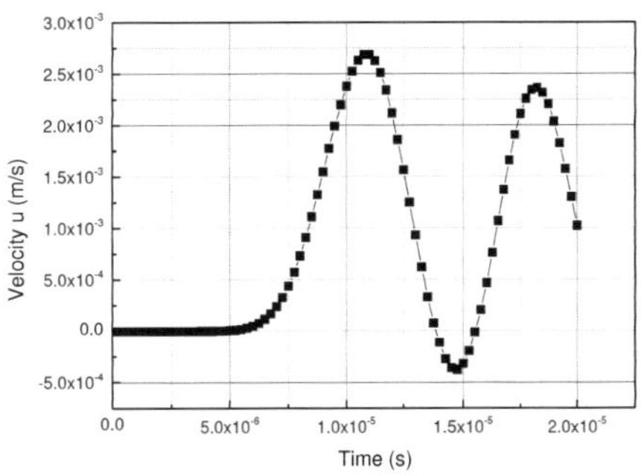

Figure 8. The fluid velocity of the outlet at different time

The velocity of the fluid reflect the intensity of the piezoelectric micro-pump, thus, we obtains the fluid velocity at the u direction at the outlet, as shown in Fig.8. We can see that the velocity curve of the fluid only has two peaks as well, which is due to the delay of the pressure propagation. Besides, the velocity also has positive and negative values, which reflect the outflow and inflow of the fluid at the outlet. As we can see that, the time for the velocity to be positive is large for the whole time, in other words, the general trend of fluid flow is the fluid outflow of the outlet. Thus, the piezoelectric micro-pump can achieve the quantitative control of fluid and can be used for 3D additive manufacturing.

4. Conclutions

Through the performance analysis of the piezoelectric micro-pump for ejecting the solidable fluid which is based on Cyanoacrylate, it is proved that the electronic device can be used for 3D additive manufacturing. Although back flow exist at the outlet part of the piezoelectric micro-pump, the general trend is that the fluid flows out of the outlet, and we can achieve high accuracy 3D printing by controlling the excitation signal. In the next step, we will establish experimental platform to carry out experimental research.

Acknowledgment

This project is supported by the National Natural Science Foundation of China (51075082, 51375107). All authors conceived and analyzed the data; all authors contributed to the writing of the paper. The authors declare no conflict of interest.

References

[1] E. Kurtoglu, A. Bilgin, M. Sesen, I. B. Misirlioglu, M. Yildiz, H. F. Y. Acar, *et al.*, "Ferrofluid actuation with varying magnetic fields for micropumping applications," *Microfluidics and Nanofluidics,* vol. 13, pp. 683-694, Oct 2012.

[2] D. Cao, Q. Jia, L. Lei, Z. Xin, and J. Mi, "Cantilever Piezoelectric Micro-pump Modeling Based

on Bond Graph Port," in *Micro-Nano Technology Xv*. vol. 609-610, F. Tang, Ed., ed, 2014, pp. 819-824.

[3] S. Kaviani, M. Bahrami, A. M. Esfahani, and B. Parsi, "A modeling and vibration analysis of a piezoelectric micro-pump diaphragm," *Comptes Rendus Mecanique,* vol. 342, pp. 692-699, Dec 2014.

[4] Z. Zhang, W. Wang, and X. Chen, "Study on the performance of piezoelectric micro pump for insulin injection," *Zhongguo yi liao qi xie za zhi = Chinese journal of medical instrumentation,* vol. 39, pp. 64-7, 2015-Jan 2015.

[5] J. H. Oh, J. N. Lim, S. S. Lee, K. J. Lim, and H. H. Kim, "A Valveless Piezoelectric Micro-Pump Using Peristaltic Motion of Two Sided Disk Type Vibrator," *Ferroelectrics,* vol. 409, pp. 128-138, 2010 2010.

[6] C. Hernandez, Y. Bernard, and A. Razek, "A global assessment of piezoelectric actuated micro-pumps," *European Physical Journal-Applied Physics,* vol. 51, Aug 2010.

[7] S. A. Wilson, R. P. Jourdain, and S. Owens, "Pre-stressed piezoelectric bimorph micro-actuators based on machined 40 mu m PZT thick films: batch scale fabrication and integration with MEMS," *Smart Materials & Structures,* vol. 19, Sep 2010.

[8] N. Okamoto, H. Hwang, Y. Morita, and E. Nakamachi, "Development of Micro-pump for Bio-MEMS by Using New Bio-compatible Piezoelectric Material MgSiO3," in *Sensors and Smart Structures Technologies for Civil, Mechanical, and Aerospace Systems 2011*. vol. 7981, M. Tomizuka, C. B. Yun, V. Giurgiutiu, and J. P. Lynch, Eds., ed, 2011.

[9] A. Benaissa and S. Belkhiat, "Performances analysis of a piezo-pump," in *Emm-Fm2011 - First Euro Mediterranean Meeting on* Functionalized *Materials*. vol. 29, A. Cheikhrouhou, Ed., ed, 2012.

[10] S. Chen, Y. Liu, Y. Shen, J. Wang, and Z. Yang, "The Structure of Wheel Check Valve Influence on Air Block Phenomenon of Piezoelectric Micro-Pump," *Micromachines,* vol. 6, pp. 1745-1754, Nov 2015.

[11] P.-H. Cazorla, O. Fuchs, M. Cochet, S. Maubert, G. Le Rhun, Y. Fouillet, *et al.*, "A low voltage silicon micro-pump based on piezoelectric thin films," *Sensors and Actuators a-Physical,* vol. 250, pp. 35-39, Oct 15 2016.

[12] R. Zhang, F. You, Z. Lv, Z. He, H. Wang, and L. Huang, "Development and Characterization a Single-Active-Chamber Piezoelectric Membrane Pump with Multiple Passive Check Valves," *Sensors,* vol. 16, Dec 2016.

[13] R. K. Haldkar, V. K. Gupta, and T. Sheorey, "Modeling and flow analysis of piezoelectric based micropump with various shapes of microneedle," *Journal of Mechanical Science and Technology,* vol. 31, pp. 2933-2941, Jun 2017.

[14] E. Gonzalez, J. Orta, C. Quero, L. Niemshik, R. Galera, D. Onay, *et al.*, "Ethyl-2-cyanoacrylate fixation of the cranial bone flap after craniotomy," *Surgical Neurology,* vol. 53, pp. 288-289, 2000.

[15] P. A. Mccarron, R. F. Donnelly, W. Marouf, and D. E. Calvert, "Anti-adherent and antifungal activities of surfactant-coated poly(ethylcyanoacrylate) nanoparticles," *International Journal of Pharmaceutics,* vol. 340, pp. 182-190, 2007.

[16] T. Landegren, M. Risling, J. K. Persson, and A. Sondén, "Cyanoacrylate in nerve repair: transient cytotoxic effect," *Int J Oral Maxillofac Surg,* vol. 39, pp. 705-712, 2010.

[17] A. G. S. B. Neto, A. M. N. Lima, C. L. Gomes, L. C. Oliveira, C. S. Moreira, and F. A. F. Tejo, "Design methodology of a self-priming PMMA valveless piezoelectric micro-pump," in *Instrumentation and Measurement Technology Conference*, 2013, pp. 1068-1073.

IOP Conf. Series: Materials Science and Engineering **409** (2018) 012024 doi:10.1088/1757-899X/409/1/012024

Electric Discharge Machining of Tungsten Carbide using Uncoated and Coated Wire Electrodes: Analysis of Cutting Speed and Workpiece Geometrical Accuracy

Rakhshanda Naveed, Nadeem Ahmad Mufti

Department of Industrial and Manufacturing Engineering, University of Engineering and Technology, Lahore, Pakistan

E-mails: rakhshanda@uet.edu.pk

Abstract. Tungsten carbide is an important tool and die material due to its excellent mechanical properties such as high hardness, good toughness, and wear resistance. Wire electric discharge machining (WEDM) is regarded as a process capable of machining complex geometries and curved features in difficult-to-cut tungsten carbide. The present research work is carried out to study the influence of wire electrode, workpiece taper angle, and WEDM parameters on cutting speed and geometrical error of curved profile machined in tungsten carbide. Experiments are performed based on Taguchi's L18 orthogonal array using two types of wires; uncoated brass wire and zinc-coated brass wire. Experimental results indicate that zinc-coated brass wire offers higher cutting speed as compared to uncoated brass wire. Main effects plots show that pulse-on time and pulse-off time are the most significant factors affecting cutting speed while taper angle is the most significant factor for geometrical error.

1. Introduction

Tungsten Carbide (WC) is a well known material for manufacturing of cutting tools, dies, molds, and other wear resistant parts due to its high hardness and strength at a wide range of temperatures [1]. High hardness of WC imparts brittleness making this material difficult to machine. The requirement of rigid machining setup and expensive tooling along with the difficulty in producing complex profiles in conventional machining of WC necessitates using non-conventional machining methods for processing of this material. Wire electric discharge machining (WEDM) is an established non-conventional process for machining of complex features and tapered profiles in very hard materials like carbides, metal matrix composites, titanium based alloys, and nickel based super alloys [2]. This process uses electrode in the form of a thin wire to erode the workpiece material. Along with other electrical and mechanical parameters, the performance of WEDM process also depends upon the properties of wire electrode. There are various types of wires that have been developed to be used as a tool for WEDM operation. As the requirements of higher productivity, better surface finish and higher precision in WEDM operation are increasing, therefore, there is a need to evaluate the machining performance of different types of wire electrodes.

Researchers have studied the performance of various types of coated and uncoated wires for electric discharge machining of titanium alloys, nickel based super alloys, and metal matrix composites. Antar et al. [3] compared the performance of two types of coated wires (copper core with CuZn50 coating and copper core with zinc rich brass coating) with uncoated brass wire for machining of aerospace alloys. They found that cutting rate of coated wires was upto 70% higher than that of brass wire. Moreover, 25% thinner recast layer was obtained with coated wires. Klocke et al. [4] evaluated the machining performance of three types of wire electrodes including uncoated brass wire, zinc-rich brass coated copper wire, and nickel coated copper wire for generation of fir tree slots in Inconel 718. The results revealed that uncoated brass wire produced best accuracy and surface integrity with almost zero recast layer but its cutting speed was lower than that of zinc-rich brass coated copper wire. Ramamurthy et al. [5] analyzed the effect of three types of wire electrodes including plain brass wire, zinc coated brass wire, and diffusion coated brass wire on machining characteristics of Ti-6Al-4V alloy. The results showed that diffusion coated brass wire produced the lowest surface roughness and minimum kerf width. Patil et al. [6] studied the effect of wire material properties on machining performance of ceramic particle reinforced aluminum matrix composites. They found that copper core CuZn50-coated wire offered higher cutting speed and lower kerf width as compared to plain brass wire but surface roughness was significantly increased by the use of coated wire.

A review of literature related to WEDM of tungsten carbide revealed that though several type of wires have been employed for machining of this material [7- 10] but the effect of wire properties on machining performance has never been studied. Also, no research work was found addressing the machining of curved profiles in taper mode in WC by using coated wire. Therefore, the present study is undertaken to fill this gap.

This research work is carried out to evaluate the machining performance of uncoated brass wire and zinc-coated brass wire in terms of cutting speed and geometrical error of curved profile in tungsten carbide. Main effects plots are drawn in order to study the influence of workpiece taper angle and WEDM parameters on performance characteristics.

2. Experimentation

Tungsten carbide workpiece was machined on CHMER G43S WEDM using de-ionized water as a dielectric fluid. The chemical composition of workpiece material is presented in Table 1. Machining setup and geometry of workpiece are shown in Fig. 1 and Fig. 2, respectively. A concave curved profile having a radius of 5 mm was generated on workpiece in taper cutting mode. Two types of wire electrodes, uncoated brass wire (Wire A) and zinc-coated brass wire (Wire B), having a diameter of 0.2 mm were used to perform experimentation.

Table 1. Chemical Composition of Tungsten Carbide

Element	W	C	Co	Ni	Fe	Ti
Weight (%)	87.6	5	6.5	0.18	0.06	0.6

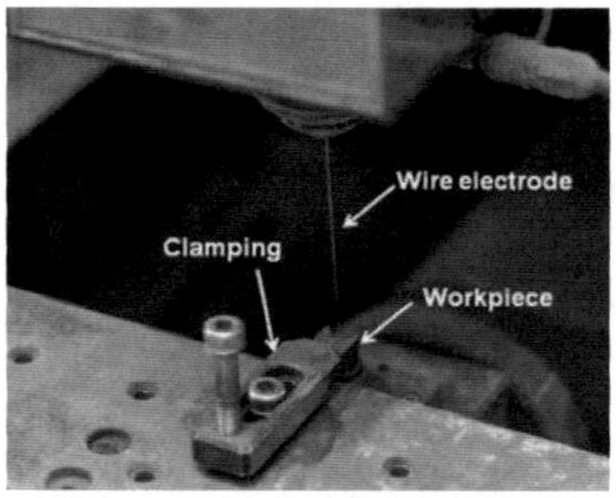

Figure 1. Machining setup

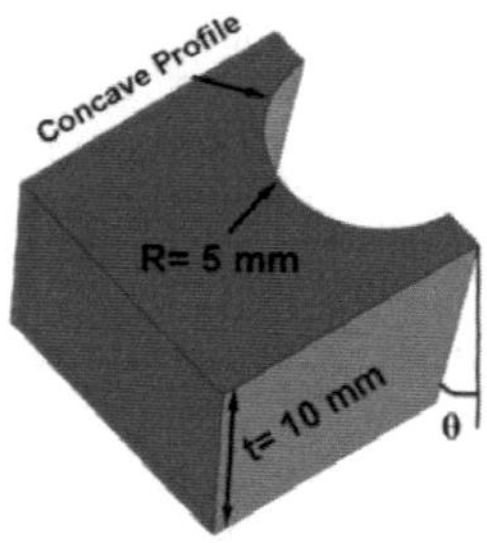

Figure 2. Geometry of machined workpiece

Experimental design was based on Taguchi's L18 orthogonal array. There were five control factors namely taper angle, pulse-on time, pulse-off time, servo voltage and wire tension. Variable control factors and their levels are presented in Table 2. Constant parameters for WEDM were open voltage =

85volt, arc-on time = 0.2 μsec, arc-off time = 20 μsec, wire feed = 140 mm/sec, and dielectric pressure (upper nozzle = 10 kgw/cm^2, lower Nozzle = 8 kgw/cm^2).

Table 2. Variable Control Factors and Their Levels

Sr. No.	Control Factor	Symbol	Unit	Level 1	Level 2	Level 3
1	Taper Angle	TA	Degree	5	10.5	---
2	Pulse-on time	ON	μsec	0.2	0.3	0.4
3	Pulse-off time	OFF	μsec	20	25	30
4	Servo voltage	SV	Volt	45	50	55
5	Wire tension	WT	g	630	730	870

There were two response variables namely cutting speed (CS) and geometrical error (GE) of machined profile. Cutting speed was directly noted from WEDM display. Radius of curved profile was measured using video camera option of co-ordinate measuring machine (Model: CHEN WEI CE-450DV). After recording the measurement values, geometrical error was calculated by Eq. (1):

$$Geometrical\ Error = Nominal\ Radius - Measured\ Radius \quad (1)$$

3. Results And Analysis

Taguchi orthogonal array L18 along with experimental results is presented in Table 3.

Table 3. Taguchi L 18 Orthogonal Array and Response Variables

Exp. No.	TA	ON	OFF	SV	WT	Cutting Speed (CS)		Geometrical Error (GE)	
	degree	μsec	μsec	volt	g	mm/min		μm	
						Wire A	Wire B	Wire A	Wire B
1	5	0.2	20	45	630	0.86	0.93	14	7
2	5	0.2	25	50	730	0.75	0.79	14	12
3	5	0.2	30	55	870	0.65	0.72	11	12
4	5	0.3	20	45	730	1.07	1.19	18	12
5	5	0.3	25	50	870	0.97	1.04	13	8
6	5	0.3	30	55	630	0.82	0.92	8	0
7	5	0.4	20	50	630	1.27	1.37	5	6
8	5	0.4	25	55	730	1.09	1.19	13	6
9	5	0.4	30	45	870	1.09	1.17	16	10
10	10.5	0.2	20	55	870	0.87	0.95	16	16
11	10.5	0.2	25	45	630	0.79	0.87	23	11
12	10.5	0.2	30	50	730	0.68	0.77	19	24
13	10.5	0.3	20	50	870	1.09	1.27	14	34
14	10.5	0.3	25	55	630	0.90	1.03	13	22
15	10.5	0.3	30	45	730	0.88	0.95	12	11
16	10.5	0.4	20	55	730	1.23	1.40	17	4
17	10.5	0.4	25	45	870	1.19	1.34	17	15
18	10.5	0.4	30	50	630	1.02	1.11	13	7

Comparison of cutting speed for wire A and wire B against each experimental condition is presented in Fig. 3. It can be seen that cutting speed of wire B is higher than that of wire A. This may be attributed to presence of zinc coating on wire B. Zinc has very low melting point (419°C) and even its boiling temperature (906°C) is lower than the melting temperature of brass (930°). Very low melting temperature of zinc helps to improve spark formation by reducing the ionization time of dielectric fluid [3]. In addition, low melting/vaporization temperature of wire coating also affects the distribution of discharge energy between wire and workpiece. Lesser energy is absorbed by zinc-coated wire electrode as compared to plain brass wire having high melting/evaporation temperature [6]. Therefore, most of the gap energy is supplied to the workpiece. This is the possible reason for obtaining higher cutting speed when machining is performed using zinc-coated wire.

Maximum improvement (16.5%) in cutting speed using wire B is obtained for Exp. No. 13. Cutting speed of wire B can be further enhanced by increasing pulse-on time but then frequent wire breakage occurs making the process unstable.

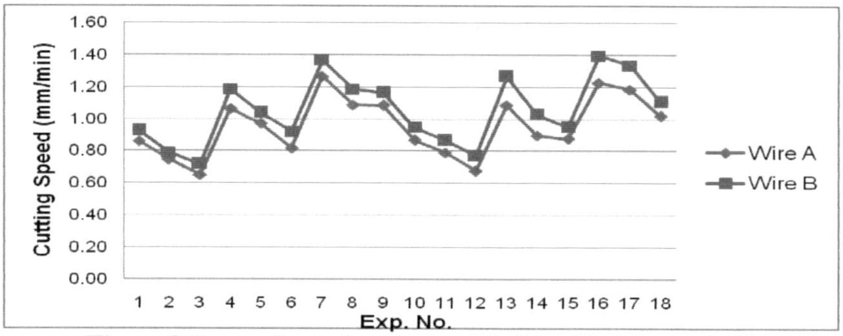

Figure 3. Comparison of cutting speeds for wire A and wire B

3.1. *Effect of Control Factors on Performance Measures*

Influence of control factors on performance measures is studied with the help of main effects plots. In order to draw the plot of a factor, the average value of response variable against each level of that factor is required. These averages are calculated by using data from Table 3.

For cutting speed, both the wires have shown similar trends as depicted in Fig. 4. Pulse-on time (ON) is the most significant factor followed by pulse-off time (OFF). Cutting speed increases with increase in pulse-on time. At higher pulse-on time, greater discharge energy is supplied to the spark gap resulting in melting and evaporation of material at a faster rate hence cutting rate increases. Cutting speed has shown a decreasing trend with increase in pulse-off time (OFF) and servo voltage (SV). Number of discharges occurring in a given time period decreases with increase in time interval between two consecutive pulses (pulse-off time) causing a reduction in cutting speed. Increase in servo voltage widens the gap between wire and workpiece, therefore, spark intensity is reduced and cutting speed becomes slow. For wire tension (WT), cutting speed is higher at level 3 (870g). The higher value of wire tension causes a reduction in amplitude of wire vibrations so the kerf width is reduced and higher cutting speed is obtained for the same value of discharge energy [11].

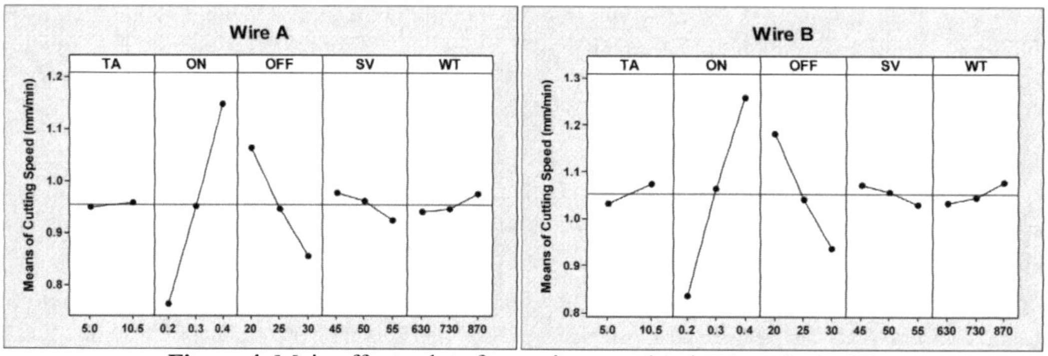

Figure 4. Main effects plots for cutting speed (wire A and wire B)

It can be seen from Table 3 that the value of geometrical error (GE) is positive for all experiments. It means that measured radius of concave curved profile is always lesser than the nominal radius of 5 mm.

Main effects plots for geometrical error in curved profile machined using wire A and wire B are presented in Fig. 5. It can be seen that taper angle (TA) is the most significant parameter. Geometrical error is higher for the workpiece that is machined at higher taper angle. This may be due to the reason that at high value of taper angle, wire cannot adapt to the geometry of the guides due to its rigidity and there is a large deviation between programmed shape and actual shape of wire [12]. When machining is performed with this deformed wire, higher value of geometrical error is experienced. Geometrical error decreases with increase in pulse-on time, pulse-off time, and servo voltage. Higher value of pulse-on time causes more erosion of workpiece material, resulting in increased radius of curved profile so the geometrical error is reduced. Increase in pulse-off time minimizes the chances of re-deposition of debris due to better flushing and higher servo voltage increases sparking gap width thereby stabilizing the WEDM process. Increase in wire tension (WT) results in increased geometrical error. This may be attributed to reduced amplitude of wire vibrations at higher WT causing a decrease in kerf width, therefore, the radius of concave curved profile decreases [13] resulting in higher value of geometrical error.

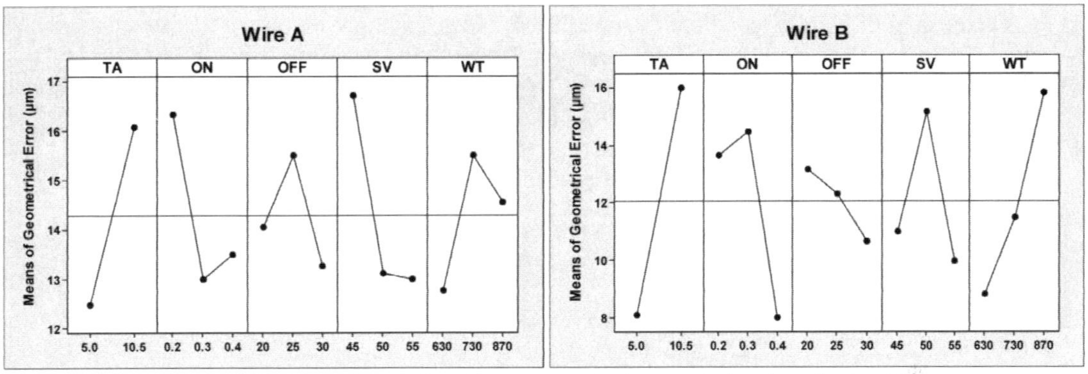

Figure 5. Main effects plots for geometrical error (wire A and wire B)

4. Conclusions

In present research work, the performance of uncoated brass wire and zinc-coated brass wire is evaluated for machining of curved profile in tungsten carbide in taper cutting mode of WEDM under various experimental conditions. Following conclusions can be drawn from experimental results and analysis:

- Cutting speed of zinc-coated brass wire is up to 16.5% higher as compared to that of uncoated brass wire.
- Pulse-on time and pulse-off time are the most significant parameters affecting cutting speed for both the wires.
- Geometrical accuracy of curved profile is significantly affected by the value of workpiece taper angle. Higher value of taper angle gives rise to geometrical error.

References

[1] Jangra K, Grover S, Chan FTS, Aggarwal A (2011) Digraph and matrix method to evaluate the machinability of tungsten carbide composite with wire EDM. Int J Adv Manuf Technol 56 :959–974
[2] Nayak BB, Mahapatra SS (2016) Optimization of WEDM process parameters using deep cryo-treated Inconel 718 as work material. Eng Sci Technol An Int J 19(1):161–170
[3] Antar MT, Soo SL, Aspinwall DK, Jones D, Perez R (2011) Productivity and workpiece surface

integrity when WEDM aerospace alloys using coated wires. Procedia Eng 19:3–8

[4] Klocke et al. (2014) Evaluation of advanced wire-EDM capabilities for the manufacture of fir tree slots in inconel 718. Procedia CIRP14:430–435

[5] Ramamurthy et al. (2015) Performance analysis of wire electrodes on machining Ti-6Al-4V alloy using electrical discharge machining process. Machining Science and Technology 19:577-592

[6] Patil NG, Brahmankar PK,Thakur DG (2016) On the effects of wire electrode and ceramic volume fraction in wire electrical discharge machining of ceramic particulate reinforced aluminium matrix composites. Procedia CIRP42(ISEM XVIII):286–291

[7] Kim C, Kruth JP (2001) Influence of the electrical conductivity of dielectric on WEDM of sintered carbide. KSME International Journal 15(12):1676–1682

[8] Jangra K, Grover S, Aggarwal A (2012) Optimization of multi machining characteristics in WEDM of WC-5.3 % Co composite using integrated approach of Taguchi , GRA and entropy method. Frontiers of Mechanical Engineering 7(3):288–299

[9] Muthuraman V, Ramakrishnan R (2012) Multi parametric optimization of WC-Co composites using desirability approach. Procedia Eng 38:3381–3390

[10] Jangra KK (2015) An experimental study for multi-pass cutting operation in wire electrical discharge machining of WC-5.3 % Co composite. Int J Adv Manuf Technol 76: 971–982

[11] Okada et al. (2008) Effect of surface quality of brass coating wire on wire EDM characteristics. Journal of Advanced Mechanical Design, Systems, and Manufacturing 2(4):735–741

[12] Selvakumar G, Jiju KB, Veerajothi R (2016) Experimental study on wire electrical discharge machining of tapered parts. Arab J Sci Eng 41:4431–4439

[13] Naveed R, Mufti NA, Mughal MP, Saleem MQ, Ahmed N (2017) Machining of curved profiles on tungsten carbide-cobalt composite using wire electric discharge process. Int J Adv Manuf Technol 93(1-4):1367-1378

IOP Conf. Series: Materials Science and Engineering **409** (2018) 012025 doi:10.1088/1757-899X/409/1/012025

Crystal Structure, Surface Topography, Surface Morphology and Optical Properties of DC Magnetron Sputtered VO_2 Thin Films using VO_2 Target

Noormariah Muslim[1,*], Muhammad Nur Syafi'ie Md Idris[1], Ying Woan Soon[2], Yuan-Fong Chou Chau[1], Chee Ming Lim[1], Nyuk Yoong Voo[1]

[1]Centre for Advanced Material and Energy Sciences, Universiti Brunei Darussalam, [2]Physical and Geological Sciences, Faculty of Science, Universiti Brunei Darussalam, Jalan Tungku Link, Gadong BE 1410, Brunei Darussalam

*E-mail: 14h8702@ubd.edu.bn, noormariah.muslim@hotmail.com

Abstract. Vanadium dioxide (VO_2) thin films were deposited at room temperature on Corning 2947 glass substrates by direct current (DC) magnetron sputtering with a high purity VO_2 target. Crystal structure, surface topography, surface morphology and optical properties of the deposited VO_2 thin films were investigated. The deposited films exhibited a single orientation of (110) with a crystallite size of 41.3 nm as confirmed by the X-ray diffraction analysis and Scherrer formula, respectively. From the surface topography analysis, the film surface had root mean square surface roughness of ~6.8 nm and consisted of round-shaped grains. Similarly, from the surface morphology analysis, spherical-like grains were observed on the surface of the deposited VO_2 thin films with estimated average grain size of 34.2 nm. The deposited thin films showed high transmittance and low reflectance in the visible and near-infrared wavelength regions at room temperature. In addition, from the optical transmittance against temperature measurements, only a few transmittance variation and a slight change in hysteresis loop were detected during heating and cooling between room temperature and 100 °C. Hence, the deposited VO_2 thin films were found to exhibit lack of phase transition.

1. Introduction

Vanadium Dioxide (VO_2) thin films have been intensively studied over the last decade to the point where the thin films were discovered as materials with distinctive characteristics such as temperature-dependent properties, for a wide range of applications for instance intelligent coatings for architectural glazing. However, accessibilities of deposition techniques to grow stoichiometric VO_2 thin films are still crucial.

Deposition techniques that have been employed to deposit VO_2 thin films include atmospheric pressure chemical vapour deposition [1,2], reactive pulsed laser deposition [3], magnetron sputtering [4-5] and solution-based techniques. Despite, among those techniques, magnetron sputtering techniques with vanadium (V) target are very popular and become the process of choice for preparations of VO_2 thin films. For instance, in 1996, direct current (DC) magnetron sputtering was chosen, and employed to fabricate stoichiometric VO_2 films at sputtering power of 250 W applied on a V target with working pressure of 1.0 Pa, substrate temperature of 400 °C, Ar gas flow of 88.9 % and O_2 gas flow of 11.1 % [6]. The films exhibited transition temperature (Tt) at 65 °C where changes in structural and optical properties of the films were observed. Additionally, Batista et al. [7] investigated several depositions of single phase of monoclinic (M) VO_2 films by using DC magnetron sputtering with substrate temperature of 450 °C and various parameters such as O_2/Ar gas flow ratio, sputtering power, DC current supply and working pressure. Moreover, according to Cui et al., VO_2 films were also successfully deposited with a V target at working pressure of 1.0 x 10^{-3} Pa, DC power of 160 W and different substrate temperature ranging from room temperature to 400 °C, as well as different O_2 partial pressure of about 9 to 20 % [8]. Mlyuka et al. reported depositions of VO_2 thin films with different film thicknesses (i.e. ranging from 25 to 120 nm) at substrate temperature of 450 °C, Ar/O_2 flow ratio of 10.7 and working pressure of 12.3 Pa [9]. The deposited films exhibited Tt at 60 °C. Meanwhile, Ji et al. studied growth of VO_2 thin films at O_2/Ar flow ratio of 0.05, working pressure of 1.2 Pa, sputtering power of 172 W and substrate temperature 380 °C with film thickness of ~80 nm [10]. Luo et al. conducted a study on depositions of nanocrystalline VO_2 thin films at low temperatures with different film thickness (i.e. 60 to 197 nm) by using DC magnetron sputtering and in-situ annealing process [11]. Furthermore, polycrystalline VO_2(M) thin films were prepared at working pressure of ~12.3 Pa, Ar/O_2 flow ratio of ~19, sputtering power of 210 W and substrate temperature of 450 °C [12]. In addition, Melnik et al. reported depositions of polycrystalline VO_2 (M) thin films at working pressure of 0.4 Pa, sputtering power of 70 W, substrate temperature of 200 °C and annealing process at 300 °C in air [13]. Besides, depositions of stoichiometric VO_2 films were also studied on various substrates. For example, Lafort et al. researched the depositions of single polycrystalline VO_2 thin films with film thickness of 100 to 200 nm on stainless steel substrates at working pressure of 1.3 x 10^{-4} Pa, DC power of 200 W, Ar flow rate of 20 sccm, O_2 flow rate of 4.5 sccm and substrate temperature of 480 °C [14]. In addition, nanostructured VO_2 films were deposited on amorphous glass slides with a V target at substrate temperature of 100 °C, DC current 0.3 A, O_2/Ar flow ratio of 3.2 % and in-situ annealing process at 350 °C in O_2 ambient for 25 min [15]. According to Zhang et al., DC sputtered VO_2 thin films were successfully deposited on K9 glass substrates and in-situ annealed at low temperature with different annealing time [16]. VO_2 thin films deposited on soda lime glass substrates were also reported at working pressure of 0.5 Pa, O_2 flow rate of 1.6 sccm, Ar flow rate of 40 sccm, substrate temperature of 200 °C and sputtering power of 110 W with negative bias voltage of -160 V [17]. Nevertheless, among those depositions, the application of DC magnetron sputtering and VO_2 target for depositions of VO_2 thin films is unexplored.

Hence, this work aims to study the structural, morphological and optical properties of VO_2 thin films deposited by using a DC magnetron sputtering and a VO_2 target at room temperature.

2. Experimental Details

2.1. Film Preparation

A VO_2 target (99.9 % purity) with a diameter of 76.2 mm and thickness of 3.0 mm was used to prepare VO_2 thin films on Corning 2947 glass substrates by using a DC magnetron sputtering technique. Before depositions, the glass substrates were cleaned in an ultrasonic bath with ethanol for 15 min, rinsed with distilled water, dried by blowing dry nitrogen gas of 99.9 % purity, and mechanically clamped to the substrate holder in the sputtering chamber. The sputtering chamber was evacuated to 8.0 x 10^{-4} Pa by a turbomolecular pump. A commercial Ar (99.9 % purity) was supplied as the sputtering gas and controlled by a mass flow controller to a flow rate of 30 sccm. Then, the VO_2 target was pre-sputtered in Ar atmosphere for 10 min to remove any impurities. The film thickness during depositions was monitored with an Inficon Model SQM-160 quartz rate/thickness meter. To ensure a uniform film thickness, the substrates were rotated at 10 rpm during film depositions. The thickness of the deposited VO_2 thin films was 100 nm. During sputtering, the working pressure and DC supply were maintained at 0.8 Pa and 160 mA, respectively with substrate temperature of room temperature.

2.2. Film Characterization

Film thicknesses were measured by a surface profilometer (Alpha-Step IQ) with a stylus force of 0.03 N (3.0 mg) and measurement accuracy of ± 1.5 nm. The structural property of the prepared films was studied by X-ray diffraction (XRD) on a Shimadzu XRD-7000. The XRD patterns were obtained using monochromatic high-intensity Cu Kα (λ = 1.54056Å) radiation operated at 40 kV and 30 mA. The patterns were recorded at a scanning rate of 2°/min in the range of 2θ = 20 – 50° and an incident angle of 1.2°. The surface topography of the films was determined by an atomic force microscopy (AFM) on a commercial Nanos (CSM Instruments). The measurements were performed in the contact tapping mode whereby the load force on the cantilever was maintained at 2.0 nN in order to apply a constant force on the samples. The surface morphology of the films was analyzed by field-emission scanning electron microscopy (FE-SEM) using a JEOL JSM-7610F field-emission scanning electron microscope. The optical properties of these films were measured using a CARY 5000 UV-Vis-NIR spectrophotometer (Agilent Technologies) equipped with a SPECAC 4000 Series^{TM} high stability temperature controller. The measurements were taken in the wavelength range of 350 – 2500 nm for the optical transmittance, and 300 – 800 nm for the optical reflectance. The scanning rate was 250 nm/min.

3. Results and Discussion

3.1. Crystal Structure of the Deposited VO_2 Films

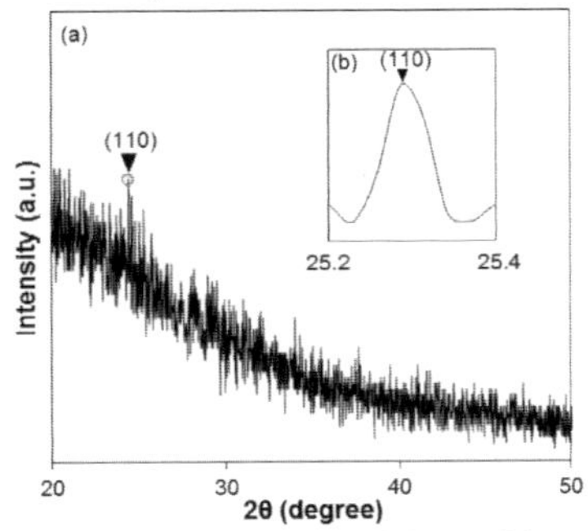

Figure 1. (a) XRD patterns of the deposited VO_2 thin films. (b) presents magnified details of (a).

Fig.1 (a) shows the XRD diffraction pattern of the VO_2 thin films deposited on glass substrates using DC magnetron sputtering and VO_2 target at room temperature. It could be seen that there is only one obvious diffraction peak at 2θ = 25.3° (as magnified in Fig.1 (b)) which is a characteristic of VO_2 phase of (110) plane, as confirmed with the standard ICDD2016: card no. 01-071-0042. The (110) crystallite orientation was used to estimate crystallite size of the deposited films by using the Scherrer formula. The estimated crystallite size of the deposited VO_2 thin films was 41.3 nm.

3.2. Surface Topography of the Deposited VO_2 Films

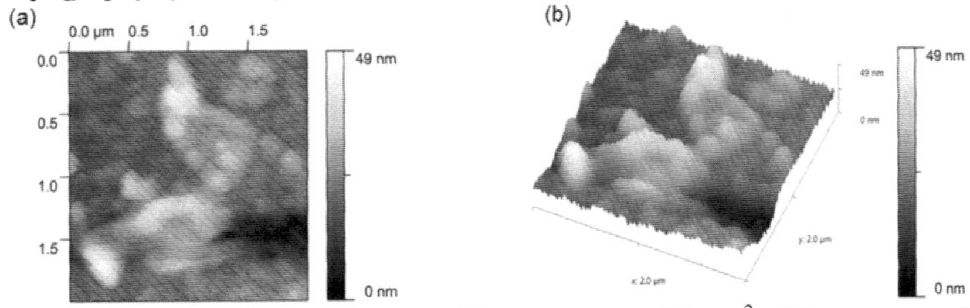

Figure 2. AFM (a) 2D- and (b) 3D- images with a scan area of 4 µm^2 of the upper surface of the deposited VO_2 thin films.

Fig.2 (a)-(b) display the AFM images of the deposited VO_2 thin films which were used to analyze surface topography and surface roughness of the films. The images were acquired using a scan area of 4 µm^2. The study of surface topography of the films from the AFM images was carried out by assuming the dark regions represent areas with zero or near zero height value along the Z scale (positive direction) and the bright regions represent higher areas i.e. the top of bulging grains. The surface of the films was composed of round-shaped grains. The root mean square surface roughness of the films was detected to be ~6.8 nm.

3.3. Surface Morphology of the Deposited VO_2 Films

Fig.3 represents the FE-SEM image of the deposited VO_2 thin films under a magnification of x 200,000 with an accelerating voltage of 2.0 kV. The surface of the films appeared smooth and uniform. VO_2 grains were grown in spherical shape with the estimated average grain size of 34.2 nm. Similar surface morphology was also observed by Zhang et al. [18] whereby nanostructured VO_2 films were prepared on glass substrates by reactive DC magnetron sputtering with a vanadium metallic target. In addition, Batista et al. [7] reported that VO_2(M) would appear as spherical-like grains whereas very flat and no particular features were detected to be in VO_2(B) phase and elongated grains were found to be in V_2O_5 phase. Hence, the surface morphology of the deposited films correspondingly featured the surface topography obtained on the AFM analysis.

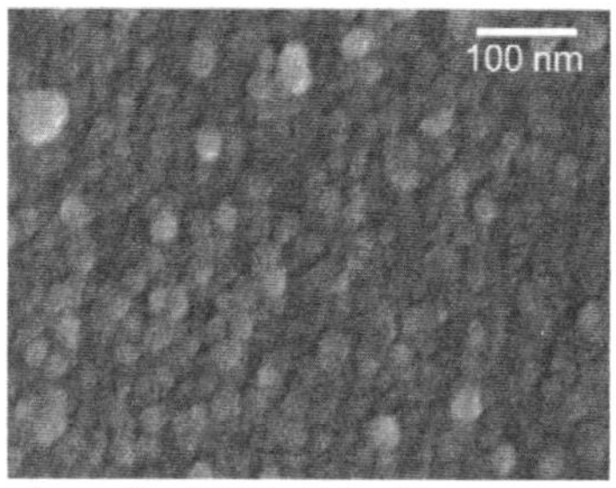

Figure 3. FE-SEM image of the deposited VO_2 thin films recorded at a magnification of x 200,000.

3.4. Optical Properties of the Deposited VO_2 Films

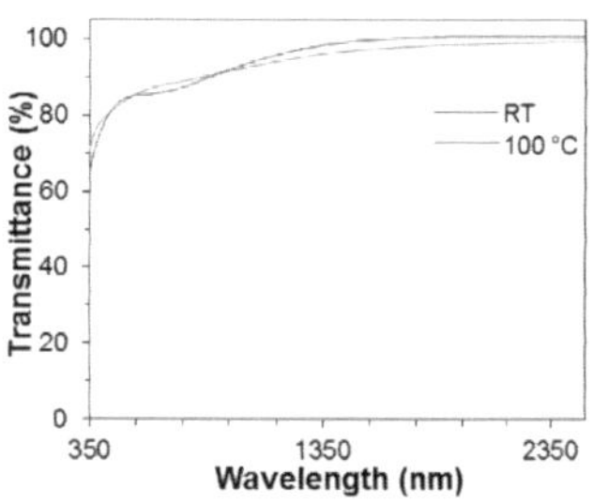

Figure 4. Optical transmittance spectra of the deposited VO_2 thin films at room temperature (RT) and 100 °C.

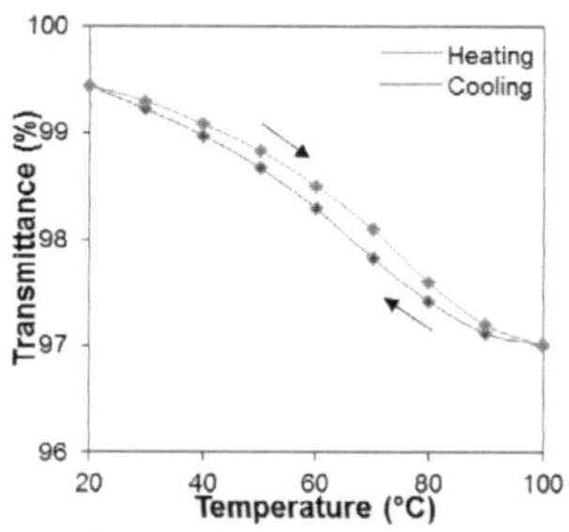

Figure 5. Optical transmittance at a fixed wavelength of 1500 nm as a function of temperature for the deposited VO_2 thin films.

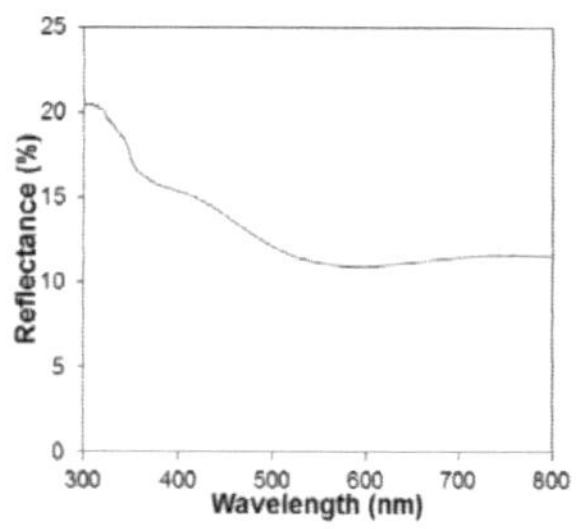

Figure 6. Optical Reflectance spectrum of the deposited VO_2 thin films measured at room temperature.

Fig.4 shows the optical transmittance spectra (in the wavelength range of 350 – 2500 nm) of the deposited VO_2 thin films at room temperature and 100 °C, i.e. on the semiconducting and metallic states, respectively. The deposited VO_2 films exhibited high transmittance in the visible and near-infrared wavelength regions. This was suggested due to the surface of the films. Cui et al. [8] reported that the VO_2 thin films prepared at room temperature showed relative smooth surface and very compact structure resulting in low surface roughness, hence leads to low light-scattering loss and high transmittance. Moreover, only a slight decrease in the transmittance of the films from room temperature to 100 °C was detected.

The curves of the optical transmittance at a fixed wavelength of 1500 nm during heating and cooling for the deposited VO_2 thin films are shown in Fig.5. Only a few transmittance variation within the temperature range could be measured. In other words, only a slight change in the hysteresis loop was observed. As a result, lack of phase transition was present [19].

Furthermore, Fig.6 represents the optical reflectance spectrum of the VO_2 thin films, which was only measured at ambient temperature in air. As observed, the optical reflectance was about 16 % at 400 nm in the visible region and around 11 % at 800 nm in the near-infrared region. The low reflectance of the films could be attributed to the smooth surface and low surface roughness of the films based on the surface morphology and surface topography analyses, respectively.

4. Conclusion

VO_2 thin films were successfully deposited on glass substrates by using DC magnetron sputtering with a VO_2 target at room temperature. X-ray diffraction pattern of the deposited VO_2 thin films displayed a single orientation of (110). In addition, with the help of the Scherrer formula, a crystallite size of about 41.3 nm was obtained. Spherical-like grains were observed on the surface of the films with the root mean square surface roughness of ~6.8 nm and the estimated average grain size of 34.2 nm obtained from the surface topography and surface morphology analyses, respectively. From optical properties analyses, the deposited thin films showed high transmittance and low reflectance in the visible and near-infrared wavelength regions at room temperature. Moreover, only a small change in the transmittance and the hysteresis loop were found during heating and cooling between room temperature and 100 °C. Subsequently, as demonstrated, DC magnetron sputtering with a high purity VO_2 target is also a good alternative technique for the fabrication of VO_2 thin films. Hence, further studies on depositions of VO_2 thin films using DC magnetron sputtering with a VO_2 target are in progress in order to improve the properties of the films, especially crystallinity and hysteresis behavior of the films.

Acknowledgment

This research was supported by the Brunei Research Council under Grant No. S&T 17.

References

[1] G. Bodurov, T. Ivanova, M. Abrashev, Z. Nenova, and K. Gesheva, "Thin film optical coatings of vanadium oxide and mixed tungsten/vanadium oxide deposited by APCVD employing precursors of vanadyl acetylacetonate and a mixture with tungsten hexacarbonyl," Physics Procedia, vol. 46, Jul. 2013, pp. 127-136, doi: https://doi.org/10.1016/j.phpro.2013.07.054.

[2] D. Vernardou, D. Louloudakis, E. Spanakis, N. Katsarakis, and E. Koudoumas, "Thermochromic amorphous VO_2 coatings grown by APCVD using a single-precursor," Solar Energy Materials and Solar Cells, vol. 128, May 2014, pp. 36-40, doi: http://dx.doi.org/10.1016/j.solmat.2014.04.033.

[3] A. Hendaoui, N. Émond, S. Dorval, M. Chaker, and E. Haddad, "Enhancement of the positive emittance-switching performance of thermochromic VO_2 films deposited on Al substrate for an efficient passive thermal control of spacecrafts," Current Applied Physics, vol. 13, Jan. 2013, pp. 875-879, doi: https://doi.org/10.1016/j.cap.2012.12.028.

[4] X. Dong, Z. Wu, X. Xu, T. Wang, and Y. Jiang, "Effects of duty cycle and oxygen flow rate on the formation and properties of vanadium oxide films deposited by pulsed reactive sputtering," Vacuum, vol. 104, Jan. 2014, pp. 97-104, doi: https://doi.org/10.1016/j.vacuum.2014.01.020.

[5] C. Batista, R.M. Ribeiro, and V. Teixeira, "Synthesis and characterization of VO_2-based thermochromic thin films for energy-efficient windows," Nanoscale Research Letters, vol. 6, Apr. 2011, pp. 301, doi: 10.1186/1556-276X-6-301.

[6] M.A. Sobhan, R.T. Kivaisi, B. Stjerna, and C.G. Granqvist, "Thermochromism of sputter deposited $W_xV_{1-x}O_2$ films," Solar Energy Materials and Solar Cells, vol. 44, Dec. 1996, pp.

451-455, doi: https://doi.org/10.1016/S0927-0248(95)00051-8.

[7] C. Batista, V. Teixeira, and J. Carneiro, "Structural and morphological characterization of magnetron sputtered nanocrystalline vanadium oxide films for thermochromic smart surfaces," Journal of Nano Research, vol. 2, Aug. 2008, pp. 21-30, doi: 10.4028/www.scientific.net/JNanoR.2.21.

[8] H.N. Cui, V. Teixeira, L.J. Meng, R. Wang, J.Y. Gao, and E. Fortunato, "Thermochromic properties of vanadium oxide films prepared by dc reactive magnetron sputtering," Thin Solid Films, vol. 516, Feb. 2008, pp. 1484-1488, doi: https://doi.org/10.1016/j.tsf.2007.03.131.

[9] N.R. Mlyuka, G.A. Niklasson, and C.G. Granqvist, "Thermochromic multilayer films of VO_2 and TiO_2 with enhanced transmittance," Solar Energy Materials and Solar Cells, vol. 93, Sept. 2009, pp. 1685-1687, doi: https://doi.org/10.1016/j.solmat.2009.03.021.

[10] Y.X. Ji, S.Y. Li, G.A. Niklasson, and C.G. Granqvist, "Durability of thermochromic VO_2 thin films under heating and humidity: effect of Al oxide top coatings," Thin Solid Films, vol. 562, Jul. 2014, pp. 568-573, doi: https://doi.org/10.1016/j.tsf.2014.03.043.

[11] Z. Luo, X. Zhou, D. Yan, D. Wang, Z. Li, C. Yang, and Y. Jiang, "Effects of thickness on the nanocrystalline structure and semiconductor-metal transition characteristics of vanadium dioxide thin films," Thin Solid Films, vol. 550, Jan. 2014, pp. 227-232, doi: https://doi.org/10.1016/j.tsf.2013.10.172.

[12] N.R. Mlyuka, G.A. Niklasson, and C.G. Granqvist, "Mg doping of thermochromic VO_2 films enhances the optical transmittance and decreases the metal-insulator transition temperature," Applied Physics Letters, vol. 95, Oct. 2009, pp. 171909, doi: https://doi.org/10.1063/1.3229949.

[13] V. Melnik, I. Khatsevych, V. Kladko, A. Kuchuk, V. Nikirin, and B. Romanyuk, "Low-temperature method for thermochromic high ordered VO_2 phase formation," Materials Letters, vol. 68, Feb. 2012, pp. 215-217, doi: https://doi.org/10.1016/j.matlet.2011.10.075.

[14] A. Lafort, H. Kebaili, S. Goumri-Said, O. Deparis, R. Cloots, J. De Coninck, M. Voué, F. Mirabella, F. Maseri, and S. Lucas, "Optical properties of thermochromic VO_2 thin films on stainless steel: experimental and theoretical studies," Thin Solid Films, vol. 519, Mar. 2011, pp. 3283-3287, doi: https://doi.org/10.1016/j.tsf.2010.12.122.

[15] Z. Luo, Z. Wu, T. Wang, X. Xu, W. Li, W. Li, and Y. Jiang, "Comparison of the optical responses of O-poor and O-rich thermochromic VO_x films during semiconductor-to-metal transition," Journal of Physics and Chemistry of Solids, vol. 73, Sep. 2012, pp. 1122-1126, doi: https://doi.org/10.1016/j.jpcs.2012.04.020.

[16] H. Zhang, Z. Wu, X. Wu, W. Yang, and Y. Jiang, "Transversal grain size effect on the phase-transition hysteresis width of vanadium dioxide films comprising spheroidal nanoparticles," Vacuum, vol. 104, Jun. 2014, pp. 47-50, doi: https://doi.org/10.1016/j.vacuum.2014.01.003.

[17] D.P. Zhang, M.D. Zhu, Y. Liu, K. Yang, G.X. Liang, Z.H. Zheng, X.M. Cai, and P. Fan, "High performance VO_2 thin films growth by DC magnetron sputtering at low temperature for smart energy efficient window application," Journal of Alloys and Compounds, vol. 659, Feb. 2016, pp. 198-202, doi: https://doi.org/10.1016/j.jallcom.2015.11.047.

[18] H. Zhang, Z. Wu, D. Yan, X. Xu, and Y. Jiang, "Tunable hysteresis in metal-insulator transition of nanostructured vanadium oxide thin films deposited by reactive direct current magnetron sputtering," Thin Solid Films, vol. 552, Feb. 2014, pp. 218-224, doi: https://doi.org/10.1016/j.tsf.2013.12.007.

[19] L.L. Fan, Y.F. Wu, C. Si, C.W. Zou, Z.M. Qi, L.B. Li, G.Q. Pan, and Z.Y. Wu, "Oxygen pressure dependent VO_2 crystal film preparation and the interfacial epitaxial growth study," Thin Solid Films, vol. 520, Jul. 2012, pp. 6124-6129, doi: https://doi.org/10.1016/j.tsf.2012.05.086.

IOP Conf. Series: Materials Science and Engineering **409** (2018) 012026 doi:10.1088/1757-899X/409/1/012026

Modeling and Experimental Validation of Deformation of Al-33Cu Droplet during Impingement on a Substrate

[1*]**Amitesh Kumar**

Department of Foundry Technology, National Institute of Foundry and Forge Technology, Ranchi, Jharkhand, India

E-mail: amitesh.iitkgp@gmail.com

Abstract Impingement of a molten Al-33Cu droplet over a substrate causes formation of a splat after spreading and solidification. Modeling of the heat transfer and fluid flow during the impingement and solidification of metal droplets on a substrate is helpful for the better understanding and control of related spray casting and thermal spray coating processes. Authors have thus developed a comprehensive mathematical model of the droplet impingement on the FLUENT 6.3.16 platform. The model takes into account all the inter-related phenomena, viz., (a) heat transfer through the droplet, solid substrate, solid-droplet interface and adjacent gaseous atmosphere, (b) fluid flow in the droplet and adjacent gas and (c) solidification. The droplet undergoes very large and rapid deformation during the impingement. The model predicts the deformation as well as the thermal and flow field within the droplet. Present work reports the experimental validation of model predictions on the droplet deformation, which was carried out using a high speed camera.

1. Introduction

The processing condition for achieving the desired splat morphology can be identified by a validated comprehensive mathematical model that predicts the flow and the thermal fields during the droplet deposition. However, due to interaction of several complex phenomena, comprehensive modeling of the droplet deposition is a challenging task. Further, the validation of the model is complicated due to the small size of the droplet as well as due to rapid deformation and cooling involved.

Mathematical models of liquid metal impingement on cold substrate have been developed to predict the deformation of impinging droplet and solidification within it. Using numerical models and experiments, Zhao et al. [1, 2] studied heat transfer and fluid dynamics during collision of a liquid droplet on a substrate. They extended the earlier model of Fukai et al. [3] to account for the relevant convection and conduction heat transfer phenomena, both in the droplet and in the substrate, in the case, when there is no solidification. Their results, therefore, are applicable to the pre-solidification stage of the impact process. Liu et al. [4] used a one- dimensional solidification model in conjunction with a two-phase flow continuum technique to track the moving liquid-solid boundary. The model, however, does not account for the convection in the liquid and conduction in the substrate. Different investigators studied the droplet deposition based on thermal resistance and temperature profile using different commercial software [5-16].

The present authors [17,24] had developed a comprehensive mathematical model that incorporates fluid flow in droplet as well as adjacent gas, heat transfer in the droplet-substrate interface and solidification. The thermal fields predicted by the models were experimentally validated using

Jackson-Hunt relationship. The present study reports the experimental validation of predictions on the rapid deformation of droplet using a high speed camera.

2. Mathematical Modeling of Droplet Impingement

Present study aims at developing a mathematical model of flow and heat transfer in a liquid metal which falls freely through gaseous medium and impinges on a cold substrate. Modeling of the phenomena is complex because it involves interactions among three phases, viz., liquid metal, solid substrate and gaseous medium. Further, as the droplet falls under the action of gravity and impinges on the substrate, its shape changes rapidly with time and its cooling is accompanied by solidification.

Figure 1 is a schematic of the impingement phenomenon. With a symmetric boundary in the middle only half of the domain (ABFCDEA) was actually computed and a very fine grid was used for the regions where impact, spreading and solidification occur. The grid for computation was finalized by grid independent test.

A droplet of Al-33wt%Cu (hereafter referred to as Al-33Cu), which is initially spherical, falls through the gaseous medium and impinges on a grade 304 stainless steel substrate. At the droplet/substrate interface a gap was created. This gap was used to model the interfacial contact resistance by prescribing an effective thermal conductivity to the gap. The thickness of the substrate was assumed to be 1mm. Since the initial shape of the droplet is assumed to be spherical, the fluid flow and the heat transfer are axisymmetric (r-z direction).

The following assumptions have been made in the mathematical model:

1. Droplet is initially having spherical geometry and it remains so just before the impingement.
2. The problem is axi-symmetric about the axis shown in Figure 1.
3. Velocity of the droplet impingement is perpendicular to the plane of the substrate and no rotation of droplet occurs along the axis.
4. The flow of metal Al-33Cu droplet and gas is laminar and incompressible. The velocity of gaseous media at domain walls is zero.
5. Walls of the domain (Figure 1) are assumed to be at 300 K.
6. The heat transfer is dominated by conduction and convection modes. Therefore, radiation is ignored.
7. To take into account the thermal contact resistance due to air gap a solid layer has been defined. The conductivity of the solid layer was assigned by a constant value.
8. Initial velocity of droplet (when it forms) is zero. It acquires velocity during the free fall. The velocity just before the impingement was $\sqrt{2gH}$, where H is the distance between the orifice of crucible and substrate. The other important information on the mathematical model like the governing equations and discretization method have been described elsewhere [20,28].

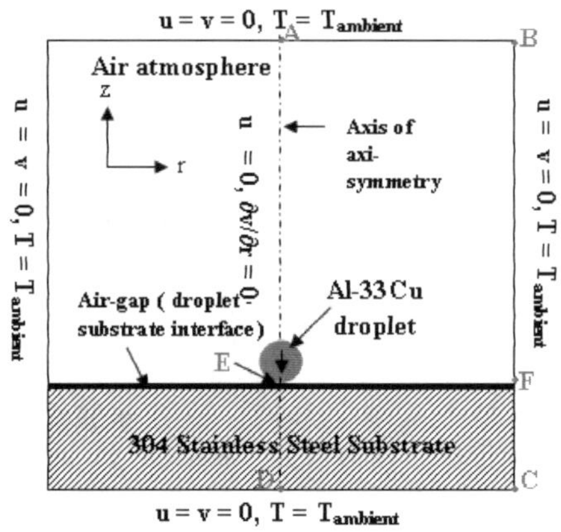

Figure 1. Schematic Representation of the Mathematical Model of Droplet Impingement.

All zones were initialized with the temperature of 300K, except the droplets. The droplets were set to a predetermined higher temperature. The initial velocities at every point inside the domain were zero, except within the droplets.

Since problem is axis-symmetric only half of the domain shown in Figure 1 was considered in the computation domain. Axis of symmetry was like a free-slip wall. The normal velocity was zero. Tangential velocity did not have normal gradient. Thus along the axis of axis-symmetry $u = 0$ and $\partial v/\partial r = 0$. At the other boundaries $u = 0$ and $v = 0$. Since these boundaries are far away from the droplet, the temperatures at these boundaries were set equal to the ambient temperature, i.e., 300 K. The above described model was simulated using FLUENT 6.3.16 plat form.

Thermo- physical properties of Al-33Cu and 304 stainless that have been used for simulation are taken from ref. [20]. The thermal contact resistance $7 \times 10^{-4}\ m^2\ Kw^{-1}$ at the interface of Al-33Cu droplet and the grade 304 stainless steel substrate was taken from the work of Kumar et al [20].

3. Experiments for Validation of model

A droplet impingement set up to validate the model was fabricated. Figure 2 shows the schematic sketch of the set-up. A vertical resistance furnace was used for melting. The furnace has a cylindrical shape having a height of 8 inch, an outer diameter of 4 inch and an inner diameter of 1 inch. The furnace was capable of rapid heating of the charge. The temperature of the melt could be controlled within $\pm 5^0$C. It was placed at the top of the droplet impingement chamber as shown in Figure 3. A quartz tube with an inner diameter of 17 mm was used as a crucible. At the bottom of the crucible there was a melt delivery nozzle. The diameter of the melt delivery nozzle was 2 mm. The inner wall of the quartz was coated with alcohol shoot, which suppressed the reaction between molten metal and quartz tube. Substrate was made of grade 304 stainless steel. The substrate was circular in shape having a diameter of 20 cm and thickness of 20 mm. It was placed inside the spray chamber. It was placed on a threaded stand, such that it could be moved up and down. Pieces of master alloy Al-33Cu were melted in the quartz crucible. The temperature of the melt was maintained at ~150°C above the melting point. For each droplet impingement experiment, 50 gram charge of the alloy was charged into the crucible where the temperature was 700^0C. The temperature of the melt was monitored by a thermocouple inserted between the quartz crucible and the furnace inner wall. By application of air pressure droplets were pushed out of the 2 mm diameter melt delivery nozzle. These droplets impacted on the substrate, deformed and finally solidified on it. For deposited splats, the size of the droplet was estimated by measuring the size of splats.

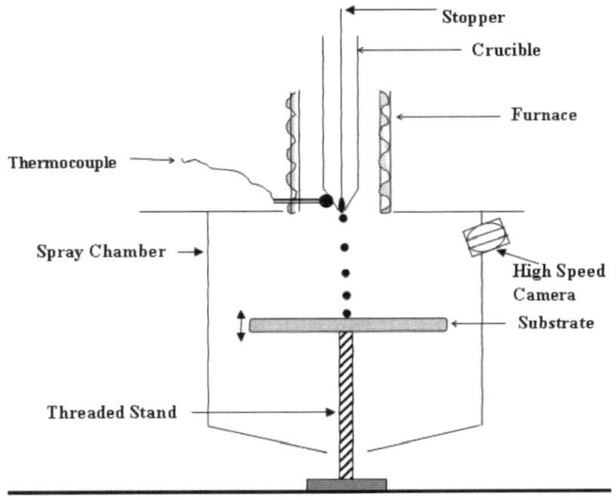

Figure 2. Schematic Sketch of Droplet Impingement Set up

4. Results and Discussions

In the present work the mathematical model of the droplet impingement assumes a thin layer of air – gap in between the impinging Al-33wt%Cu (Al-33Cu) droplet and the stainless steel substrate (Figure

1). $7x10^{-4}m^2KW^{-1}$ was taken as the value of the thermal contact resistance and 1.43 $Wm^{-2}K^{-1}$ as the value of the conductivity of the interface air-gap layer in the droplet impingement model (Shown in Figure 1).

5. I. Deformation of Al-33Cu Droplet

Numerical simulations and experiments were carried out for the impingement of Al-33wt%Cu droplets on the grade 304 stainless steel substrate. In the current study simulation and experimental studies were performed for shape evaluation of impinging molten 6.2 mm diameter droplet with velocity 3.13 m/s on to the stainless steel substrate.

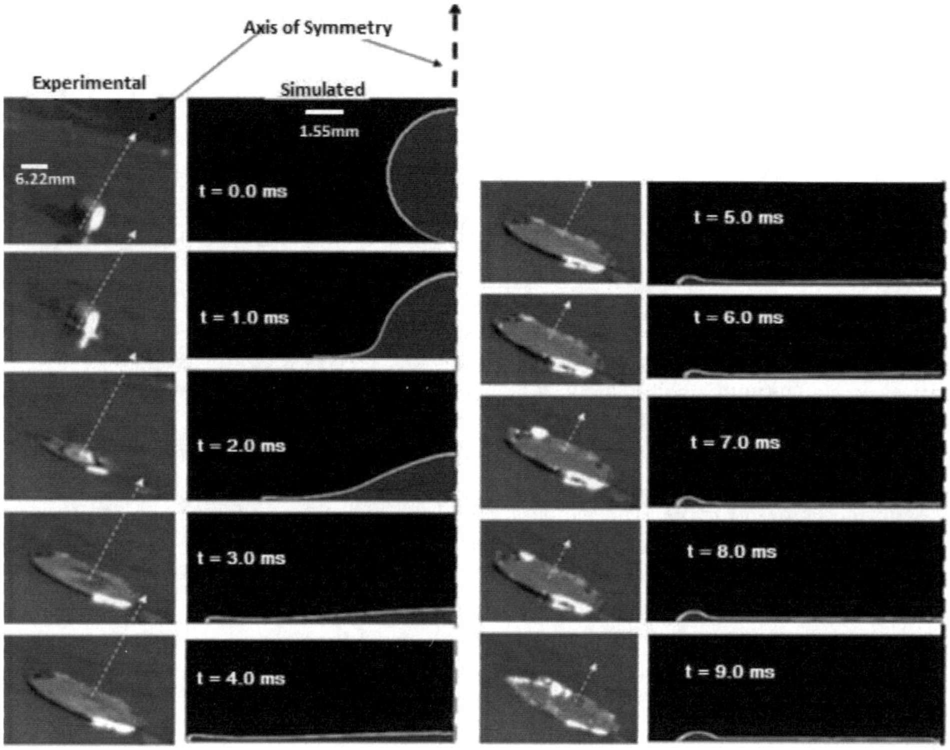

Figure 3. Comparisons of Photographs and Simulated Images a 6.2 mm Diameter Al-33Cu droplet.

At the moment of impingement, the high impingement velocity creates a high pressure region under the droplet and forced the liquid metal to flow over the contact surface while the effect of surface tension was comparatively less; the maximum flow velocity was achieved at the edge where all the momentum was transferred to the radial direction, and the cross-section area was small. The sequence of deforming droplet was recorded using a high speed camera (FASTCAM- Super 10 KC, Photron USA, San Diego, CA), which can capture the photographs at the speed of 1000 frame per seconds. Figure 3 presents the result of simulation of impingement of a Al-33 Cu droplet having a diameter of 6.2 mm and compares the computed shape of the droplet for different durations of time after the droplet comes in contact with substrate, with that recorded by high speed video camera. It is clear that the simulated Al-33wt%Cu droplet profile is in good agreement with that from the experiment as shown in the Figure 3. The forces which cause the deformation of droplets are gravitational force and the force due to high velocity of impingement. The forces due to viscosity and surface tension oppose the flow of molten metal. Droplet spreads radially after impingement on the substrate along the surface of the substrate as shown in the Figure 3. With the spreading and solidification progressed, the flow of liquid metal reduced and the effect of surface tension became significant, which generated high pressure regions at the edges and finally reversed the flow back towards the centre as shown in the 4.0 ms time frame plot in Figure 3. This resulted into the formation of bump at the edge of the splat. When a solidification layer grows, the lower part of the liquid splat loses momentum as it solidifies to become a part of solidified layer. Solidification may take place during spreading, but the solidified layer growth velocity is slow compared to when solidification starts after complete spread. Therefore, a splat retains radial momentum during spreading and reaches its maximum diameter while solidifying. The simulated evolution of droplet shape has been reported earlier paper [8]. The bump near the edge has been observed even when there is no cooling [22]. The physical reasons for the bump formation are not fully understood although surface tension and the existence of the bump because of the maximum cooling near the edge (Figure 3) [8] could be suggested as two of the reasons. The shape of the splat remains same after 9.0 ms till the complete solidification. Droplet becomes solid after 67.7ms. The solidification of droplet starts from the periphery of the splat because outer edge of the droplet remains in continuous colder part of the substrate, which causes faster heat transfer.

The evolution of the shape of liquid droplet impinging a solid substrate is a complex phenomenon. However, this has been studied by many investigators [8, 22-25]. An important parameter which decides upon the evolution of shape of the liquid droplet is Weber number, i.e., density, viscosity, impingement velocity and surface tension. Depending on the Weber number different types of shape evolution are possible. In numerical method the shape of the splat depends on contact angle, cooling and solidification characteristics. Cooling and solidification characteristics of molten metal depend on thermal contact resistance between molten droplet and the substrate. The value of contact angle for which simulation results were in good agreement with experimental results was 58^0,although simulations have been carried out for different value of contact angle and found different morphologies of the splat.

They have calculated by inverse method between the grade 304 stainless steel and molten Al-33wt%Cu.In experiment droplet impinged the substrate normally, but due to inclined camera arrangement in the set up, the droplet appears to be hitting the substrate at some angle or the substrate appears to have some inclination with horizontal as shown in the Figure 3. In fact substrate was horizontal and direction of droplet impingement was perpendicular to the substrate.

The sequence of impingement as shown in the Figure 3 shows that the droplet deforms on to the substrate into the shape of a cap and cone shape with flat edge symmetric around the axis of symmetry up to 2.0ms followed by formation of bump at the periphery of the impacting droplet after 3.0ms. Theoretical predictions [20, 28] have shown that a molten Al-33wt%Cu droplet impacting on a 304 stainless steel substrate starts freezing first at the outer most edge, where the substrate temperature is lowest. Therefore, solidification restricts the further spread of the splat and surface tension becomes predominant at the lower temperature and tries to pull back the liquid metal inwards. Therefore there is formation of bump at the periphery of the splat as shown in the Figure 3. Therefore, the final shape of solidified splat is like a circular disc as obtained. As shown in Figure 3 the numerical results are in qualitative as well as quantitative agreement with the experiments. So, it is one of the way in which the impingement model can be validated is by comparing the simulated deformation of droplet with the video photographed image. The same has been reported in this study. Figure 3 shows the actual deformation of a droplet having diameter of 6.2 mm (photographed using high speed video camera) along with the simulated deformation. The initial temperature of the droplet was 973 K and the velocity of impingement was 3.13 ms^{-1}.

6. Conclusion

In the present study, the impingement of a molten metal droplet (Al-33wt%Cu) is simulated using a commercial computational fluid dynamics software, FLUENT 6.3.16, in which a two dimensional axis-symmetric domain for deformation considering the free surface deformation and solidification have been developed by using the VOF technique with a enthalpy porosity method. The VOF technique in combination with solidification model was used to simulate the impingement of Al-

33wt%Cu droplets on the 304 stainless steel substrate. The model was validated by comparing the predictions with experimental data. The outcomes of the present investigation are listed below.

1. Fluent 6.3.16, finite volume method based code has the capability of obtaining a convergent solution of the equations established in simulation of impingement of a molten droplet impacted at a high velocity.

2. The deformation behavior of the droplet was characterized by the spreading and evolution of shape of the splat, the predicted results were in good match with the experimental.

3. A comprehensive model has been developed that accounts the effect of surface tension as well as solidification.

4. Model can predict the formation of bumps accurately during droplet deposition based processes.

References:

[1] Z. Zhao, D.Poulikakos, J. Fukai: Int. Journal of Heat and Mass Transfer, Vol. 39, pp. 2771-2789, 1996.
[2] Z. Zhao, D. Poulikakos, J. Fukai: Int. Journal of Heat and Mass Transfer, vol. 39, pp. 2791-2802, 1996.
[3] J. Fukai, Z. Zhao, D. Poulikakos, C.M. Megaridis, O. Miyatake: Physics of Fluids A, Vol. 5, pp. 2588 – 2599,1993.
[4] H. Liu, E. J. Lavernia, R. H. Rangel: Journal of Physics D: Applied Physics, Vol. 26, pp 1900-1908, 1993.
[5] G. Trapaga, E.F. Matthys, J.J. Valencia, J. Szekely: Metall. Mater. Trans. B, Vol. 23, 701 – 718,1992.
[6] J.M. Sicilian, C.W. Hirt, R.P. Harper: FLOW-3D, Report No. FSI-88-00-1, Flow Science Inc., Los Alamos, NM, Vol. 1-4,1988,.
[7] M. Bertagnolli, M. Marchese, G. Jacucci, St. Doltsinis, I. Noelting: Journal of Computational Physics, Vol. 133, No. 2, pp. 205-221,1997.
[8] J. M. *Waldvogel*, D. Poulikakos: Int. Journal of Heat and Mass transfer, vol. 40, No. 2, pp. 295-309,1997.
[9] H. Liu, E.J. Lavernia, R.H. Rangel: Materials Science and Engineering A, Vol. 191, pp. 171 – 184, 1995.
[10] G.X. Wang, E.F. Matthys: Journal of Heat Transfer, Vol. 118, pp. 157 – 163, 1996.
[11] M. Pasandideh-Fard, J. Mostaghimi: Plasma Chem. Plasma Process, vol.16, pp 83S-98S, 1996.
[12] M. Pasandideh-Fard, R. Bhola, S. Chandra, J. Mostaghimi:, Int. J. Heat Mass Trans., vol. 41, pp 2929-2945, 1998.
[13] M. Chung, R. H. Rangel: Numerical Heat Transfer, Part A, Vol. 37, No. 3, pp. 201- 226, 2000.
[14] S. Kamnis, S. Gu: Journal of Physics D: Applied Physics, Vol. 38, pp. 3664 – 3673, 2005.
[15] S. Shakeri, S. Chandra: Int. Journal of Heat and Mass Transfer, Vol. 45, pp. 4561 – 4575, 2002.
[16] D. Sivakumar , H. Nishiyama: Journal of Heat Transfer, Vol. 126, No. 6, pp. 1014 – 1022, *2004.*
[17] Korol'kov *A. M*, Casting Properties of Metals and Alloys, Consultants Bureau, New York, NY, 1960, pp 64-66, (1960).
[18] Lang G., Effect of addition elements on the surface tension of liquid very high Purity aluminum, Aluminium, Vol. 50, No. 11, pp. 731 – *734*,1974.
[19] Yang Y.-S., Kim H.Y., Chun J.H., Spreading and solidification of a molten microdrop in the solder jet bumping process, IEEE *Transactions* on Components and Packaging Technologies, Vol. 26, No, 1, pp. 215 – 221,2003.
[20] A. Kumar, S. Ghosh, B.K. Dhindaw: Acta Materialia, Vol. 58, pp. 122-133,2010.
[21] W.I. Geldrop, R. Rioboo, S. Jakirlic, S. Muzaferija, C. Tropea: Eighth International Conference on Liquid Atomization and Spray Systems, Pasadena, CA, USA, July 2000.
[22] S. Sikalo, E. N. Ganic: Experimental Thermal and Fluid Science, Vol. 31, pp. 97-110, 2006.
[23] S. Kamnis, S. Gu: Journal of Physics D: Applied Physics, Vol. 38, pp. 3664 – 3673,2005.

[24] M. Fukumoto, E. Nishioka, T. Matsubara: Surface and Coatings Technology, Vol. 120 – 121, pp. 131 – 137,1999.
[25] H. *Zhang*: Int. Journal of Heat and Mass Transfer, Vol. 42, pp. 2499-2508,1999.
[26] Mo *C*hung, R.H. Rangel: International Journal of Heat and Mass Transfer,Vol. 44, Issue 3, pp. 605-618, 2001.
[27] M. Pasandideh-Fard , S. Chandra, J. Mostaghimi: International Journal of Heat and Mass Transfer, Vol. 45, pp. 2229 – 2242, 2002.
[28] A. Kumar, S. Ghosh, B.K. Dhindaw: Metallurgical Transaction B**,** Vol. 42 B, pp. 269 – 273, 2011.

IOP Conf. Series: Materials Science and Engineering **409** (2018) 012027 doi:10.1088/1757-899X/409/1/012027

The role of $CaCO_3$ in the extraction of valuable metals from low-nickel matte by calcified roasting—acid leaching process

Yong Zhao[1], Shuhua Geng[1], Hongwei Cheng[1], Guangshi Li[1], Xionggang Lu[1], and Qian Xu[1]

[1]School of Materials Science and Engineering, Shanghai University, Shanghai 200444, People's Republic of China

E-mails: hwcheng@shu.edu.cn; luxg@shu.edu.cn

Abstract. The extraction of valuable metals from low-nickel matte via a calcified roasting-acid leaching process was investigated. The influences of process parameters on the extraction of metals were studied, including the roasting temperature, roasting time and the addition of $CaCO_3$. Under the optimum condition, 92.4% of Ni, 97.1% of Cu, 95.1% of Fe, and 91.2% of Co were recovered. In addition, the removal of iron via goethite precipitation process was studied. The results show that 99.8% of Fe could be removed by goethite process. The reaction mechanism of calcified roasting of low-nickel matte was also studied by XRD and XRF analysis. It was shown that the formation of ferrite can be effectively inhibited by $CaCO_3$ addition, leading to enhancement of the metal recovery.

1. Introduction

Nickel (Ni) is the twenty fourth most common elements in the earth's crust; it naturally occurs as sulfides or laterites (oxides), in which almost 60% of the nickel production is via processing sulfide ores, while only 30% of the nickel reserves exist in sulfide deposits [1-2]. As a strategic metal, Ni is widely used in stainless steel, alloys, batteries, and catalysts [3]. With the increasing discrepancy between the growing industrial demand and the continuous diminution of nickel sulfide ores, alternative methods applicable to complex sulfide ores have become an important research topic.

Conventionally, the floating-smelting-converting process, which is of high efficiency and low cost, is used to treat sulfide ores [4]. This process includes the following steps. Firstly, Nickel sulfide minerals are amenable to concentration by milling and flotation, with rejection of a high proportion of ore rock and pyrrhotite prior to smelting. Secondly, the nickel sulfide concentrate is smelted in a flash furnace or electric furnace, separating gangue and resulting in low-nickel matte. During this process, Ni, Cu, and Co are enriched and the loss of valuable metals is rare. Thirdly, the low-nickel matte is treated by converter blowing to decrease the contents of Fe and S, resulting in high-nickel matte which is further enriched in Ni, Cu, and Co. However, in this process, part of valuable metals, especially a large amount of cobalt, might be over oxidized and translated into slag. Finally, for refining high-nickel matte, different refining processes such as floatation-magnetic separation, selective leaching and layered melting processing have been used to eventually obtain Ni, Cu, and Co metals or compounds [5]. Thus, to reduce the loss of metals during the converter blowing process and shorten the traditional complex treating process, we selected low-nickel matte as the subject of our study to elucidate the metals extraction process and mechanism.

Recovery of Ni, Cu, and Co from the low-nickel matte has been studied by several researchers via a hydrometallurgical process, such as atmospheric pressure acid leaching, ammonia leaching, oxygen pressure acid leaching and oxidative leaching [6,7]. Among these processes only the oxygen pressure acid leaching method could achieve a high recovery of metals, while it still encounters some problems, such as high requirement for specialized facilities and complex operation.

The combined of pyrometallurgy and hydrometallurgy process, which is of less energy consumption and less pollution, has strong adaptability to raw materials and gradually becomes a hotspot of metallurgy research [8]. Among the alternative processes, calcified roasting followed by acid leaching is an important technique in the recovery of many nonferrous metals from minerals or concentrates. Calcified roasting is generally used for vanadium extraction from vanadium ore, which has been replaced sodium roasting that may cause corrosion of equipment and environment pollution [9-12]. The treatment of sulfide ores via calcified roasting was firstly reported in the extraction of copper from flotation copper concentrates, which can achieve a high recovery of Cu and is friendly to the environment [13-15]. Researchers in the Central South University used the calcification roasting, sulphation roasting and water leaching process to treat the carbonaceous shale containing Ni and Mo and studied the recovery rate of metals from the ore. The results show that via this process the leaching rate of Ni and Mo can reach a high level and the experiments confirm that $CaCO_3$ can decompose $NiFe_2O_4$ and iron silicates during roasting [16-18].

In this study, we investigated the feasibility of calcified roasting of low-nickel matte and the application of the subsequent iron-removing method through a goethite process. The effects of the roasting temperature, the addition of $CaCO_3$ and the roasting time on the extraction behavior of metals were studied. In addition, the role of $CaCO_3$ in the roasting process and the reaction mechanisms were analyzed and discussed.

2. Experimental

2.1. Materials

The low-nickel matte sample used in this study was from Jinchuan Co., Ltd in Gansu Province, China. In this study, the low-nickel matte was ground into particles smaller than 75 μm. The composition of the sample was analyzed by X-ray fluorescence (XRF) and the result is shown in Table 1. It is shown that the concentrations of nickel, copper and iron in the low-nickel matte is higher, while the cobalt content is very low, only 0.5 wt%. Fig. 1 presents the X-ray diffraction (XRD) pattern of material, of which the main phases are pentlandite ($(Ni,Fe)_9S_8$), chalcocite (Cu_2S), magnetite (Fe_3O_4), trevorite (Ni_3Fe), and quartet pyrite (FeS). The diffraction line of silicon (SiO_2) was not detected by XRD, due to its low content.

All chemical reagents used in this study were analytical grade and purchased from Sinopharm Chemical Reagent Co., Ltd., China. The acid used for this study is H_2SO_4 (98 wt%) diluted with deionized water to 2 mol/L.

Table 1. Chemical Composition Of The Low-Nickel Matte (Wt%).

Fe	Ni	Cu	Co	S	Si	O
29.90	27.24	16.21	0.50	20.91	0.18	4.94

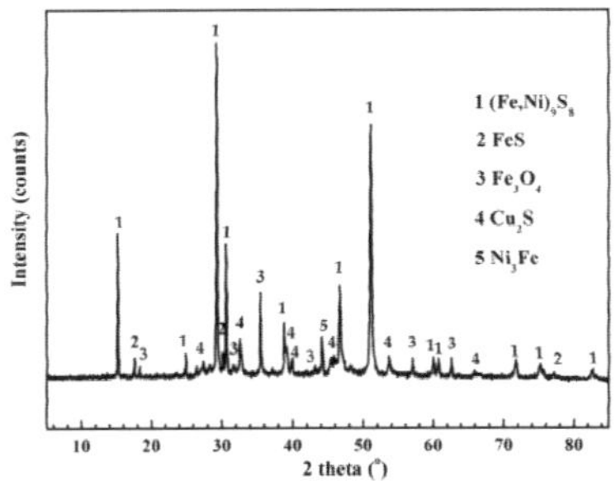

Figure 1.XRD pattern of the low-nickel matte.

2.2. Methods

2.2.1. Calcified roasting and acid leaching: In this experiment, 10 g of low-nickel matte was mixed thoroughly with a certain amount of $CaCO_3$, and the mixture was transferred into a 100 mL crucible. The crucible was then placed into a vertical furnace containing a quartz reaction tube, held for a required time after the furnace had been preheated to the designated temperature. After cooling to room temperature, the roasted sample was removed from the crucible, and then leached in 2 mol/L dilute H_2SO_4 at 90 °C with a liquid-to-solid ratio of 25:1 ml/g for 60 min to extract valuable metals. The experimental procedure is illustrated in Fig. 2. The leaching solution was then filtered and transferred to a 1000 mL volumetric flask. The extractions of Ni, Cu, Fe and Co (η) were calculated according to the following equation:

$$\eta = \frac{CV}{1000m\omega}$$

where C is the content of metals from the leach liquor, mg/L; V is the volume of the leach liquor, L; m is the mass of the low-nickel matte used in the experiments, g; and ω is the percent content of a certain metal in the low-nickel matte. Parallel experiments were conducted thrice, and the average extractions were obtained to ensure the reproducibility of the results.

Removal of irons from the leaching liquor was carried out in a 2 L water-heated reaction kettle. The leaching liquor was first diluted to a concentration of irons less than 1 g/L, and then added to the reaction kettle at 85 °C and stirred at 80 r/min. The pH value of the solution were adjusted using $NH_3{\bullet}H_2O$ by adding slowly to maintain the appropriate pH values as shown in Fig. 2. The experiment duration was 60 min.

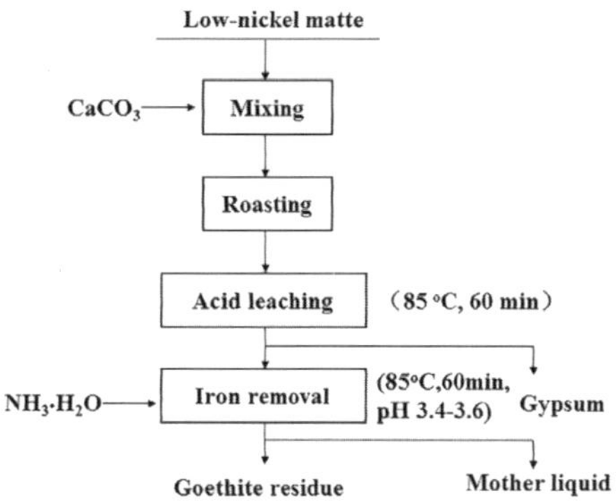

Figure 2. The flow diagram of the calcified roasting—acid leaching process.

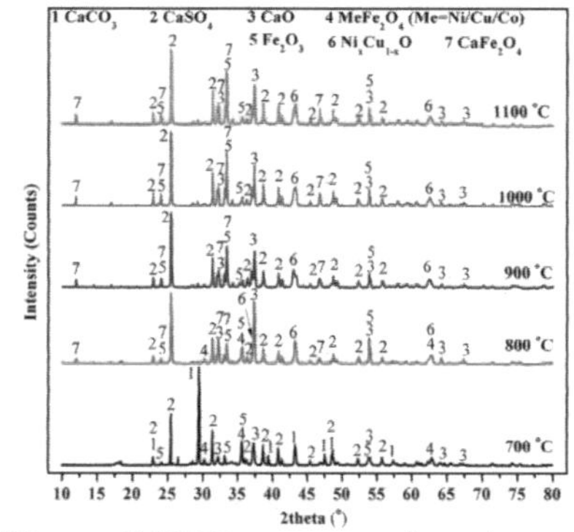

Figure 3.XRD patterns of products after roasting at different temperature

2.2.2. Characterization: The chemical compositions of the samples were determined by X-ray Fluorescence (XRF, XRF-1800, Shimadzu, Kyoto, Japan). The contents of nickel, copper, cobalt and iron in the leach liquor were determined by ICP-OES (Perkin-Elmer 7300DV, USA). The pH values of the leach liquor were measured using a pH meter (pH, Shanghai INESA, PHS-3C). The phases of the low-nickel matte and roasted products were characterized by X-ray diffraction analysis (Bruker AXS, D8 Advance, Cu radiation, 40kV/40mA).

3. Results And Discussion

3.1. Phase changes during calcified roasting

To identify the reaction process of calcified roasting, a series of roasting experiments were carried out between 700 ~ 1100 °C for 2 h with the addition of $CaCO_3$. The experimental results are presented in Fig. 3. The main products roasting at 700 °C are $CaCO_3$, $CaSO_4$, CaO, $MeFe_2O_4$ (Me=Ni/Cu/Co), and

Fe_2O_3. As the temperature increased to 800 °C, $CaCO_3$ is completely converted to CaO, indicating that the decomposition temperature of $CaCO_3$ is 700 ~ 800 °C. Compared to 700 °C, an obvious difference is that the peak intensity of $MeFe_2O_4$ decreased and the peaks of $CaFe_2O_4$ and $Ni_xCu_{1-x}O$ started to appear. When the temperature reached 900 °C, the peaks of $MeFe_2O_4$ disappeared and the main phases are $CaSO_4$, CaO, Fe_2O_3, $Ni_xCu_{1-x}O$ and $CaFe_2O_4$. As the temperature continues to increase, the main phases have no change.

Possible chemical reactions during the roasting process are as follows [19,20]:

$$4(Fe,Ni)_9S_8+29O_2 \rightarrow 18Fe_2O_3+4NiO+32NiS$$

$$CaCO_3 \rightarrow CaO+CO_2\uparrow$$

$$NiS+2O_2+CaO \rightarrow NiO+CaSO_4$$

$$2Cu_2S+5O_2+2CaO \rightarrow 4CuO+2CaSO_4$$

$$MeO+Fe_2O_3 \rightarrow MeFe_2O_4 \text{ (Me=Ni/ Cu/ Co)}$$

$$MeFe_2O_4+CaO \rightarrow CaO\bullet Fe_2O_3+MeO \text{ (Me=Ni/ Cu/ Co)}$$

$$xNiO+(1-x)CuO \rightarrow Ni_xCu_{1-x}O$$

3.2. *Effect of roasting temperature*

A series of calcified roasting experiments were carried out between 700 and 1100 °C with the addition of $CaCO_3$ and with the reaction time kept constant. The extraction of metals was determined by the methods described in section II. The effect of roasting temperature on the extraction of metals is shown in Fig. 4. A significant improvement was observed in the extraction of nickel, copper, iron, and cobalt as the temperature was increased to 1100 °C; further raising the roasting temperature will increase energy consumption, thus for subsequent tests the roasting temperature was kept to 1100 °C.

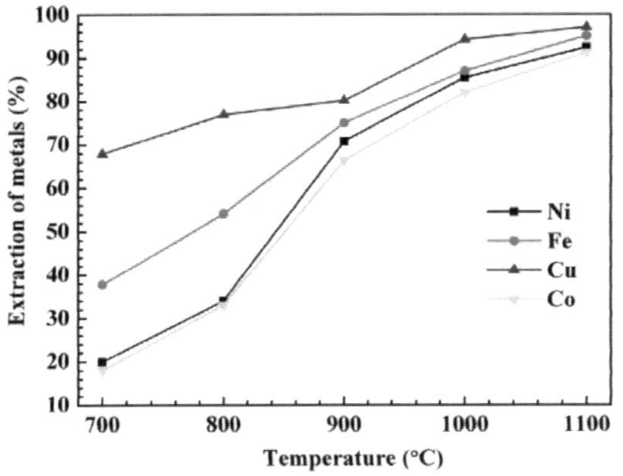

Figure 4. Effect of roasting temperature on the extraction of metals

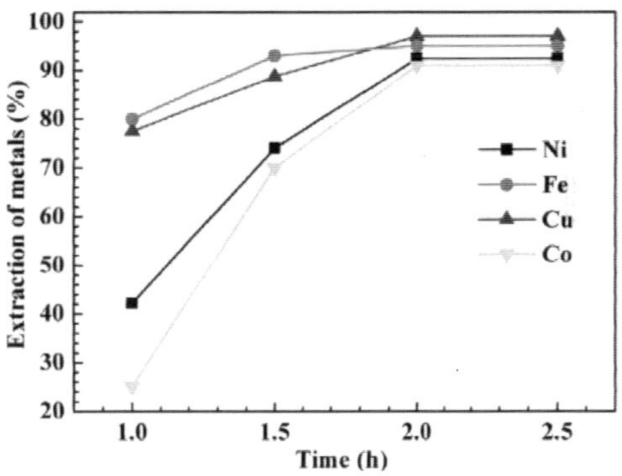

Figure 5. Effect of roasting time on the extraction of metals

3.3. *Effect of roasting time*

The effect of roasting time on the extraction of valuable metals was investigated. Fig. 5 presents the extraction of metals as a function of roasting time at 1100 °C with addition of $CaCO_3$ (1:1.8 in mass ratio of low-nickel matte and $CaCO_3$). It is shown that the extraction of the metals increased rapidly as the roasting time increased to 2 h, which maintained almost constant as the roasting time continued to increase. It is indicated that after 2 h, the reactions have been basically completed. Hence, 2 h was taken as the optimum roasting time for the rest of the experiments.

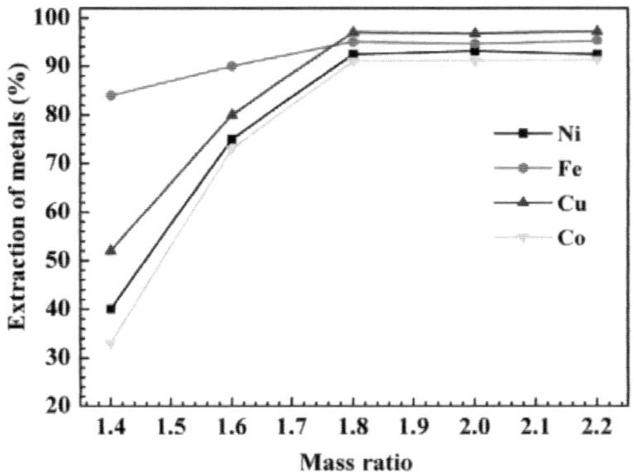

Figure 6. Effect of $CaCO_3$ addition on the extraction of metals

3.4. *Effect of $CaCO_3$ addition*

To save the material costs and enhance the extraction of metals, an appropriate amount of $CaCO_3$ should be added. Fig. 6 shows the relationship between the metals extraction and the amount of $CaCO_3$ added at 1100 °C for 2 h. The extraction of these metals increased with the increased amount of $CaCO_3$ addition when the mass ratio of $CaCO_3$ to low-nickel matte is lower than 1:1.8, which maintained almost constant as the continued addition of $CaCO_3$. Therefore, the optimum mass ratio of low-nickel matte to $CaCO_3$ should be 1:1.8.

3.5. *Iron Removal Process*

As the experimental results shown above, under the optimum conditions (i.e, 1:1.8 mass ratio of the low-nickel matte to $CaCO_3$, 2 h, 1100 °C), about 95.1% of Fe was transferred into the leach liquor. And so, an iron-removal step (goethite process) is adopted to separate iron from the solution. The essential feature of the goethite process is to keep low concentrations of ferric iron (<1 g/L) in the reactor during precipitation. This requirement can be met by either reducing all ferric ions to the ferrous state (V.M. method) or by adding the concentrated pressure leaching solution into a large precipitation vessel at the same rate as goethite precipitation (E.Z. method) [21,22]. Considering the main iron ions in leach liquor are trivalent and the concentration is not high, here we chose the E.Z. method. The reaction equation can be written as

$$Fe^{3+}+2H_2O=FeOOH+3H^+$$

The precipitation process will generate acid, which decreases the removal rate of iron. A neutralizing agent is added at a fixed rate to adjust the pH within the range of 3.4–3.6. The experiment duration was 60 min.

The results after the iron-removal process are shown in Table 2. It indicates that 99.8% of Fe can be removed via the goethite process. Although about 2.4% of Ni, 0.9% of Cu and 2.6% of Co may be lost, due to the adsorption and co-precipitation with iron particles in the goethite process, a high recovery of these metals can still be achieved in the final mother liquor.

Table 2. The Extraction Of Metals After Leaching Process, The Metal Losses During The Iron-Removal Process And The Total Recovery Of The Metals.

Metals	Ni	Cu	Co	Fe
Metals extraction/%	92.4	97.1	91.2	95.1
Metal losses /%	2.4	0.9	2.6	99.8
Total recovery /%	90.1	96.2	88.8	

4. Conclusion

The extraction of valuable metals (Ni, Cu and Co) from low-nickel matte can be performed in a three-step procedure consisting of calcified roasting, acid leaching and an iron-removal process. The results showed that the formation of ferrite can be effectively inhibited by $CaCO_3$ when calcified roasting low-nickel matte at 1100 °C. The main reaction products after roasting were confirmed as calcium ferrite ($CaFe_2O_4$) for Fe and oxides for Ni, Cu and Co. All of them can be easily dissolved by sulphuric acid. Under the optimum experimental conditions (the weight ratio of low-nickel matte to $CaCO_3$ reached 1:1.8; calcified roasting at 1100 °C for 2 h; leaching for 60 min at 85 °C with 2 mol/L dilute H_2SO_4), the maximum extraction of Ni, Cu, Fe and Co achieved 92.4%, 97.1%, 95.1% and 91.2%, respectively. And iron was effectively precipitated in the form of goethite using a goethite process with temperature and pH carefully controlled. Though, a small amount of Ni, Cu and Co (2.4% of Ni, 0.9% of Cu and 2.6% of Co) may be lost, ~99.8% of Fe could be effectively removed from the leach liquor after the iron-removal step.

Acknowledgment
This work was financially supported by the Grant 2014CB643403 from the National Key Program for Basic Research of 973 Program.

References

[1] C.Y. Lu, X.G. Lu, X.L. Zou, H.W. Cheng, and Q. Xu, "Current situation and utilization technology of nickel ore in China," Chin. J. Nat., vol.37, pp.269-277, April 2015.
[2] S.L. Chen, X.Y. Guo, W.T. Shi, and D Li, "Extraction of valuable metals from low-grade nickeliferous laterite ore by reduction roasting-ammonia leaching method," J. Cent. South. Univ., vol.17, pp.765-769, 2010.
[3] H.R. Watling, "The bioleaching of nickel-copper sulfides," Hydrometallurgy, vol.91, pp.70-88, March 2008.
[4] C. Xu, H.W. Cheng, G.S. Li, C.Y. Lu, X.G. Lu, X.L. Zou, and Q. Xu, "Extraction of metals from complex sulfide nickel concentrates by low-temperature chlorination roasting and water leaching," Int. J. Miner. Metall. Mater., vol.24, pp.377-385, April 2017.
[5] G.J. Chen, J.M. Gao, M. Zhang, and M. G, "Efficient and selective recovery of Ni, Cu, and Co from low-nickel matte via a hydrometallurgical process," Int. J. Miner. Metall. Mater., vol.24, pp.249-256, March 2017.
[6] Y. Liu, "Influence of different pretreatment methods on extracting of nickel matte by sulfate roasting," Nonferrous Min. Metall., vol.27, pp.23-25, 2011.
[7] F. Yin, Z.W. Wang, C.Y. Wang, and P.H. Jiang, "Research on pessure leaching process for low nickel matte," Min. & Metall., vol.18, pp.35-37, 2009.
[8] O.G. Olvera, M. Rebolledo, and E. Asselin, "Atmospheric ferric sulfate leaching of chalcopyrite: Thermodynamics, kinetics and electrochemistry," Hydrometallurgy, vol.165, pp.148-158, October 2016.
[9] Z.L. Cai, Y.M. Zhang, T Liu, and J. Huang, "Mechanisms of vanadium recovery from stone coal by novel $BaCO_3$/CaO composite additive roasting and acid leaching technology," Minerals, vol.6, pp.26, 2016.
[10] M.Y. Wang, X.W. Wang, J.F Shen, and R.N. Wu, "Extraction of vanadium from stone coal by modified salt-roasting process," J. Cent. South. Univ., vol.18, pp.1940-1944, 2011.
[11] B. Wang, T. Liu, Y.M. Zhang, and J. Huang, "Effect of CaF_2/CaO composite additive on roasting of vanadium-bearing stone coal and acid leaching kinetics," Minerals, vol.7, pp.43, 2017.
[12] T.Y. Wang, L.J. Xu, C.L. Liu, and Z.D. Zhang, "Calcified roasting-acid leaching process of vanadium from low-grade vanadium-containing stone coal," Cin. J. Geochem., vol.33, pp.163-167, June 2014.
[13] H.H. Haung and R.W. Bartlett, "Oxidation kinetics of a lime-copper concentrate pellet," Metall.

Trans. B, vol.7, pp.369-374, September 1976.

[14] Y.X. Hua, C.J. Cai, and Y. Cui, "Microwave-enhanced roasting of copper sulfide concentrate in the presence of $CaCO_3$," Sep. Purif. Technol., vol.50, pp.22-29, June 2016.

[15] R. Gabriel, M Tanai, and P. Carlos, "Lime-concentrate roasting studies—effect of activated limestone," Miner. Eng., vol.17, pp.469-471, March 2004.

[16] Y.L. Liao, J. Zhou, F.R. Huang, and Y.Y. Wang, "Leaching kinetics of calcification roasting calcinate from multimetallic sulfide copper concentrate containing high content of lead and iron," Sep. Purif. Technol., vol.149, pp.190-196, July 2015.

[17] M.Y. Wang and X.W. Wang, "Extraction of molybdenum and nickel from carbonaceous shale by oxidation roasting, sulphation roasting and water leaching," Hydrometallurgy, vol.102, pp.50-54, April 2010.

[18] X.W. Wang, J. Peng, M.Y. Wang, P.H. Ye, and Y. Xiao, "The role of CaO in the extraction of Ni and Mo from carbonaceous shale by calcification roasting, sulphation roasting and water leaching," Int. J. Miner. Process., vol.100, pp.130-135, 2011.

[19] J.G. Dunn, "The oxidation of sulphide minerals," Thermochim. Acta., vol.300, pp.127-139, October 1997.

[20] S. Prasad and B.D. Pandey, "Alternative processes for treatment of chalcopyrite —A review," Miner. Eng., vol.11, pp.763-781, August 1998.

[21] T. Yue, H.S. Han, W. Sun, Y.H. Hu, P. Chen, and R.Q. Liu. "Low-pH mediated goethite precipitation and nickel loss in nickel hydrometallurgy," Hydrometallurgy, vol.165, pp.238-243, 2016.

[22] P.S.R. Prasad, K. Shiva Prasad, V. Krishna Chaitanya, E.V.S.S.K. Babu, B. Sreedhar, and S. Ramana Muthy. "In situ FTIR study on the dehydration of natural goethite," J. Asian. Earth. Sci., vol.27, pp.503-511, 2006.

Research on Microstructure and Wear Resistance of Fe-Based Hardfacing Layer Using Plasma Surfacing Technology

SONG Lingling[1], GUO Jian[2], MA Xiaolei[2], ZHAO Jingnan[2]

[1]Mechanical and Material School, Tianjin Sino-German University of Applied Sciences, Tianjin 300350, China ;

[2]College of Mechanical Engineering, Tianjin University of Science & Technology, Tianjin 300222, China

E-mail: jingnanzhao@tust.edu.cn

Abstract. In order to enhance the wear resistance of material surface, the hardfacing alloy is usually deposited on its surface by using plasma surfacing technology (PST). The technology has the advantages of the low dilution rate, small deformation, smooth welding path, food metallurgical combination, and no major defects. In this paper, a Fe-based alloy coating had been prepared by PST, and the structure and properties of surfacing layer had been studied. The results showed that the hardness of surfacing layer was significantly greater than that of the substrate. And the coating layer had a snowflake-like structure. Its Lorraine hardness was up to about 60HRC, and the friction coefficient was about half of the 65Mn after quenching. With the increase of the material hardness, the coating layer has excellent abrasion resistance.

1. Introduction

With the rapid development of modern industry, mechanical parts often work under extremely complicated and demanding conditions, which scrapped the equipment parts easily due to wear, corrosion or abrasion. This requires the materials of equipment parts with good wear resistance, corrosion resistance, high temperature resistance and oxidation resistance under extreme temperature, high pressure and other complex conditions. Therefore, it is important to pay more attention to surface technology and develop surface modification technology (1, 2).

Plasma Surfacing Technology (PST) was developed in the 1960s. Its application area is then further expanded from surface repair to manufacture industry in following 20 years (3, 4). Ever since the 1990s, further researches had been taken on plasma surfacing machines with the rapid development of advanced manufacturing technology. Plasma arc welding is a welding method that takes combined or transferred plasma arc as the heat source and alloy powder or wire as filler metal. In order to obtain desired properties of the surfacing layer, the welding metal and the surface of the substrate need to be melted together into the molten pool, and the condensing to room temperature. The difference between the plasma arc and general mainly lies in that of the general arc is the free arc, while plasma arc compressed in the nozzle and has more characteristics as thermal compression effect, mechanical compression effect, and electromagnetic compression effect (5-8).

Surface coating plays an important role on improving surface properties and prolonging the service life of metal parts (9). The close combination, smooth surface, and superb formation can be obtained by using PST, and macroscopic defects such as cracks are avoided through the this technology. In order to improve the surface wear resistance of materials, the hardfacing alloy tends to deposit on the

IOP Conf. Series: Materials Science and Engineering **409** (2018) 012028 doi:10.1088/1757-899X/409/1/012028

surface. Extra special performance such as high-stress wear property can be obtained by adding cemented carbide particles (10).

PST is a kind of welding method to obtain high quality surfacing layer, which uses argon transfer arc as the main heat source to fill metal alloy powder and form a dissolution bath on the surface of the workpiece. According to the different types of the power supply, the plasma arc can be divided into the transfer arc and the non-transfer arc. The transfer arc burns between the electrode and the workpiece, while the non-transferred arc between the electrode and the nozzle. The main difference between them is that the voltaic arc is generally free arc, while the plasma arc is compressed arc. And the plasma arc is influenced by hot compression, mechanical compression and electromagnetic compression in the nozzle, by the influence of which the arc column temperature and energy density greatly increased.

2. Method

The main parameters of powder plasma arc surfacing include welding voltage, current, swing parameter, protective gas flow rate, welding speed and working distance, etc.. Welding voltage has a correlation of working distance: smaller the distance, lower the voltage. The welding current determines the heat input. There won't be a metallurgical combination of good quality if the welding current is too low since that porosity and micro-cracks will easily happen on the materials. Swing parameters determine the width of the Surfacing path. Protection gas flow will influence the protective effect of the molten pool. Welding speed has a significant influence on the weld thickness and penetration (1, 4).

2.1. Experiment devices and methods

The PST was conducted by using the plasma welding machine of DML-V02BD. In order to increase the bonding strength of the surfacing layer and the substrate, the surface roughness of the material surface was increased through surface sandblasting via the long-empty gp-1 type dry sandblasting machine. The substrate is 65Mn plate, The Fe base alloy powder consist of 80-85%Fe, 8-10%Cr, 2.8-3.2%B, 2-3%Si, 0.9% C.

In order to identify the difference of the material hardness on substrate and hardfacing layer, Rockwell hardness tests were taken on both of the materials. Indentation hardness tests were performed using a electronic MH-6 micro-hardness tester, which can be accurate to 0.01 HRC. The load was set at 100 N and the test duration was set at 10 s. There are 3 tests on each material, and then the average value of the tests was calculated as the final experimental data. The tested specimen size is 20 mm x 8 mm x 8 mm.

Before the surfacing test, the base material of 65Mn material was pretreated. Sand blasting machine was first used to give sand blasting on the substrate surface, in order to remove oil pollution and impurities, meanwhile it increases roughness and bonding strength. The pretreated base material was then overlaying welded with Fe-based powder. Heat treatment was given after welding, and then wire cutting followed after air-cooling in order to test the hardness and observe the morphology of the tissue.

TESCAN VEGA3-SCAN electron microscope and EDS spectroscopy analyzer were used to test the microstructure morphology of the coating and the composition of various phases. The abrasion resistance analysis was used to test the friction coefficient and wear loss of the composite coating and 65Mn materials using the HRS-2M reciprocating friction wear tester. The grinding wheel has a rotational speed of 250 r/min, the wear test is 180min, and the quality loss of the material is measured in real time. The measurement can be accurate to 0.1 mg.

2.2. Maintaining the Integrity of the Specifications

The template is used to format your paper and style the text. All margins, column widths, line spaces, and text fonts are prescribed; please do not alter them. You may note peculiarities. For example, the head margin in this template measures proportionately more than is customary. This measurement and others are deliberate, using specifications that anticipate your paper as one part of the entire proceedings, and not as an independent document. Please do not revise any of the current designations.

3. Results And Discussion

Plasma arc welding technology was used to prepare Fe-based coating on the 65Mn substrate. Side observation was given by electron microscopy after wire cutting, as shown in figure 1. It can be clearly seen that the surfacing layer and the substrate formed quite well metallurgical combination.

Figure 1. The metallurgical combination of matrix and coating

3.1. Hardness analysis

The Rockwell hardness tests of substrate and hardfacing layer were conducted, as shown in figure 2. Results of the 3 groups of substrate and hardfacing layer were 42.10, 38.50, 43.4053, 65, 62.40, and 65.73 HRC, respectively, as shown in table1. The average hardness of the hardfacing layer is 60.59 HRC, while the average hardness of the substrate is only 41.3HRC. It is obvious that the hardness of the surfacing layer is better than the substrate.

Table 1. Hardness Test Result

Hardness	Coating layer (HRC)	Matrix layer (HRC)
1	53.65	`42.10
2	62.40	38.50
3	65.73	43.40
Average	60.59	41.3

Figure 2. The hardness test of the surfacing layer

The coating microstructure and morphology were observed by the electron microscope, as shown in Figure 3(a, b). Figure 3(a) shows the microstructure of the substrate and figure 3(b) shows the morphology of the surfacing layer. It can be seen that the coating is mainly composed of a grey matrix, hexagonal particles, lath-shaped particles, oxide, and pore structure. Among that the dark gray hexagonal particles and plate-like particles deposited on the gray substrate, the black hole is pore and part of the black dots for oxide impurities.

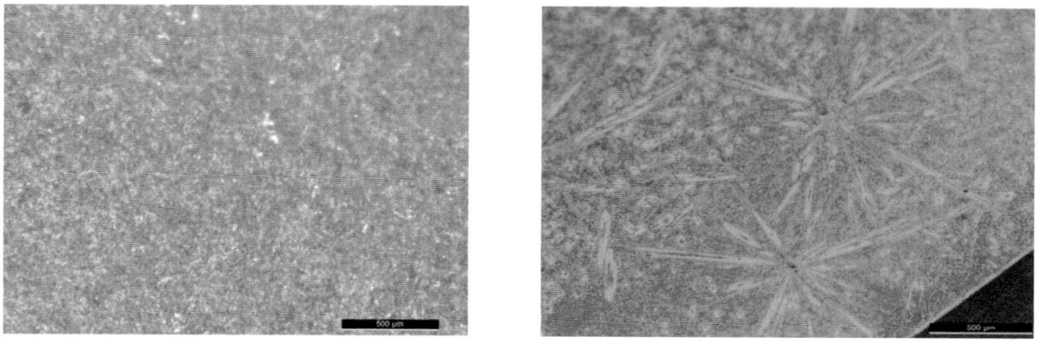

(a) (b)

Figure 3. The structure and morphology of the matrix and the coating (a) morphology of substrate material. (b) Microstructure of surfacing layer.

3.2. Wear resistance analysis

Figure 4 shows the friction coefficient test on the substrate and the coating. The figure is a time-dependent contrast line graph about friction coefficient of the 65Mn substrate and that of Fe coating after quenching, Plasma surfacing Fe powder coating friction coefficient is significantly lower than that of 65Mn. It was almost half of the 65Mn friction coefficient after quenching. It means that the prepared coating layer has greatly enhanced the wear resistance of the substrate.

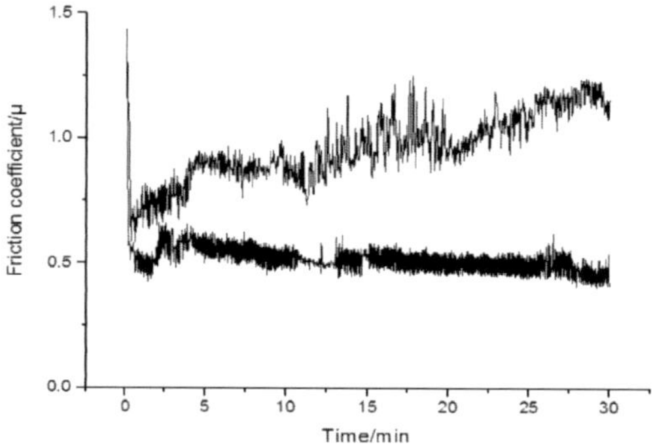

Figure 4. Comparison chart of friction coefficient

The wear loss of the surfacing layer was measured by HSR-2M reciprocating friction and wear test machine. The change law of the sample wear rate and time was shown in Tab.2 and figure 5. As shown by the graph, the wear rate increases with time. The wear rate was relatively high of 65Mn, it increased linearly with the increase of wear time. Compared with 65Mn, the wear volume was about 1mg of plasma surfacing coating at 30min. Then it increased in the amplitude of miniature 2mg~4mg in 45-90 min. However after 120min, the wear volume increased straightly and rapidly. If the sample surface contains more CrC hard phase, the better wear resistance and smaller changes of wear loss it will be. With the extension of time, the wear scar grew deeper to the coating layer. Then the hard particle content would reduce, and the wear resistance became poorer so that the wear rate increased significantly.

Table 2. Comparison Table Of Wear And Loss

Wear time（min）	**30**	**60**	**90**	**120**	**150**	**180**
65Mn (mg)	1.1	3	4.1	6.05	9	11
surfacing layer（mg）	1.1	2.8	3	4.05	5.9	7.9

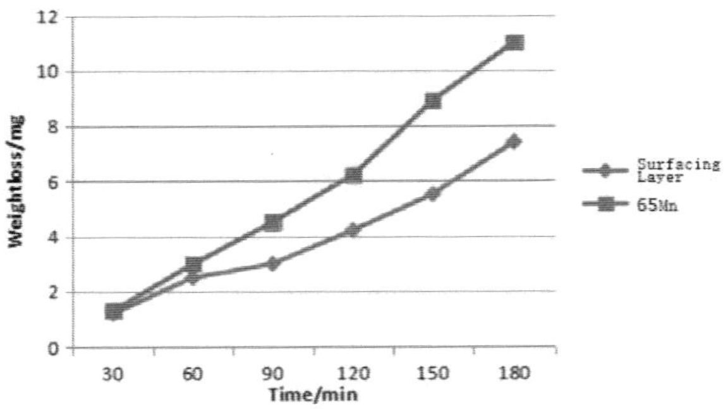

Figure 5. Comparison chart of wear and loss

4. Conclusion

In this paper, plasma welding technology was utilized to prepare Fe-based composite coatings on 65Mn substrate. The specimen coating formation looked smooth and without major defects. The welding layer was uniform and had a compact structure. The surfacing layer and the substrate formed an excellent metallurgical combination.

The hardness of the surfacing layer reached 60HRC, much higher than the hardness of the 65Mn base material, which is about 40HRC. And the friction coefficient was approximately half of the 65Mn after quenching. The average wear amount was reduced by 1/3, and the coating had better wear resistance much better.

The PST is a simple and feasible way to effectively improve the wear resistance of the mechanical parts and prolong its service life. Therefore, it could be applied to strengthening the structure surface and improve the quality of work in industry.

Acknowledgment

The authors would like to gratefully acknowledge the financial support provided by Tianjin Key Laboratory of Integrated Design and On-line Monitoring for Light Industry & Food Machinery and Equipment, China under the Grant 2017LIMFE01, Tianjin Municipal Education Commission, China under the Grant 2017KDYB20, and Tianjin Natural Science Foundation Grant 17JCYBJC42400.

References

[1] Hou QY, Gao JS, and Zhou F.“Microstructure and wear characteristics of cobalt-based alloy deposited by plasma transferred arc weld surfacing”J.Surface and Coatings Technology, vol.194, pp.238-243, 2005.

[2] Liu YF, Mu JS, Xu XY, and Yang SZ, “Microstructure and dry-sliding wear properties of TiC-reinforced composite coating prepared by plasma-transferred arc weld-surfacing process”, Materials Science and Engineering: A, vol.458, pp.366-370, 2007.

[3] Dong Lihong, Zhu Sheng, Xu Binshi, and Du Zeyu, “Development of wear and corrosion resisting powder Plasma-arc surfacing,” Weld, pp.6-9, July 2004.

[4] M. Young, The Technical Writer’s Handbook. “Optimization design of composite powder composition of plasma arc surfacing” Journal of Materials Protection, vol.37, pp.7-8, July 2004.

[5] YANG Kun-yu, “Based on the working process of welding methods and equipment curriculum design”. Journal of Mechanical professional education , pp. 57–59, 2012.

[6] KANG Jian-xin, GUO Li-jie, and GAO Fa-ming, “ HAZOP analysis of gas fractionation unit based on process simulation”. Journal of Yanshan University, vol. 38, pp.189–196, May 2014.

[7] LU Jian-bo,YAO Sun,LOU Song-nian,DU Ze-yu,and LI shao-qing, “Coat Components Optimization and Wearability for Plasma Transferred Arc Processing,”Materials for

Mechanical Engineering, vol.30, pp.35–37, Mar 2006.

[8] DENG Yuan-jiang, PANHou-hong, and LUO Tao, "Comparison between Two Methods of Plasma Arc Powder Automatic Surfacing on Valve," Journal of Chongqing Institute of Technology, vol. 21, Nov 2007

[9] Zhao J, Wong KS, Shrotriya P. Hybrid CO2 laser waterjet heat (LWH) treatment of bindered boron nitride composites with hardness improvement. Ceramics International. 2017;43(11):8031-9.

[10] Zhao J, Shrotriya P., "Increase the hardness of polycrystalline cubic/wurtzite boron nitride composite through hybrid laser/waterjet heat (LWH) treatment" Advances in Applied Ceramics , vol. 1-8, May 2017

IOP Conf. Series: Materials Science and Engineering **409** (2018) 012029 doi:10.1088/1757-899X/409/1/012029

Effect of Air Entraining Agent on Uniaxial Tensile Properties of PVA-ECC

Jincai Zhu[1,a], Zigeng Wang[2,b]*, Yue Li[3,c], Zhanguo Li[4,d]

The Key Laboratory of Urban Security and Disaster Engineering, MOE, Beijing Key Lab of Earthquake Engineering and Structural Retrofit, Beijing University of Technology, Beijing 100124, PR China.
Email: [a]zjc00288@126.com, [b]zigengw@bjut.edu.cn, [c]liyue@bjut.edu.cn, [d]lizg@bjut.edu.cn
Abstract: In this paper, polyvinyl alcohol (PVA) fiber reinforced ECC (Engineered Cementitious Composites) mortar was selected as the research object. The effect of air entraining agent on the tensile properties of PVA-ECC was studied by uniaxial tensile test with different dosages. The results showed that the tensile strength reduced by adding air entraining agent. When the amount of fly ash increased with adding a large amount of air entraining, the tensile strain reduced. In addition, the tensile strain of the cementitious material with no fly ash increased when incorporating air entraining agent.

1. Introduction

In order to overcome the disadvantages of low tensile strength, poor toughness and crack width control, high toughness cementitious composite (Engineered Cementitious Composite, ECC) was invented. This kind of material reinforced by short fiber was designed on micro mechanics and fracture mechanics principle with significant strain hardening and multiple cracking characteristics. The content did not exceed 2.5% of the total volume. The ductility and fracture had high control ability with the ultimate tensile strain of more than 3% [1-5]. Some scholars [6,7], in order to increase the ultimate tensile strain and tensile strength of the concrete, treated the surface of PVA fiber to reduce the chemical bonding force of the fiber substrate interface. In this paper, a large number of uniform, stable and closed microbubbles were introduced into the ECC with different proportioning ratios to change the bonding interface between the fiber and the matrix. The effect of air entraining agent on the tensile properties of PVA-ECC was discussed.

2. Experiment profiles

2.1 Materials and mix proportion

The cement was P·O 42.5 cement. The mesh of quartz sand was 200. The density of II grade fly ash was 2.31 g/cm^3. The air entraining agent was FLOTAGE AE-1 produced by a company in China. PVA fiber was K-II REC15 fiber produced by Japanese Kuraray Company. The mechanical and geometric properties of PVA fiber are listed in Table 1. The fiber content was 1.25% of the material by mass and the water binder ratio was 0.33. The mix proportion is shown in Table 1.

Table 1. Mechanical properties of PVA fiber.

Fiber	Tensile strength(MPa)	Length (mm)	Diameter (μm)	Density (g·cm^{-3})	Elasticity modulus (GPa)
PVA	1600	12	39	1.3	42.8

Table 2. Mixture proportions of fiber cement-based materials.

sample	Cement	Fly ash/ Cement	Silica sand/B	Air entraining agent
A	386	2.2	0.36	0
B	386	2.2	0.36	0.62（0.05%）
C	386	2.2	0.36	1.3（0.1%）
D	1233	0	0.36	0
E	1233	0	0.36	0.62（0.05%）
F	1233	0	0.36	1.3（0.1%）
G	1113	0	0.51	0
H	1113	0	0.51	0.62（0.05%）
I	1113	0	0.51	1.3（0.1%）

B – Binder i.e. cement and fly ash

2.2 Testing methods

(1) The forming process of the specimen: cement, fly ash, quartz sand and air entraining agent were well weighed, stirred for 2min in the mixer and adding water stirred for 3min. Then PVA fiber was added, continuing to stir for 5min at high speed to uniform, demolded after pouring in 24h. Curing conditions: (20 + 2) oC, humidity of 95%, curing age of 7d, 28d, 90d.

(2) The axial tensile: the specimen was dumbbell-shaped and the thickness was 13mm. The axial load was measured by a 10kN sensor. The axial deformation was measured by an electronic extensometer with the maximum deformation amount of 10% and the loading speed was 0.1mm/min. The size of specimen is shown in Figure. 1.

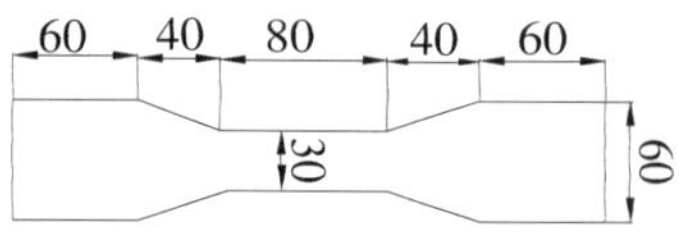

Figure 1. Uniaxial tensile specimen size

3. Experimental results and Discussion

The tensile stress-strain curve of the specimen at different ages is shown in Figure. 2. It can be seen from the specimens A, B and C with fly ash that the tensile stress fluctuation is small and the crack distribution of the specimen is fine. The uniaxial tensile strength decreases with the increase of the amount of air entraining agent. Compared with the specimen A, the tensile strain of the specimen B with air entraining agent content of 0.05% increases. The tensile strain of the specimen C decreases with air entraining agent of 0.1%. At 7d, 28d and 90d, the stresses of B decrease by 5%, 22% and 21%, and the strains increase by 39%, 12.5% and 33%, respectively. The stresses of C decrease by 31%, 31% and 18%, and the strains decrease by 42%, 35% and 35%, respectively. It can be seen from D, E and F that there is no fly ash in the cementitious material with the sand / cement of 0.36. At 7d and 28d, the tensile stress of the specimen with air entraining agent is almost unchanged, and the tensile stress reduces by 18% at 90 d. The tensile strain increases with the increase of the amount of air entraining agent. At 7d, 28d and 90d, the tensile strains of E increase by 2.7, 0.96, 1 time, respectively. The tensile strains of F increase by 4, 1.3 and 2.7 times, respectively. From G, H and I, it can be seen that when the cementing material does not contain fly ash with sand / cement of 0.51, the tensile stress decreases and the tensile strain increases with the increase of air entraining agent content. At 7d, 28d and 90d, the tensile stresses of I decrease by 29.8%, 6.5%, 33.9, respectively, and the tensile strains increase by 1.9, 3 and 2.1 times respectively.

The reasons of the results above: When a large amount of fly ash is added, the inert fine particles of fly ash adhered to the surface of the fiber to form a layer of film ball, avoiding the direct contact of fiber and matrix of hydration products, reducing the effect of matrix to fiber the chemical bond. Additionally, The resultants of fly ash reacting with cement hydration products are spherical particles with dense smooth surface, which can play ball bearing lubrication. Therefore, the tensile strain hardening phenomenon is obvious. When there is no fly ash in cementitious material, the bond force between fiber and matrix is large, and the fiber is broken when pulling out from the matrix. The bridging effect of fiber is not fully functional, and the tensile stress fluctuates greatly. When air entraining agent is mixed into mortar, a large number of uniform and stable micro bubbles are introduced to reduce the contact area between fiber and mortar matrix, thus reducing the bond force between fiber and matrix. Under tensile load, the fiber cannot break, but the bonding force between the fiber and the matrix would be overcome when the fiber is pulled out, which is greater than the cracking stress of the matrix. The fiber plays a bridging effect, the tensile strain increases to produce strain hardening.

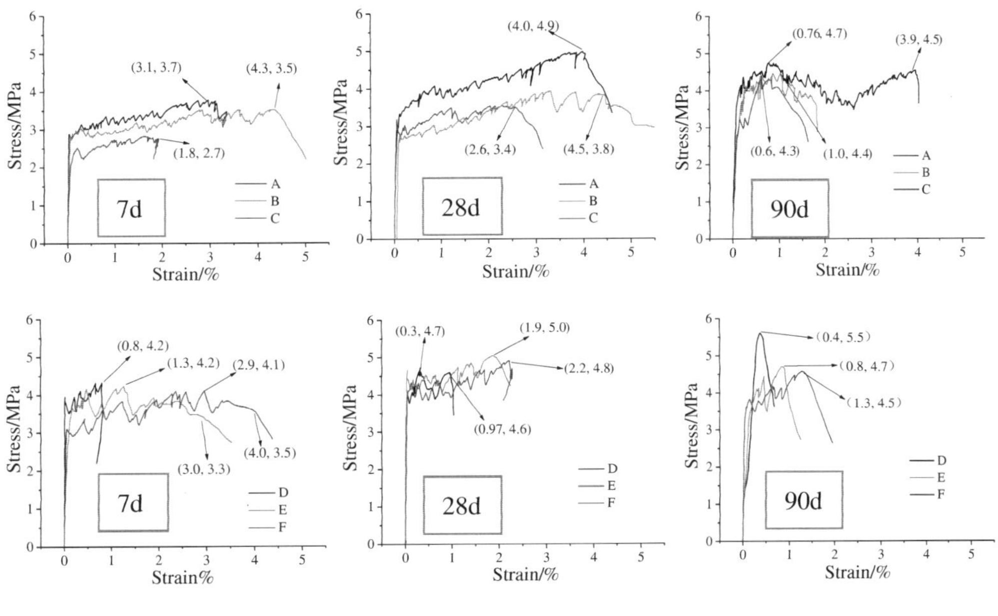

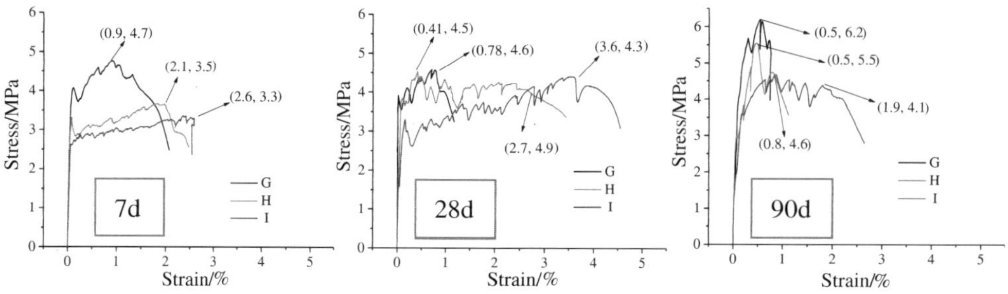

Figure 2. Tensile stress-strain curves at different ages

4. Conclusions

(1) The uniaxial tensile strength decreases with the increase of the amount of air entraining agent.

(2) When the amount of fly ash is higher and a large amount of air entraining agent are added, the tensile strain reduces. When there is no fly ash in the cementitious material, the tensile strain increases with the incorporation of air entraining agent.

References

[1] Nurdeen M Altwair M A M J 2012 *Flexural performance of green engineered cementitious composites containing high volume of palm oil fuel ash*. Construction and Building Materials pp18-525.

[2] Xiaoyan Huang R R Q Z 2013 *Mechanical and thermal properties of green lightweight engineered cementitious composites*. Construction and Building Materials pp 954-960.

[3] Zhitao Chen, Yingzi Yang, Yan Yao. 2013 *Quasi-static and dynamic compressive mechanical properties of engineered cementitious composite incorporating ground granulated blast furnace slag*. Materials and Design pp 500-508.

[4] En-Hua Yang V C L. 2014 *Strain-rate effects on the tensile behavior of strain-hardening cementitious composites*. Construction and Building Materials pp 96-104.

[5] Li Yue Zhu Jincai P. 2018 *The Research on Mechanics Properties of PVA fiber Cementitions Composites*. Journal of Beijing University of Technology pp 121-128.

[6] Zhang Jun S H L V. 2001 *Crack bridging model for fiber reinforced concrete under fatigue tension*. International Journal of Fatigue pp 655-670.

[7] Li V C W C W S. 2002 *Interface tailoring for strain-hardening polyvinyl alcohol-engineered cementitious composite (PVA-ECC)*. ACI Materials Journal pp 463-472.

Study on Axial Compression Test of Corroded Reinforced Concrete Columns Reinforced by MPC Bonding CFRP

Chang Zhao[1,a], Yue Li[2,b*], Zigeng Wang[3,c] and Zhanguo Li[4,d]

[1]The Key Laboratory of Urban Security and Disaster Engineering, MOE, Beijing Key Lab of Earthquake Engineering and Structural Retrofit, Beijing University of Technology, Beijing 100124, PR China

E-mails:[a]zhaochang13@126.com,[b]liyue@bjut.edu.cn,[c]zigengw@bjut.edu.cn,[d]lizg@bjut.edu.cn

Abstract. In this paper, reinforced concrete was first reinforced by magnesium phosphate cement (MPC) bonding carbon fiber sheets (CFRP). The axial compression test was conducted to evaluate the property of the MPC-CFRP reinforcement corroded reinforced concrete column. The results showed that the MPC-CFRP reinforcement corroded reinforced concrete columns had better structural bearing capacity and ductility in the process of compression failure.

1. Introduction

Structural adhesives have special advantages in construction and bonding effect. But there are also problems such as poor durability, poor fire resistance, environmental protection, and uneconomic issue. In this paper, the authors put forward the formation of MPC-CFRP by using magnesium phosphate cement (MPC) instead of structural adhesive to bond CFRP. MPC has the advantages of good conductivity, promising permeability, high bond strength, rapid solidification and hardening, superior corrosion resistance to steel and so on[1-3]. It was found that the effective binding coefficient of MPC bonding CFRP was higher than that of epoxy resin, and the strength model of MPC binding CFRP confined concrete column was in conformity with the existing theoretical model[4]. Based on the above research results, MPC bonding CFRP was innovatively used to reinforce the corroded reinforced concrete and the reinforcement effect was studied.

2. Test scheme

2.1. Raw materials

The used cement was P.O 42.5. The stone was 5~20mm continuous graded macadam. The sand was river sand with density of 2.7g/cm3 and fineness modulus of 2.56. Water reducing agent was Polycarboxylic Water Reducing Agent with reduction rate of 23%. The sodium chloride was industrial

IOP Conf. Series: Materials Science and Engineering **409** (2018) 012030 doi:10.1088/1757-899X/409/1/012030

purity with the NaCl content of more than 98%. The content of NaCl in concrete was controlled as 3% of cement by mass.

The density of MgO used in MPC was 3460kg/m3 with specific surface area of 806kg/m3 and the compositions and physical properties shown in Table 1. The potassium dihydrogen phosphate was industrial grade with purity of 98%. The purity of the retarder borax was 99.5%.

Carbon fiber sheets(CFRP) was offered by Japan torayn grade I with type of UT70-30. The thickness was 0.19mm and the surface density standard value was 300g/m2. The standard value of tensile strength, tensile modulus and fracture strain were 4200g/m2, 210GPa and 0.02.

Table 1. The compositions and physical properties of MgO.

sample	MgO	CaO	SiO2	Al2O3	Fe2O3
magnesium oxide	91.7%	1.6%	4%	1.4%	1.3%

2.2. Mix proportion design

The concrete mix design and mechanical properties of C30 concrete are shown in Table 2.

Table 2. Concrete mix ratio and mechanical properties.

W/C	Cement (kg/m3)	Water (kg/m3)	fine aggregate (kg/m3)	coarse aggregate (kg/m3)	water reducer (g)	NaCl (kg/m3)	compressive strength (MPa)	Slumps (mm)
0.57	325	185	662	1228	1.625	9.75	30	190

The MPC mix ratio and mechanical properties used in the test are shown in Table 3.

Table 3. Concrete mix ratio and mechanical properties.

MgO(kg)	KH2PO4(kg)	Borax(kg)	Water(kg)	compressive strength(MPa)			
				3h	1d	7d	28d
1	0.75	0.05	0.28	22	30	42	53

2.3. Sample preparation

Concrete size was Φ100mm×200mm. The center of the specimen was inserted with a Φ10mm steel. The overall length of the steel bar was 250mm with 150mm inside the specimen and 100mm outside the specimen, shown in figure 1. After demolded, the specimen was maintained for 28d in curing room with temperature of 20±1oC and humidity of 50±2%. The corrosion of the specimen was accelerated by electrochemical method with device shown in figure 2. The steel bar is the anode, the stainless steel is the cathode. The solution is tap water. The current for accelerating the corrosion process was 100mA. According to Faraday's law[5], the time for corroding the steel bars is calculated when reaching to 5%, 15% and 30% of corrosion ratio. After all the specimens were corroded, parceled by CFRP with MPC. The length of the lap part of CFRP was 100mm, shown in figure 3.

2.4. Axial compression test

Figure 4. shows the standard t axial compression test device. The load rate was controlled as 0.5mm/min. The detection range was the length of 100mm in the middle of concrete.

IOP Conf. Series: Materials Science and Engineering **409** (2018) 012030 doi:10.1088/1757-899X/409/1/012030

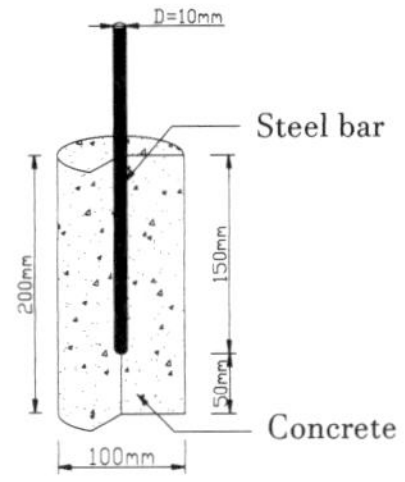

Figure 1. Schematic of reinforced concret

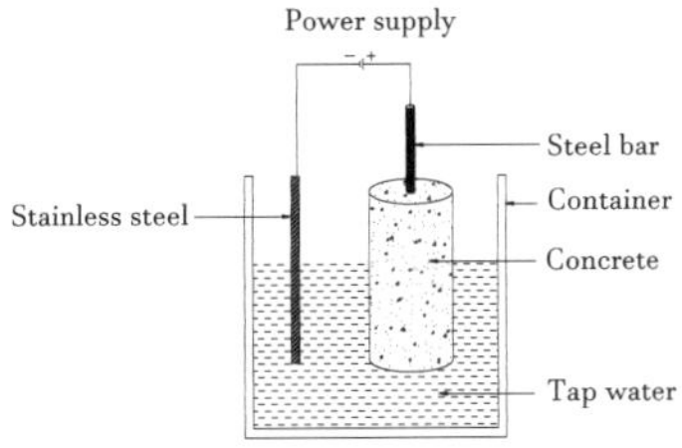

Figure 2. Electrochemical accelerated corrosion

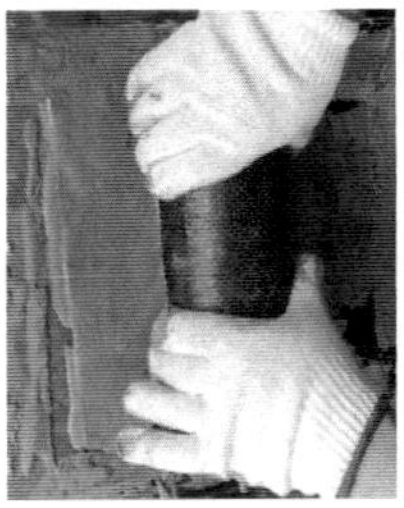

Figure 3. Schematic of package

Figure 4. Axial compression test

3. Results and discussion

After corrosion of the specimen was accelerated, the accumulation of steel corrosion caused the additional tensile stress in the concrete around the steel bar, eventually leading to a smooth expansion of the concrete protection layer, as shown in figure 5. After Corroded reinforced concrete was compressed, the vertical corrosion cracks along the upper expanded rapidly. As the load continued to increase, the brittle failure of cylindrical concrete appeared and the load reached to the peak.

The damage of MPC-CFRP confined reinforced concrete columns was due to the rupture of CFRP, indicating that the length of the CFRP lap was sufficient. Figure 6. shows the CFRP fracture failure in the axial compression test.

(a) (b) (c)

Figure 5. Different degree of corrosion of specimens: (a) Corrosion ratio 5%; (b) Corrosion ratio 15%; (c) Corrosion ratio 30%

Figure 6. Fracture failure of CFRP

The axial load-displacement curves of the reinforced concrete columns with natural corrosion and corrosion ratio of 5%, 15% and 30% are shown in figure 7.

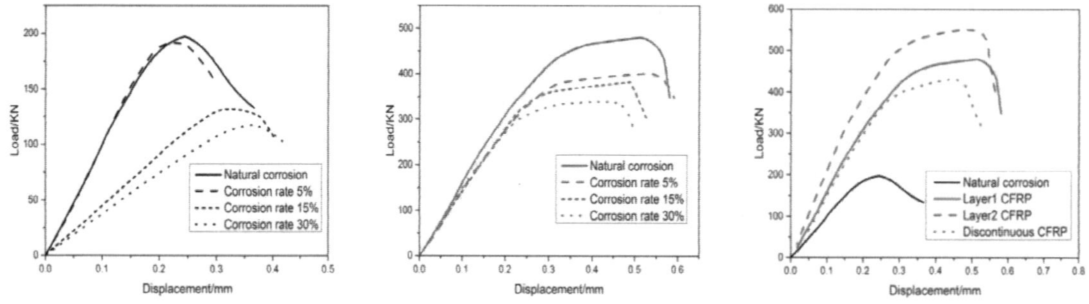

(a) (b) (c)

Figure 7. Load-displacement curves of reinforced concrete columns: (a) unconfined concrete columns; (b) confined concrete columns; (c) concrete columns with different constraints

As seen from Fig. 7(a), the ultimate bearing capacity of the unconstrained reinforced concrete columns decreases with the increase of corrosion ratio. Compared with the natural corroded concrete column, the ultimate strengths of reinforced concrete columns with corrosion ratios of 5%, 15% and 30% decrease by 8.15%, 32.61% and 42.53%, respectively. In Fig. 7(b), the ultimate bearing capacity of MPC-CFRP restrained corroded concrete columns decreases with the increase of corrosion ratio. Compared with the natural corroded concrete column, the ultimate strengths of reinforced concrete columns with corrosion ratios of 5%, 15% and 30% decrease by 13.14%, 17.53% and 30.04%, respectively. Comparison with Fig. 7(a), the ultimate strengths of natural corrosion column and columns with corrosion ratios of 5%, 15% and 30% increase by 57.4%, 58.27%, 65.9%, 66.21%, respectively. Therefore, the MPC-CFRP can obviously improve the ultimate strength and ultimate displacement of the concrete column. It is because MPC-CFRP effectively constrains the expansion of the compressed concrete. Compared with the unreinforced concrete columns, the ultimate strength and ultimate displacement are obviously improved. As you can see in Fig. 7(c), the discontinuous package has a lower strength and ductility than the whole package. The compressive strength of the two-layer whole package is obviously greater than that of the one-layer whole package. The ratios of the compressive strengths of discontinuous one-layer wrapped concrete, one-layer whole wrapped concrete, two-layer whole wrapped concrete to natural corroded concrete are 2.32, 2.35, and 3.01 respectively.

4. Conclusion

(1) The ultimate bearing capacity and ultimate displacement of corroded reinforced concrete columns can be obviously improved by MPC-CFRP reinforcement.

(2) The increase rate of the ultimate strength of reinforced concrete columns strengthened with MPC-CFRP decreases with the increase of corrosion ratio.

(3) The compressive strength of all wrapped reinforced concrete is greater than that of discontinuous wrapped reinforced concrete, which is lower than that of two-layer wrapped reinforced concrete.

Acknowledgement

The authors would like to acknowledge the financial support provided by National Key R&D Program of China (2017YFB0310100); National Natural Science Foundation of China (51678011).

References

[1] Li Yue, Li Yaqiang, Shi Tongfei, Li Jiaqi 2015 J Experimental study on mechanical properties and fracture toughness of magnesium phosphate cemnet. Construction and Building Materials. 96:346-352.

[2] Li Yue, Sun Jia, Li Jiaqi, Shi Tongfei 2014 J Effects of fly ash, retarder and calcination of magnesia on properties of magnesia phosphate cement. Advances in Cenment Research. 12:1-8.

[3] Li Yue, Shi Tongfei, Chen Bing Li Yaqiang 2015 J Performance of magnesium phosphate cement at elevated temperatures. Construction and Building Materials. 91:126-132.

[4] Li Yue, Bai Weiliang, Shi Tongfei, Liu Xiongfei 2017 J Research on Axial Compression Test of CFRP Confined Concrete Column With MPC. Journal of Beijing university of technology. 43 (3): 467-472.

[5] A.H.Al-Saidy, A.S.Al-Harthy, K.S.Al-Jabri 2010 J Structural performance of corroded RC beams repaired with CFRP sheets. Composite Structures. 92:1931–1938.

IOP Conf. Series: Materials Science and Engineering **409** (2018) 012031 doi:10.1088/1757-899X/409/1/012031

Study on the improvement of water resistance of magnesium phosphate cement by steel slag

Mengyang Xie[1], Zhanguo Li[1*], Yue Li[1], Zigeng Wang[1]

1 The Key Laboratory of Urban Security and Disaster Engineering, MOE, Beijing Key Lab of Earthquake Engineering and Structural Retrofit, Beijing University of Technology, Beijing 100124, PR China.

E-mail:[1]xiemengyang1991@163.com, [2]lizg@bjut.edu.cn,
[3]liyue@bjut.edu.cn,[4]zigengw@bjut.edu.cn

Abstract: Steel slag powder was used as a mixed material of magnesium phosphate cement (MPC) to improve its mechanical water resistance. The mechanism of steel slag on MPC was studied through the composition and microstructure test of MPC hydration products. The results showed that the strength of magnesium phosphate cement gradually decreased with the increase of immersion time after adding steel slag. Adding 15% steel slag could improve the water resistance of MPC, make the structure compact and reduce the strength loss rate.

1. Introduction

Magnesium phosphate cement (MPC) is a new type of rapid hardening cement cementitious materials, mainly composed of burned MgO, water soluble phosphate and retarder, which is prepared according to appropriate proportion. The hydration products of MPC depends on the used phosphate, for example, when the phosphate is potassium dihydrogen phosphate, the main hydration product of MPC is MgKPO4·6H2O (MKP) [1]. Compared with the traditional Portland cement, MPC has a rapid hardening, high early strength, good bonding properties, abrasion resistance, dry shrinkage, frost resistance, energy saving and environmental protection advantages [2-6]. However, many studies have shown that the water resistance of MPC is poor [7, 8]. After soaking the samples of MPCBM natural curing 28d and above, Seehra found that the strength was reduced by about 15% [9]. With the studies going on, many researchers found that the mineral admixture could improve the mechanical properties and water resistance of MPC [10, 11, 12]. Zheng found that the strength retention ratios of MKPCs with the combination of fly ash and silica fume at 56 days are found to be significantly higher than those without fly ash and silica fume [13]. Chen studied the latex powder, silica fume and fly ash on the water resistance of MPC and found that latex powder was better than that of silica fume and fly ash on the water resistance improvement of MPC, but its modification effect was limited, the shrinkage of MPC was too large [14]. The poor water resistance of MPC has bad effect on its durability, which greatly limits the widely application of MPC. Therefore, how to improve the water resistance of MPC is one of the most difficult problems to be solved. In this paper, steel slag was used to the modification of MPC. In addition, according to the analysis of strength loss, XRD and SEM to evaluate the effect of steel slag on the water resistance property of MPC.

2. Test situation

2.1 Raw material

MgO is calcined by magnesite at 1600 ℃ and its chemical components and physical properties are shown in Table 1. Potassium dihydrogen phosphate and borax are pure chemical with a content of 98% and 99.5% respectively. Steel slag finely grounded, and its chemical composition physical properties are shown in Table 2.

Tab.1 Physical and chemical properties of MgO

Materials	MgO[%]	SiO_2[%]	Al_2O_3[%]	CaO[%]	Fe_2O_3[%]	Specific surface [m2/kg]	Density[g/cm3]
MgO	92	4	1.4	1.6	1.3	805.9	3.46

Tab.2 Physical and chemical properties of steel slag

Materials	CaO [%]	Fe_2O_3 [%]	SiO_2 [%]	Al_2O_3 [%]	MgO [%]	MnO [%]	P_2O_5 [%]	Loss [%]	Specific surface [m^2/kg]	Density [g/cm^3]
Slag	38.2	24.9	18.3	6.87	5.05	2.49	1.39	2.8	657	3.62

2.2 Mix proportion

The mix ratio of MPC is shown in Table 3, where P/M is the molar ratio of phosphate to magnesia, the dosage of retarder and steel slag is the mass fraction of magnesia, W/B is the water to binder ratio.

Tab.3 The mix of MPC

Sample	P/M	Borax /[%]	W/B	steel slag /[%]
A	1/4	5	0.14	0
B	1/4	5	0.14	15

2.3Test method

2.3.1 Strength test method

The specimens were cured in air (20 ± 2℃ with a relative humidity of 70%) for 28 d, then soaked in water respectively for different ages. The compressive strengths of the specimens at 30 days, 60 days, and 90 days were tested according to Method of Testing Cements-Determination of Strength (GB/T17671-1999)

2.3.2 Micro test

The samples for micro tests were taken from the broken specimens after the compressive strength test. The hydration product morphologies and chemical components of the samples were analyzed by QUANTA FEG 250 environmental scanning electron microscope (SEM) and D8 ADVANCE X ray diffract meter (XRD) respectively.

3. Results and discussion

3.1. The influence of steel slag on the strength of MPC after soaking in water.

The compressive strength of samples A and B immersed in water for different ages was tested. The results are shown in Fig 1. And the strength loss rate of samples are shown in Fig 2.

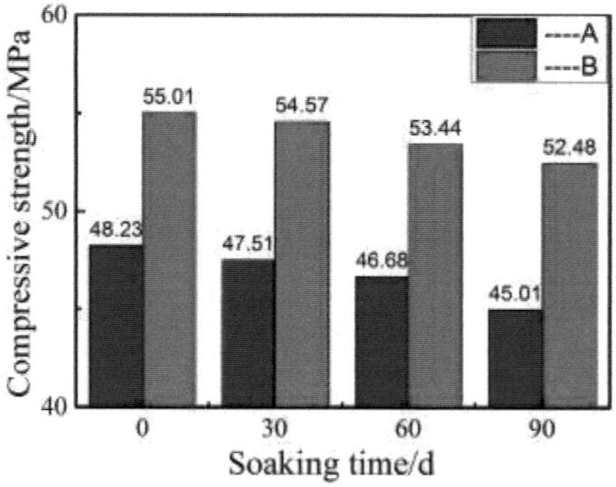

Fig 1. The Compressive strength of MPC samples soaked in water at different ages

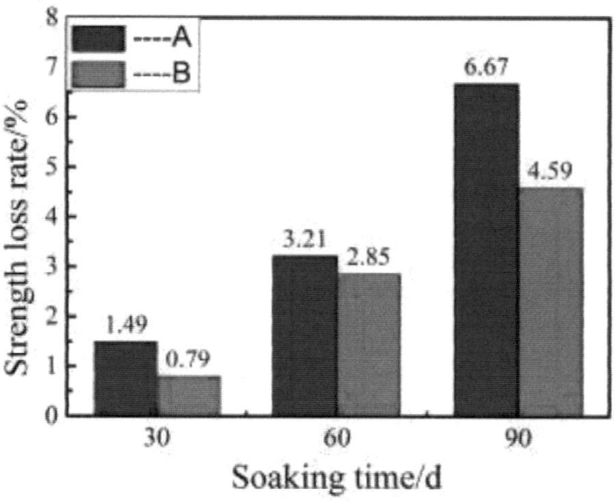

Fig 2 . Strength loss rate of sample in water for different time

Fig. 1 shows that, before soaking, the compressive strength of sample B is larger than sample A, which indicated that steel slag has a good effect on the compressive strength of MPC due to the micro-aggregate effect and higher elastic modulus of steel slag. With the increasing of soaking time, the compressive strength of sample A and B decreased. Compared with the control sample, after soaking 90d, the compressive strength of sample with slag just decreased 2.53MPa, and the strength was 4.59%, which is lower than the sample without slag. The reason is mainly due to the dissolution of the hydration products and unreacted phosphate in MPC, which will result in poor compactness of MPC and the decrease of compressive strength. While the adding of steel slag can reduce the pore content and improve the matrix density, which can reduce the ion dissolution rate and improve the water resistance of MPC. At the same time, C2S and C3S in steel slag hydration may generate C-S-H and Ca $(OH)_2$, the continuous accumulation of Ca $(OH)_2$ makes the environmental alkaline increase, and the dissolution of MKP crystals and gels is weakened. Besides, $Ca(OH)_2$ can further react with SiO_2 in steel slag to form C-S-H gel, which makes the structure compact and promotes the development of compressive strength.

3.2. XRD analysis

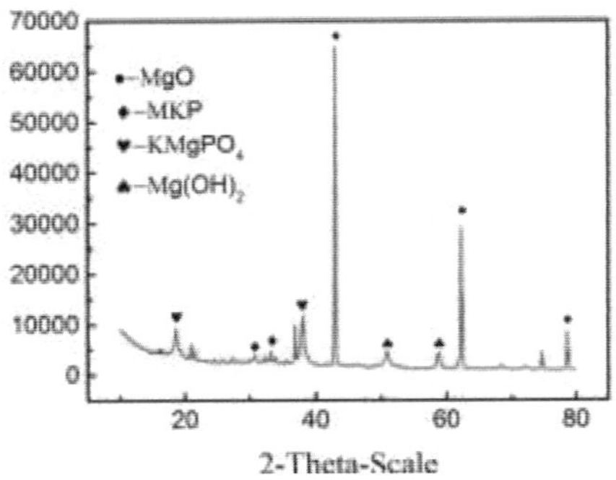

Fig 3.XRD Atlas of sample A

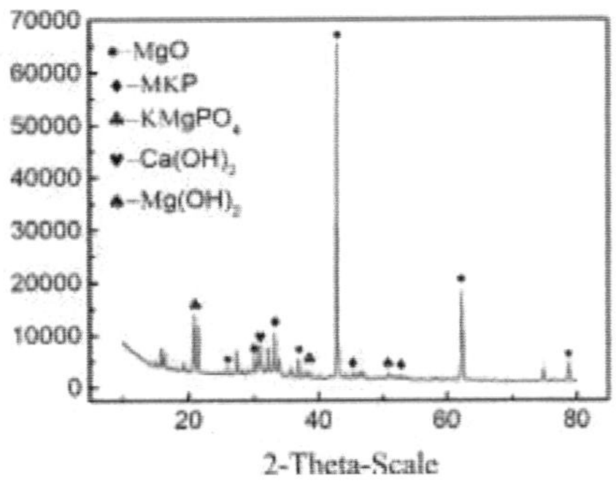

Fig 4.XRD Atlas of sample B

The sample A and B were soaked in water for 30d, and then taken out, dried and lapping to the specified fineness for XRD test. The results were shown in Figure 3, 4. It can be seen that after soaking, all the samples have a large number of $MgKPO_4 \cdot 6H_2O$（MKP）, $KMgPO_4$, unreaction MgO, and a small amount of $Mg(OH)_2$. By contrast, it can find that the diffraction maximum intensity of MKP and $KMgPO_4$ is obviously improved after adding steel slag, and there is a small amount of $Ca(OH)_2$ generation. This is mainly because the addition of steel slag increased the nucleation efficiency of MPC hydration products, reduced the nucleation barrier in the reaction process, and promoted the reaction degree of MPC matrix. Therefore, after adding steel slag, the content of MKP and $KMgPO_4$ increases, the structure of MPC matrix is compact and the strength is increased. Figure 4 shows that there was Ca $(OH)_2$ generation in MPC after adding steel slag, it may be due to the occurrence of CaO hydration reaction in water, it may also be due to the hydration of C2S and C3S in steel slag to produce Ca $(OH)_2$ and C-S-H gel. The hydration reaction increased the bonding properties of the MPC matrix, made the structure more compact and enhanced and improved the water resistance.

3.3. SEM analysis

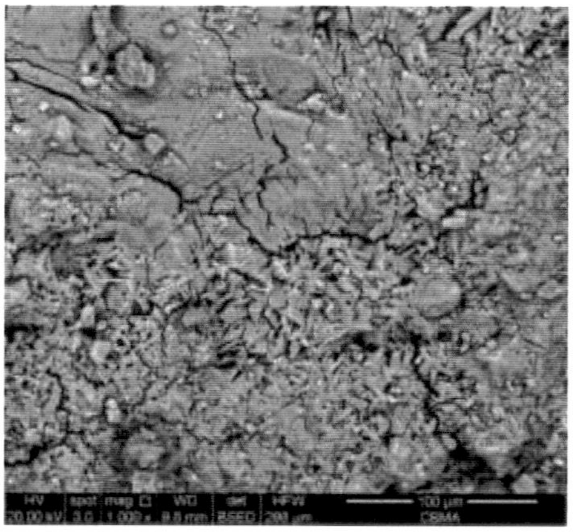

Fig 5.SEM Atlas of sample A

Fig 6.SEM Atlas of sample B

The SEM result of samples A and B after soaking 30 days is shown in Fig 5&6. By contrast it can be seen that there are many cracks in the MPC sample without slag and the matrix is loose, which indicated that the adding of steel slag could optimize the internal pore structure of MPC and improve its water resistance. This result is consistent with the analysis of section 3.1&3.2.

4. Conclusion

(1) With the increase of soaking time, the compressive strength of MPC gradually decreased.

(2) Under the same soaking time, the loss rate of steel slag-MPC strength is lower than that without slag.

(3) The main hydration products of MPC are MKP, but when steel slag is added to MPC, there is a small amount of Ca $(OH)_2$ generation.

(4) Steel slag can improve the water resistance of MPC. When soaked in water at the same time, the steel slag -MPC is weakly eroded and the crack is smaller.

References

[1] Mestres G, Aguilera F S, Manzanares 2014 Magnesium phosphate cements for endodontic applications with improved long - term sealing ability. International endodontic journal pp 127-139.

[2] Ding Z, Li Z 2005 High-early-strength magnesium phosphate cement with fly ash. Materials Journal pp 375-381.

[3] Soudée E, Péra J 2000Mechanism of setting reaction in magnesia-phosphate cements.Cement and Concrete Research pp 315-321.

[4] Iyengar S R, Al-Tabbaa A 2007 Developmental study of low-pH magnesium phosphate cement for environmental applications. Environmental technology pp1387-1401.

[5] Li Z, Ding Z, Zhang Y 2004 Development of sustainable cementations materials. Proceedings of international workshop on sustainable development and concrete technology pp 55-76.

[6] Li Y, Chen B 2013Factors that affect the properties of magnesium phosphate cements. Construction and building materials pp 977-983.

[7] Sarkar A K 1990 Phosphate cement-based fast-setting binders. American Ceramic Society Bulletin pp 234-238.

[8] Yang Q, Zhu B, Wu X 2000 Characteristics and durability test of magnesium phosphate cement-based material for rapid repair of concrete. Materials and Structures pp 229-234.

[9] Seehra S S, Gupta S, Kumar S 1993 Rapid setting magnesium phosphate cement for quick repair of concrete pavements—characterisation and durability aspects. Cement and Concrete Research pp 254-266.

[10] Jiang Z, Qian C, Chen Q 2017 Experimental investigation on the volume stability of magnesium phosphate cement with different types of mineral admixtures. Construction and Building Materials pp 10-17.

[11] Xu B, Ma H, Shao H 2017 Influence of fly ash on compressive strength and micro-characteristics of magnesium potassium phosphate cement mortars. Cement and Concrete Research pp 86-94.

[12] LIN Wei, SUN Wei, LI Zong-jin 2010 Study on the Effects of Fly Ash in Magnesium Phosphate Cement . Journal of Building Materials pp 716-721. (In Chinese)

[13] Zheng D D, Ji T, Wang C Q 2016 Effect of the combination of fly ash and silica fume on water resistance of Magnesium–Potassium Phosphate Cement. Construction and Building Materials pp 415-421.

[14] Chen Bing, Wang Jing 2010 Experimental study on the properties of magnesium phosphate cement. Cement pp 14-18. (In Chinese)

IOP Conf. Series: Materials Science and Engineering **409** (2018) 012032 doi:10.1088/1757-899X/409/1/012032

Effects of Substrate on Thermal and Optical Characteristics of High-Power ThinGaN White LED

Muna E. Raypah, Mutharasu Devarajan, and Fauziah Sulaiman

School of physics, Universiti Sains Malaysia (USM), Penang 11800, Malaysia

E-mail: muna.ezzi@gmail.com

Abstract. Thin Gallium Nitride (ThinGaN)-based light-emitting diodes (LEDs) demonstrate high extraction efficiency and brightness. Their slim designs make them appropriate for lighting applications that demand high luminous flux, small space, and stability. Thermal resistance and junction temperature values under different boundary conditions can help in understanding thermal management of ThinGaN-based LEDs. In this paper, thermal and optical characteristics of ThinGaN (UX:3) white LED mounted on Metal-Core Printed Circuit Board (MCPCB) and a SinkPad MCPCB are investigated. Thermal and optical parameters of the LED are measured by thermal transient tester (T3Ster) and spectrometer devices. Results show that the LED attached to the SinkPad MCPCB obtained lower thermal resistance and junction temperature than MCPCB.

1. Introduction

LEDs have a significant impact in lighting applications such as displays, indicators, signaling, and public lighting technology [1-3]. Advantages of LEDs include high brightness, long lifetime, reliability, and low power consumption [4, 5]. White LEDs are evolving rapidly and widely used due to their high luminous efficacy. Recently, LED manufacturers are designing more reliable and efficient white LEDs than before. The new type of white LED is of the highly efficient UX:3 chip that is based on ThinGaN technology. A metallic mirror below the active layer and scattering surface are employed in this technology to optimize the light extraction. In addition, the efficiency is increased by interring the n-type contact inside the chip that may reduce Auger recombination. The ThinGaN basic idea is confining the light to the GaN layer through a reflective mirror on the carrier side and extracting the light by a structured surface at the top side of LED. Therefore, the factors for increasing light extraction are mirror with high reflectivity, improved surface structure, and diminished the absorption within or on the GaN layer [6, 7]. This type of LED is used in applications that demand high brightness, such as automotive headlights, general lighting, or pocket projectors. High injection current yields high brightness, but causes a decline in the LED efficiency that is attributed to the increase in LED junction temperature. This will lower the reliability and efficiency of the device [8-11]. Proper thermal management would cause a reduction in junction temperature that affects the overall LED's performance. Lower junction temperature increases LED brightness, lifetime, light output, and reliability. To reduce the junction temperature, the heat at the LED chip ought to dissipate to the substrate and then to the atmosphere. One of the solutions to dissipate the heat generated on the chip is using a good heat spreader such as metal base board. A SinkPad is a MCPCB technology that allowing a direct and more efficient thermal path between the LED system and the ambience. The SinkPad can eliminate the thermal resistance introduced by the dielectric layer in MCPCB.

In this paper, the thermal and optical performance of ThinGaN (UX:3)-based white LED attached to the MCPCB and a SinkPad are investigated. The thickness of MCPCB and SinkPad is 1.57 mm and the metal base is aluminium (Al 6063). The thermal transient technique is used via T3Ster system for determining the thermal parameters of the LED. Furthermore, the LED optical properties are measured by using the spectrometer.

2. Experiment Details

Cool white LED (6500 K) of power 4.5 W (OSLON Black Flat) from Osram Opto Semiconductors was used in this work. The LED was attached to the boards which were MCPCB and SinkPad using lead-free solder paste (SAC 305, Heraeus solder paste) by surface mount technology (SMT). After soldering process, the package was fixed on a cold plate of T3Ster machine using a thermal pad of 3.0 W/m.K thermal conductivity and 0.35 mm thickness. Firstly, the calibration measurement was done to determine the calibration factor (K-Factor) of the LED. In a calibration process, a small sensor current of 1.0 mA (to ignore the generated heat in the junction) and temperature range of 20-65 °C with a step of 5 °C were employed. Secondly, the LED on each board was heated for a period of 900s and cooled down for another 900s to reach thermal equilibrium with the surrounding temperature at various injection currents and temperatures. The LED was operated at different injection currents and the temperature of cold plate was fixed at 25 °C. In addition, the LED was driven at 700 mA for various ambient temperatures. The temperature was controlled through the thermostat of the T3Ster system. The cooling curves were evaluated using T3Ster Master Software to attain the LED junction temperature and thermal resistance. The optical properties such as emission spectrum, correlated color temperature (CCT), and color coordinates of the LED package at same operating conditions was obtained using UPRtek MK350 Spectrometer.

3. Results and Discussions

The LED temperature-sensitive parameter (TSP) which is the reciprocal of K-Factor that shows the relationship between the change in a forward voltage (ΔV_f) and ambient temperature (ΔT_A). The TSP of the LED under study equals to -1.59 mV/°C.

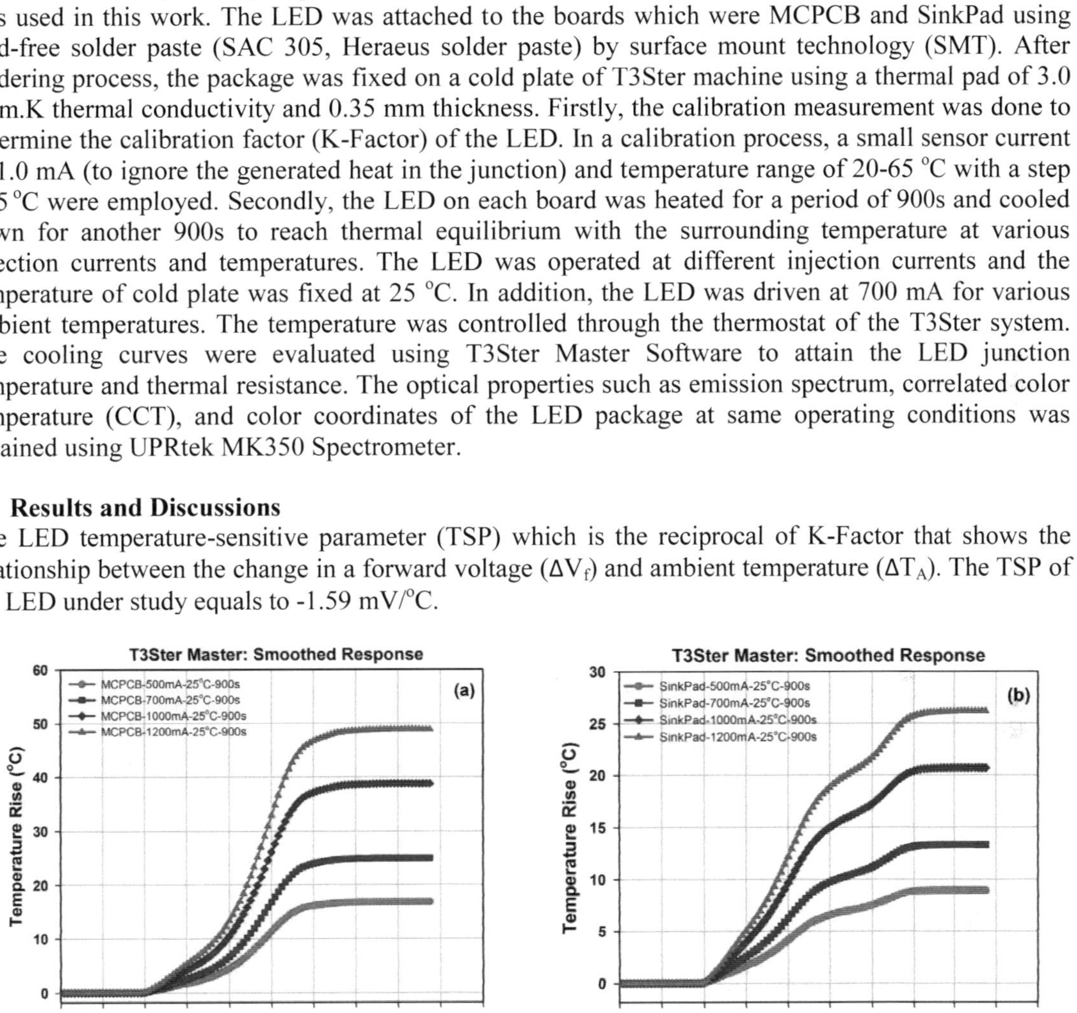

Figure 1. Junction temperature transients of high-power ThinGaN LED mounted on (a) MCPCB and (b) SinkPad at different injection currents.

The thermal transient technique was used to evaluate the thermal performance of the LED under consideration [12-14]. Structure functions and smoothed response curves were utilized to determine the LED thermal resistance (R_{th}) and junction temperature (T_J). Figure 1 represents the heating curves of the LED at different injection currents (I_f). In the beginning of heat flow, the heat is transferred from the LED chip to the board. Obviously, the temperature rising rate of SinkPad is slower than MCPCB. This means that the heat dissipation is more efficient in SinkPad. As a result, the heat accumulation will be avoided. Once the heat dissipated to the environment is equal to the heat generated in the LED chip, the device will stabilize, and the temperature rise will decrease to zero for either MCPCB or SinkPad. The steady state at 1200 mA is 49.5 °C for MCPCB, while being 26.4 °C for the SinkPad. The results also show that there is no overheat during the experiment, which means the SinkPad perform well as the current varies from 700 mA to 1200 mA.

Based on the K-Factor value, T_J of the LED on MCPCB and SinkPad at different I_f and T_A is determined and shown in Figure 2. It is apparent that the T_J increases with I_f and T_A. Moreover, the T_J of the SinkPad is always lower than that of MCPCB. T_J of the LED on MCPCB is 74 °C when I_f is 1200 mA and 51.5 °C for SinkPad. In addition, the T_J of LED on MCPCB is 113 °C when T_A is 85 °C and 102 °C for SinkPad at 700 mA. Hence, the SinkPad demonstrates better thermal transfer performance than MCPCB at different operating conditions.

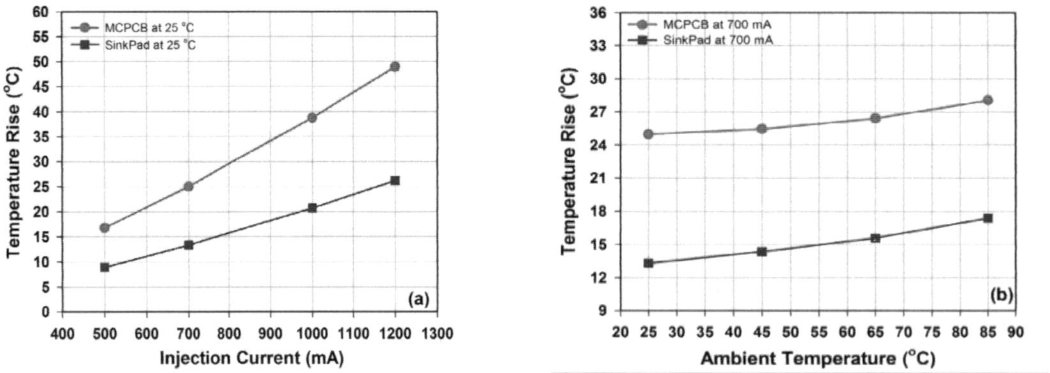

Figure 2. Temperature rise of high-power ThinGaN LED at various (a) injection currents and (b) ambient temperatures.

Figure 3 reveals the cumulative structure functions of MCPCB and SinkPad at different I_f and 25 °C. It is noticeable that the R_{th} increases with increase in I_f. However, R_{th} due to MCPCB is more increases with I_f than SinkPad. R_{th} from junction to ambient ($R_{th\text{-}JA}$) is called total thermal resistance that equals 12.6 K/W and 6.5 K/W at 1200 mA for MCPCB and SinkPad, respectively. Hence, the reduction in $R_{th\text{-}JA}$ due to SinkPad is about 48%.

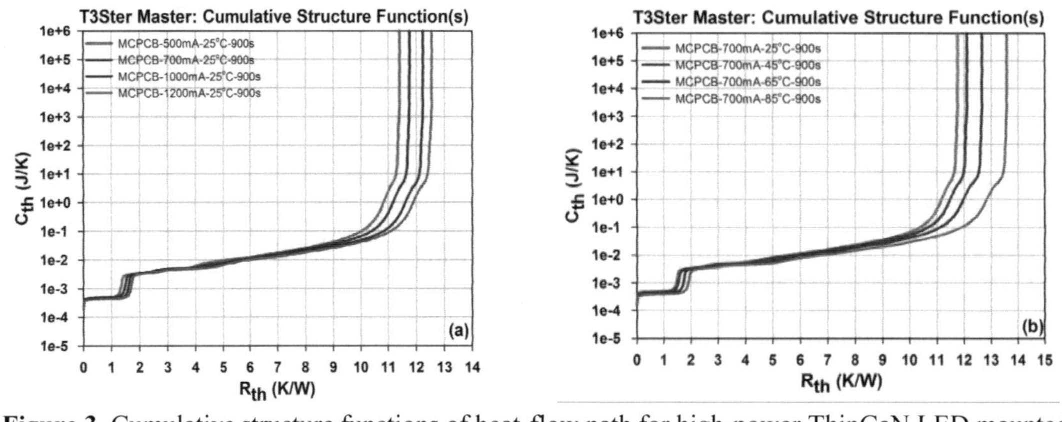

Figure 3. Cumulative structure functions of heat-flow path for high-power ThinGaN LED mounted on MCPCB at different injection currents.

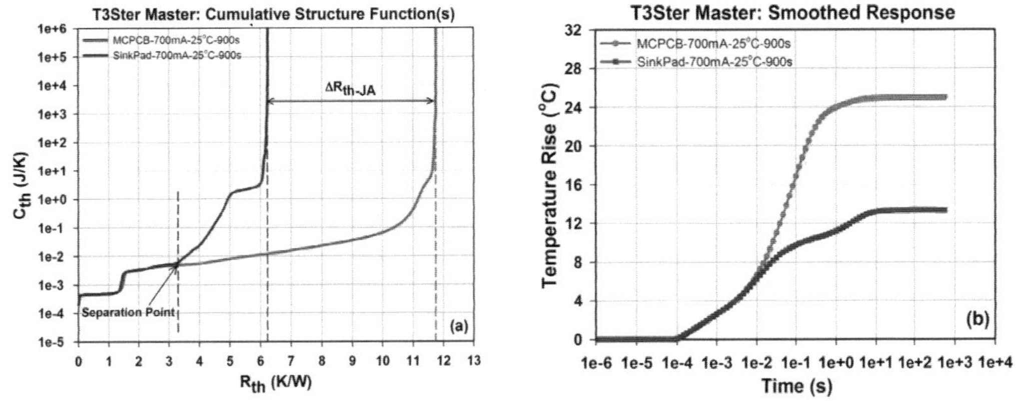

Figure 4. (a) Cumulative structure functions and (b) junction temperature transient of high-power ThinGaN LED mounted on MCPCB and SinkPad at 700mA and 25°C.

Figure 4 shows the cumulative structure functions of MCPCB and SinkPad at 700 mA and 25 °C. It is noticeable that there is a separation point between the two curves. This point means that the heat flow is entering different layers with different thermal resistance. The two samples have identical heat flow paths on the left of separation point. This region represents the R_{th} of the LED package (R_{th-JS}) which equals to 3.3 K/W. After the separation point, the heat flow will pass in the surface of the circuit board because it is the first different layer between the two thermal paths. R_{th} from the separation point to the end of a curve is for the board and TIM. It can be observed that the major R_{th} of MCPCB comes from the dielectric layer. It is noticed that the structure functions shift to the left hand when the substrate changed from MCPCB to SinkPad. This shifting demonstrates that the SinkPad has higher thermal conductivity than MCPCB. Therefore, the heat transfer from the bottom side of the LED to ambience is faster which contributes much to decrease R_{th-JA}. Also, the large R_{th} indicates a serious block of the insulating layer in the heat flow from the LED junction to the aluminum layer. Thus, the improvement of the heat conduction path has been extremely minimized in the case of MCPCB [15]. The R_{th-JA} can be divided into three parts which are partial thermal resistance from: junction to solder point (R_{th-JS}), solder point to the board (R_{th-SB}), and the board to ambient (R_{th-BA}). Therefore, R_{th-SB} significantly affects the heat conduction path and can improve the heat dissipation properties. Due to increase in LED T_J with I_f and T_A, the thermal resistance also increases as follows:

$$R_{th-JA} = \frac{\Delta T_J}{P_{heat}} = \frac{T_J - T_A}{P_{el} - P_{opt}} \qquad (1)$$

where P_{heat} is the heat dissipation power and given by the subtraction of optical output power (P_{opt}) from the electrical input power ($P_{el}= I_f x V_f$) [16, 17]. P_{heat} of SinkPad is higher than MCPCB which causes lower R_{th-JA}. Figure 5 represents the R_{th-JA} as a function of I_f and T_A. The increase in R_{th-JA} with increase in I_f can be attributed to internal series resistance, current crowding phenomenon, and the changes in conductivity of materials caused by increasing heat [18-20]. It is obvious that the heat dissipation properties improved with a good design of the substrate which can help in cooling the components of the system and increase the performance of LED device. As a result, the thermal resistance and junction temperature will be decreased that will expand the LED lifetime.

Figure 6(a) displays the LED spectra measured at 700 mA and 1200 mA. It is obvious that the LED on MCPCB and SinkPad exhibit almost similar spectra behavior at lower I_f. Nevertheless, as the I_f increases, the effectiveness of the board on the LED spectrum is noticeable. At 1200 mA, the peak wavelength of the LED on SinkPad and MCPCB are 444 nm and 446 nm, respectively. Additionally, the intensity drop is of 4 mW/nm between SinkPad and MCPCB. This is attributed to higher T_J with

MCPCB which causes shrinking in band gap energy. As a result, the peak wavelength shifts towards longer wavelength (red shift) [21]. In this case, the quantum efficiency reduces and leads to more non-radiative recombination in LED active region [22]. Ratio of the radiant power within blue emission (B) to the total radiant power (W) of the LED is defined as B/W. The total radiant power (W) is the blue emission (B) and phosphor emission (Y). The B/W increases with the input current (Figure 6(a)) which means the blue emission part enhances the total radiant power. The B/W ratio varies between MCPCB and SinkPad which is mainly because of their different heat transfer performance. Higher junction temperature decreases both of the blue emission of the LED chip and phosphor efficiency [23]. Also, it can be observed that there is a degradation in the energy of phosphor converted light which is due to the thermal quenching effect of phosphor or the silicone carbonization [24, 25]. This means that the phosphor temperature is also an important factor that characterizes the thermal performance of white LEDs besides the junction temperature [26]. Variation in the spectrum causes corresponding change in the chromaticity coordinates. Figure 6(b) exhibits the x- and y-color coordinates variation with I_f at 25 °C. Apparently, with increasing I_f, the white point shifts to higher correlated color temperature (CCT) for both MCPCB and SinkPad. The CCT of MCPCB and SinkPad samples is almost changes linearly with increasing I_f. This is due to that the blue light emission from the LED chip increases with I_f and leads to shifting in the color coordinates to the blue portion in the chromaticity diagram that known as a blue shift mechanism [15]. Moreover, higher CCT leads to more blue light to convert to yellow light which results in increase in the overall efficiency of the LED device [27]. As the current varies from 400 mA to 1200 mA, the chromaticity change $(\sqrt{\Delta x^2 + \Delta y^2})$ is 0.0206 and 0.0396 for MCPCB and SinkPad, respectively. The higher color change in case of using SinkPad is because of its bigger shift to the blue region in the chromaticity diagram. The chromaticity shift can be further explained in terms of the emission spectra. It is known that the blue emission increases with increasing I_f. As a result, the overall emission of LED light and the chromaticity coordinates of the two samples shift toward higher CCT. The maximum intensity of the blue emission (BE) is 6.77 and 7.12 mW/nm at 700 mA for MCPCB and SinkPad, respectively. Furthermore, the BE at 1200 mA is 7.52 and 10.9 mW/nm for MCPCB and SinkPad, respectively. This variation is due to the difference in thermal characterization between MCPCB and SinkPad. Higher junction temperature reduces the blue light efficiency and subsequently LED total efficiency. It is shown from the results that the color shift and CCT values for the LED on SinkPad is as expected because its higher thermal conductivity than MCPCB. Therefore, the efficiency of the LED on SinkPad is higher and more blue light is produced by the LED chip that results in higher blue shift.

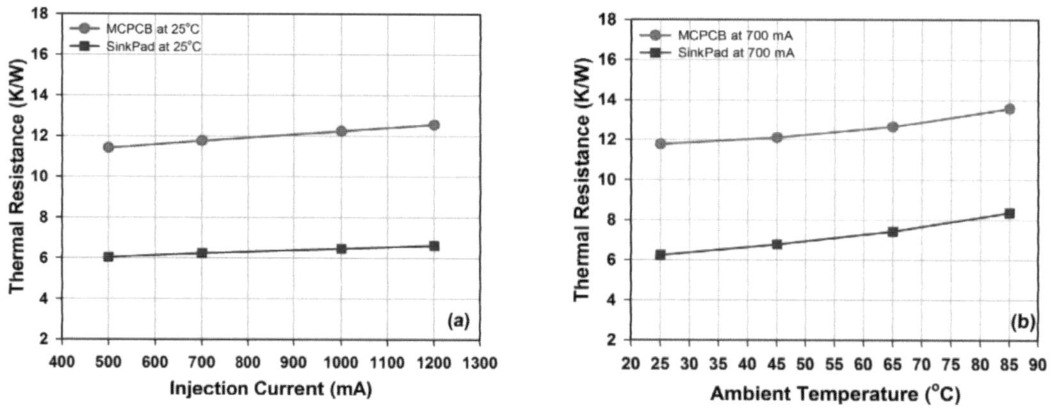

Figure 5. Thermal resistance of high-power ThinGaN LED at various (a) injection currents and (b) ambient temperatures.

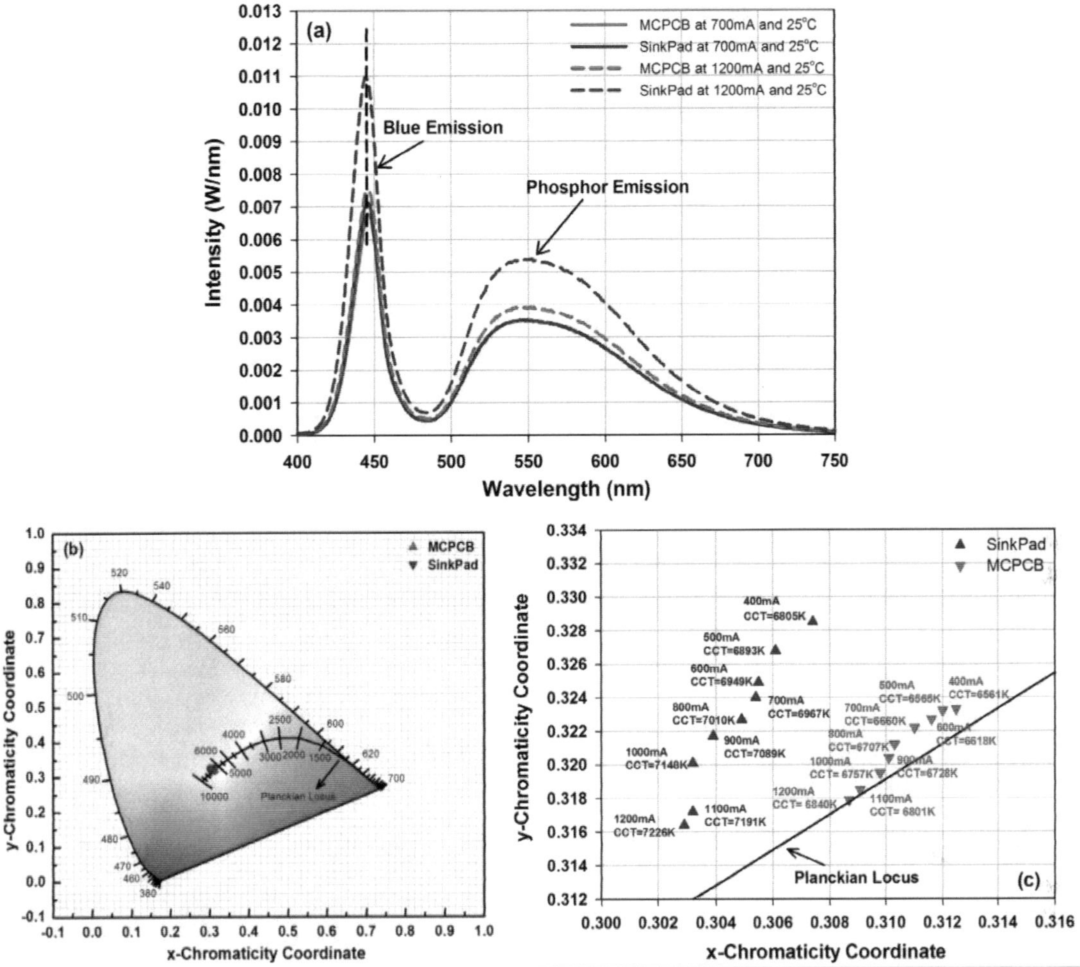

Figure 6. (a) Emitting spectra and (b) x-y chromaticity space and (c) chromaticity coordinates of high-power ThinGaN LED at different currents.

4. Conclusion

In this paper, thermal and optical characteristics of high-power ThinGaN white LED soldered on MCPCB and SinkPad are investigated. The thermal and optical parameters of the LED are measured via T3Ster and spectrometer devices. The thermal resistance and junction temperature are determined using thermal transient method through T3Ster system. It is found that at injection current of 1200 mA, the junction temperature and total thermal resistance of SinkPad is lower by 23 °C and 6 K/W than of MCPCB, respectively. In addition, the SinkPad has a better chromaticity shift control at high injection currents than MCPCB.

Acknowledgment

Author would like to thank Institute of Postgraduate Studies (IPS), Universiti Sains Malaysia (USM) for financial support through USM Fellowship.

References

[1] Chiu H-J, Lo Y-K, Chen J-T, Cheng S-J, Lin C-Y and Mou S-C 2010 *Industrial Electronics, IEEE Transactions on* **57** 735-43.
[2] Gu L, Ruan X, Xu M and Yao K 2009 *Power Electronics, IEEE Transactions on* **24** 1399-408.
[3] Gacio D, Alonso J M, Calleja A J, Garcia J and Rico-Secades M 2011 *Industrial Electronics, IEEE Transactions on* **58** 589-99.
[4] Cheng T, Luo X, Huang S and Liu S 2010 *International Journal of Thermal Sciences* **49** 196-201.
[5] Hu J, Yang L and Shin M W 2008 *Journal of Physics D: Applied Physics* **41** 035107.
[6] Baur J, Baumann F, Peter M, Engl K, Zehnder U, Off J, Kuemmler V, Kirsch M, Strauss J and Wirth R 2009 *physica status solidi (c)* **6**.
[7] Laubsch A, Sabathil M, Baur J, Peter M and Hahn B 2010 *IEEE transactions on electron devices* **57** 79-87.
[8] Chitnis A, Sun J, Mandavilli V, Pachipulusu R, Wu S, Gaevski M, Adivarahan V, Zhang J, Khan M A and Sarua A 2002 Applied Physics Letters 81 3491-3.
[9] Cao X, LeBoeuf S, Rowland L, Yan C and Liu H 2003 *Applied Physics Letters* **82** 3614-6.
[10] Pavesi M, Manfredi M, Salviati G, Armani N, Rossi F, Meneghesso G, Levada S, Zanoni E, Du S and Eliashevich I 2004 *Applied* physics *letters* **84** 3403-5.
[11] Egawa T, Ishikawa H, Jimbo T and Umeno M 1996 *Applied physics letters* **69** 830-2.
[12] Farkas G, vanVoorstVader Q, Poppe A and Bognár G 2005 *IEEE transactions on components and packaging technologies* **28** 45-50.
[13] Choi J H and Shin M W 2012 *Microelectronics Reliability* **52** 830-5.
[14] Kim L and Shin M W 2007 IEEE Transactions on Components and Packaging Technologies 30 632-6.
[15] Ding X, Tang Y, Li Z, Chen Q, Li Y and He Y 2015 *International Communications in Heat and Mass Transfer* **66** 32-9.
[16] Alexander H, Maximilian S, Liu E and Gordon E 2015 *Chinese Physics B* **24** 068105.
[17] Siegal B 2003 *Laser Focus World* **39** S12-S4.
[18] Yang L, Hu J, Kim L and Shin M W 2006 *physica status solidi (c)* **3** 2187-90.
[19] Jayasinghe L, Dong T and Narendran N 2007 In: *Optical Engineering+ Applications*: International Society for Optics and Photonics) pp 666911--6.
[20] Ji P F and Moon C-H 2013 *Solid-State Electronics* **85** 1-5.
[21] Peng L-H, Chuang C-W and Lou L-H 1999 *Applied physics letters* **74** 795-7.
[22] Liu J, Tam W-S, Wong H and Filip V 2009 *Microelectronics Reliability* **49** 38-41.
[23] Tang Y, Ding X, Yu B, Li Z and Liu B 2014 *Applied thermal engineering* **66** 632-9.
[24] Smet P F, Parmentier A B and Poelman D 2011 *Journal of the electrochemical society* **158** R37-R54.
[25] Luo X, Fu X, Chen F and Zheng H 2013 *International Journal of Heat and Mass Transfer* **58** 276-81.
[26] Ma Y, Hu R, Yu X, Shu W and Luo X 2017 International Journal of Heat and Mass Transfer 106 1-6.
[27] Ye H, Koh S, Yuan C and Zhang G 2012 In: *Electronic Packaging Technology and High Density Packaging (ICEPT-HDP), 2012 13th International Conference on*: IEEE) pp 1535-9.

IOP Conf. Series: Materials Science and Engineering **409** (2018) 012033 doi:10.1088/1757-899X/409/1/012033

Investigation on energy absorption properties of Al-foam, foam filled and empty MS tube under 3-point loading condition at room temperature

[1*]Sanchit Yogesh Joshi, [2]V. A. Kolhe

Department of Mechanical Engineering, Sandip Institute of Technology and Research Centre, Nashik- 422213, MH- India

* E-mails: sanchityj@icloud.com; kolhe.vikram@gmail.com

Abstract. This paper aims to investigate the comparison of energy absorption of metal foam (AlSi10Mg), foam filled mild steel tube and empty mild steel tube under dynamic loading condition experimentally. The metal foam used is aluminum-silicon alloy foam and the material of steel tube is mild steel. The applications and properties of metal foam were studied. The area, volume and density of all three specimens was studied. Al-Si foam has novel properties like light weight, high energy absorption, vibration dampening and it has prevalent use in applications such as ship building, automobile industries, aeronautical industries, etc. The energy absorption capacity and deformation behavior of metal foam, foam filled mild steel tube and empty mild steel tube was studied under 3 point bending test at room temperature. The graph results showed that Al-Si foam filled mild steel tube absorbed more energy than the foam and empty mild steel tube.

1. Introduction

In passenger cars, light duty vehicles and heavy-duty vehicles the bumper beams are used at the front and back. The bumper beam is the first component in vehicle which fails after the impact. The function of a bumper beam is to absorb the impact load and save the vehicle and passengers form shock. But in most of the cases it begs to differ. Generally hollow aluminum bumper beams are used in light duty vehicles and hollow mild steel beams are used in heavy duty vehicles [1-6]. These hollow beams have low absorption capacity and immediately fail after the impact thus penetrating shock to the passengers which is fatal [7-14]. In recent times, aluminum foam has shown some novel properties like light weight due to porous structure, high energy absorption, vibration dampening, lower draught and hence it has prevalent use in applications such as automotive industries, aviation, ship building, etc. [14-15]. So, hollow bumper beams can be filled with metal foam and can be tested for endurance.

In present work, the mild steel tubes are cut as per 3-point bending test requirement. The aluminum-silicon foam is cut to exact dimensions as the tubes by electrical discharge machining. The foam is then filled in the tube. Three specimens are made viz. empty mild steel tube, Al-Si foam filled mild steel tube and Al-Si foam. These samples are tested on 3 point bending test setup. The results have shown that Al-Si foam filled MS tube absorbed more energy than other two samples. This indicates that foam filled tubes have more energy absorption capacity and can be used as bumper beams in light duty and heavy duty vehicles. The loads versus displacement graphs of the test are studied and the conclusion is drawn. Hollow bumper beams have low density and low energy absorption capacity and they fail early, solid bumper beams have very high density and they absorb very less amount of shock so there is threat of bounce back of chassis after the impact, which is more dangerous. So, it is desirable to use foam filled bumper beams having density greater than hollow beams but lower than solid beams which makes it optimum for use.

2. Experimentation

2.1. Sample preparation

In this paper, three specimens are considered and tested. A hollow section of mild steel 400mm x 25mm x 75.5mm of 1.5mm thickness was cut to be prepared as first sample. Another same section of same material was cut for foam to be filled inside it as second sample. The specimen after cutting was machined to remove its burr smoothly. The same was done for the second specimen. Now for the third specimen two sections of Al-Si foam 400mm x 21mm x 71.5mm was cut by electrical discharge machining (EDM) without damaging the pores of the foam, for fine and smooth cutting of the foam. A chemical adhesive aero light was applied to another cut section of the foam and inserted in the mild steel tube. Adhesive adheres to the foam so well and helps to keep the foam inside hollow mild steel tube well enact.

2.2. Test Rig

The 3-point bending test was carried out on all three specimens. The test setup consists of the two supports at the bottom separated by the span length, third support which acts as the load. Fig 1 (a) shows the line diagram of 3-point bending test rig. The test is carried out on Universal Testing Machine (UTM). The span length between two supports is adjusted according to the length of the specimen. For the test of above mentioned three specimens the span length was adjusted to just less than 400mm so that the specimens of length 400mm rest properly on the supports. Then, the specimen is placed on the supports and the third support is brought near to the specimen. It then starts applying load gradually and the connected computer shows the load versus displacement graph. A point of maximum deflection occurs depending on the load which a material can sustain, and the test stops. Then, the third support is brought upwards and the specimen undergoing deformation is then finally removed. Fig 1 (b) shows the 3-Point bending test setup.

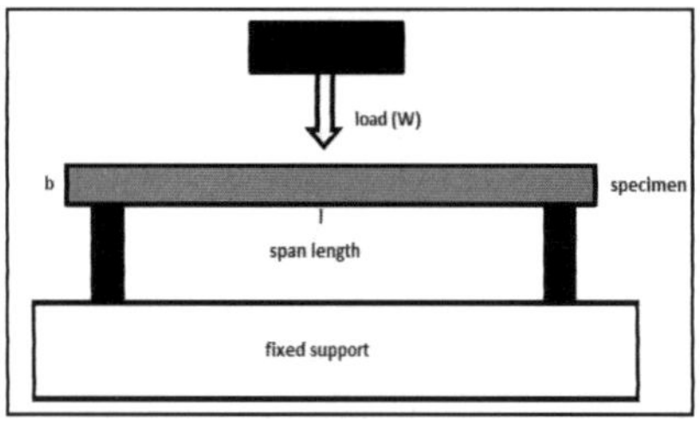

(a)

(b)

Figure 1. (a) 3-Point Bending Test Rig (b) 3-Point Bending Test Setup

2.3. Macrostructure and Microstructure

Macrostructure is the large scale overall structure of the material. Fig 2(a) shows the macrostructure of metal foam. Microstructure is the very small-scale structure of a prepared surface of al foam sample. Microstructure of metallic materials can influence properties like strength, toughness, ductility, hardness, corrosion resistance, temperature behavior, wear resistance, etc. Fig 2(b) shows the microstructure of metal foam. By using field emission scanning electron microscope test (FESEM), microstructure of a material can be well studied up to 100X. Also, by FESEM test high resolution imaging can be performed with very low accelerating voltage. FESEM machine uses a fine electron beam which is to be shot on the surface whose microstructural properties to be determined.

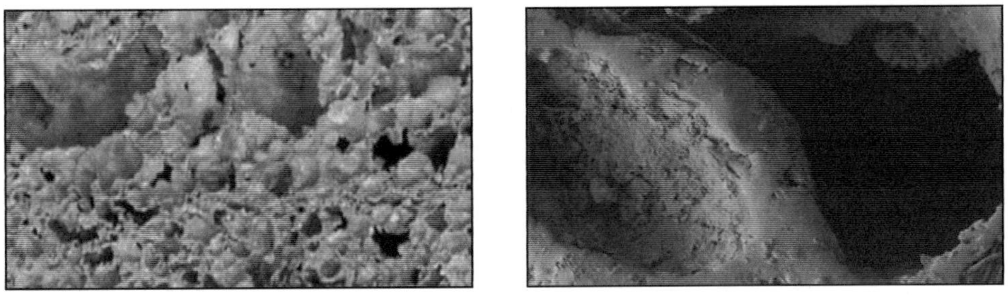

(a) (b)

Figure 2. (a) Macrostructure (b) Microstructure

2.4. Analytical and Mathematical Modeling of Samples

Table 1. Characteristics Of Prepared Samples

Sr. No.	Sample	Height (mm)	Width (mm)	Length (mm)	Thickness (mm)	Mass (g)	Cross-Sectional area (mm^2)	Volume (mm^3)	Density (g/cm^3)
1	Foam Beam	25	75	400	-	190	1875	750000	0.253
2		25	75	400	-	230	1875	750000	0.306
3		25	75	400	-	240	1875	750000	0.320
4	Hollow MS Beam	25	75	400	1.5	610	291	116400	5.240
5		25	75	400	1.5	590	291	116400	5.068
6		25	75	400	1.5	650	291	116400	5.584
7	Foam Filled MS Beam	25	75	400	-	810	1875	750000	1.080
8		25	75	400	-	803	1875	750000	1.070
9		25	75	400	-	790	1875	750000	1.053

The density is by far the most important parameter as low density would absorb more energy which is seen in porous metal foam as shown in table 1. Higher the density lesser is the energy absorbed which is seen in hollow MS beam. When such beam is used as bumper beam for vehicles, it creates the threat of bounce back. Foam filled beam shows the most optimum results of density absorbing more energy. Fig 3(a) signifies hollow MS tube. Fig 3(b) signifies Al-Si foam and Fig 3(c) signifies foam filled tube.

(a) (b) (c)

Figure 3. Specimen for 3 point bending (a): Hollow Beam, (b): Foam Beam, (c): Foam Filled Beam

2.5. 3-Point Bending Test

In this paper, three specimens in each set were tested and comparison between them is studied. The 3-point bending test was carried on three specimen viz. 1. AlSi10Mg foam filled MS tube 2. Empty MS tube 3. AlSi10Mg foam beam. The tests were conducted one by one on UTM. The span length between two supports was reduced to just less than 400mm since the length of all three specimen is 400mm. The first specimen was put forward for the test and the load was brought to the specimen.

IOP Conf. Series: Materials Science and Engineering **409** (2018) 012033 doi:10.1088/1757-899X/409/1/012033

Gradually the load is applied and the graph is made on the data acquisition system. The specimen may break or deflect as per the material. Then the load was brought upward and the specimen was removed. Fig 4(a) signifies the bending evolution of hollow MS beam. Fig 4(b) signifies the bending evolution of AlSi10Mg foam. Fig 4(c) signifies the bending evolution foam filled beam. As the load goes on increasing the displacement from the zero position goes on increasing.

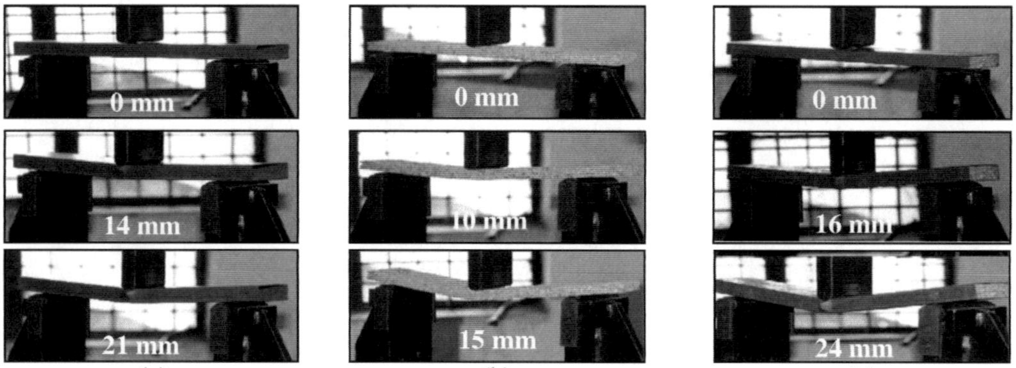

(a) (b) (c)

Figure 4. Crack Propagation in the specimens (a): Hollow Beam, (b): Foam Beam, (c): Foam Filled Beam

3. Result And Discussion

3.1. Point Bending Test Result

The results of the 3-point bending test showed that hollow MS tube sustained 8.08 KN load which is depicted in Fig 5(a), Al-Si foam sustained 0.60 KN load which is depicted in Fig 5(b), where Al-Si foam filled MS tube sustained 10.04 KN load which is shown in Fig 5(c). This shows that foam filled tube had greater load absorption capacity, greater strength, and higher density than that of hollow MS tube [13-15]. The deflection from the initial position in the hollow MS tube is 21 mm, the deflection in the foam is 15 mm and the deflection in the foam filled MS tube was observed to be 24 mm. This shows foam filled MS tube absorbed more load and is beneficial to be used as the bumper beam.

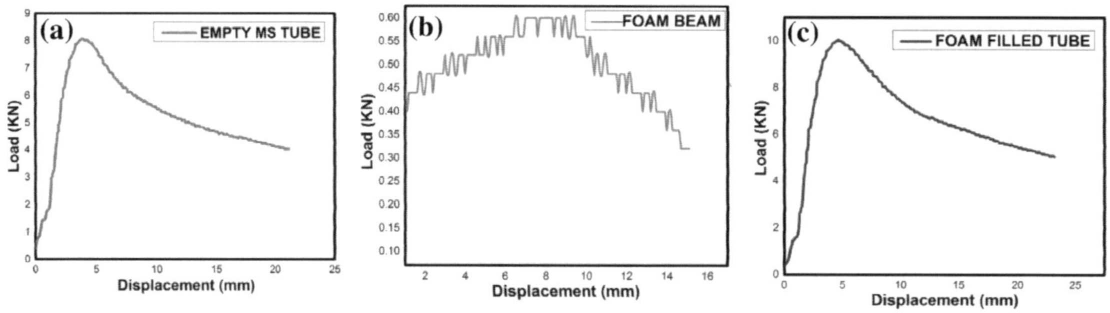

Figure 5. (a) Empty MS tube (b) Foam beam (c) Foam filled MS tube

4. Conclusion

The results and graphs of all three specimens show that only Al-Si foam has least load absorbing capacity among the three specimens, hollow MS tube has moderate load absorbing capacity and the foam filled tube has the highest load absorbing capacity. This indicates that if foam filled tubes are used as bumper beams, the safety of the passengers increases with the high energy absorption of the foam filled beam. The foam filled beam will safeguard the passengers and vehicle as well.

- Al-Si foam has comparatively least absorbing capacity 0.60 KN
- Hollow MS tube has moderate absorbing capacity 8.08 KN
- Foam filled tube has the highest energy absorption capacity 10.04 KN

Acknowledgment
One of the authors, Sanchit Yogesh Joshi, gratefully acknowledges to Dr. D. K. Rajak, Assistant Professor, Department of Mechanical Engineering, SITRC, Nashik, for going through the manuscript and their valuable comments, suggestions and appreciation to improve this manuscript.

References

[1] D. K. Rajak, L.A. Kumaraswamidhas, and S. Das, Energy Absorption Capacity of Empty and Foam Filled Mild Steel Tube Under Low Strain Rate at Room Temperature, Adv. Mater. Lett., 2015, 6(6), p 548–553
[2] J. Banhart and J. Baumeister, Deformation Characteristics of Metal Foams, J. Mater. Sci., 1998, 33, p 1431–1440
[3] S. Santosa, J. Banhart, and T. Wierzbicki, Experimental and Numerical Analyses of Bending of Foam-Filled Sections, Acta Mech., 2001, 148, p 199–213
[4] I.W. Hall, M. Guden, and C. J. Yu, Crushing of Aluminum Closed Cell Foams: Density and Strain Rate Effects, Scr. Mater., 2000, 43, p 515– 521
[5] T. Mukai, H. Kanahashi, T. Miyoshi, M. Mabuchi, T.G. Nieh, and K. Higashi, Experimental Study of Energy Absorption in a Close-Celled Aluminum Foam under Dynamic Loading, Scr. Mater., 1999, 40,p 921–927
[6] D.K. Rajak, L.A. Kumaraswamidhas, and S. Das, An Energy Absorption Behaviour of Foam Filled Structures, Proc. Mater. Sci., 2014, 5, p 164–172
[7] O. Mohammadiha and H. Ghariblu, Crush Behavior Optimization of Multi-Tubes Filled by Functionally Graded Foam, Thin Walled Struct., 2016, 98, p 627–639
[8] V.S. Deshpande and N.A. Fleck, High Strain Rate Compressive Behavior of Aluminum Alloy Foams, Int. J. Impact Eng., 2000, 24, p 277–29
[9] L. Peroni, M. Scapin, C. Fichera, D. Lehmhus, J. Weise, J. Baumeister and M. Avalle: 'Investigation of the Mechanical Behaviour of AISI 316L stainless steel syntactic foams at different strainrates', Composites Part B, 2014, 66, p 430-442.
[10] N. Babcsan, D. Leitlmier and J. Banhart: 'Metal Foams-high Temperature Colloids Part I. Ex situ analysis of metal Foams', Colloids Surf. A, 2005, 261, (1–3) , p123–130.
[11] W. J Witteman, "Improved Vehicle Crashworthiness Design by Control of the Energy Absorption for Different Collision Situations", Doctoral dissertation, Eindhoven University of Technology, 2000.
[12] Hosseinzadeh RM, Shokrieh M, and Lessard LB,"Parametric study on Automotive Composite Bumper Beams Subjected to Low-Velocity Impacts", J. Composite Stuct., 2005, 68, p 419-427.
[13] J. Marzbanrad, M. Alijanpour and M. S. Kiasat, "Design and Analysis of an Automotive Bumper Beam in Low-Speed Frontal Crashes", thin-Walled Structures, 2009, Vol. 47 902–911.
[14] Q. H. Ma., C.Y. Zhang. S.Y. Han and Z. T. Qin, "Research on the Crash Safety of The Bumper Base on the Different Standard", International Journal of Security and Its Applications, 2013.7(6), p147-154.
[15] T. Mukai, H. Kanahashi, T. Miyoshi, M. Mabuchi, T.G. Nieh, and K. Higashi, Experimental Study of Energy Absorption in a Close-Celled Aluminum Foam under Dynamic Loading, Scr. Mater., 1999, 40, p921–927

AUTHOR INDEX

Abbasi, Roohallah 82
Abu, Aminudin 114
Ahmad, H. 88
Ahmed, Naser Mahmoud 38
Alaaeddin, M. H. 127
Alhiassah, Saleh I. 38
Al-Oqla, Faris M. 127
Alshammari, Anoud 38
Asavavisithchai, Seksak 1
Bakar, Maria Abu 76
Balochi, Mehdi 82
Chau, Yuan-Fong Chou 61, 67, 153
Chee, Pei Song 7
Chen, Wei-Shan 141
Cheng, Hongwei 166
Cheong, Hau Ran 7
Chia, Min Yan 13
Chiang, Hai-Pang 67
Chin, Yik Heng 44
Devarajan, Mutharasu 194
Dong, Yong-Gang 134
Duskiardi 20
Fansuri, A. F. H. 88
Far, Ali Karimi 82
Fong, Sean Yang 50
Geng, Shuhua 166
Gul, Waheed 121
Guo, Jian 173
Guo, Zhenjiang 26
Hak, Chen Hong 44
Halim, Mohd Mahadi 38
Hee, Lim Meng 114
Idris, Muhammad Nur Syafi'Ie Md 61, 67, 153
Irmayani 20
Ismail, Ariffin 76
Ismail, Norliza 76
Jalar, Azman 76
Jamali, Peyman 82
Joshi, Sanchit Yogesh 201
Julisman, Erwin 20
Karami, Ahmad 82
Kaus, Noor Haida Mohd 38
Khodadadi, Mahmoud 82
Kolhe, V. A. 201
Koo, Chai Hoon 13
Kumar, Amitesh 159
Kwong, Yam Fong 38
Lai, Koon Chun 7
Leong, Kah Hon 44, 50
Leong, Loong Kong 13
Li, Guangshi 166
Li, Kai 141
Li, Qiulin 26
Li, Yue 179, 183, 188
Li, Zhanguo 179, 183, 188
Lim, Chee Ming 61, 67, 94, 153
Lim, Ping Feng 50
Liu, Dongji 34
Liu, Jun-Kao 141
Liu, Wei 26
Liu, Z L 56
Lu, Xionggang 166
Mahadi, Abdul Hanif 67
Mahmoudi, Hossin 82
Manoochehri, Ali 82
Marlaiwong, Teeranan 1
Mufti, Nadeem Ahmad 147
Mulyanef 20
Muslim, Noormariah 61, 67, 94, 153
Naveed, Rakhshanda 147
Nik, N. M. Z Mohamed 88
Nor, Siti Maryam Md 108
Nuttayasakul, Nuthaporn 1
Quen, Lee Kee 114
Rashid, M. F. F Ab 88
Raypah, Muna E. 194
Razali, Muhamad Farid Bin 108
Rose, A. N. M 88
Safee, Nur Shafiqa 76
Salleh, Emee Marina 76
San Thiam, Hui 13
San, Lim Hwee 38
Sapuan, S. M. 127
Saravanan, Pichiah 44, 50
Shu, Guogang 26
Sim, Lan Ching 44, 50
Song, Jian-Feng 134
Song, Lingling 173
Soon, Ying Woan 94, 153
Sudin, Mohd Nizam 108
Suherman, Hendra 20
Sulaiman, Fauziah 194
Unuh, M Hishamuddin 114
Voo, Nyuk Yoong 61, 67, 94, 153
Wang, Y M 102
Wang, Zigeng 179, 183, 188
Waziralilah, N. Fathiah 114
Xiaolei, Ma 173
Xie, Mengyang 188
Xu, Qian 166
Yang, H P 102
Yang, Yuling 34
Yu, Hui 134
Yusoff, Wan Yusmawati Wan 76
Zainudin, E. S 127
Zhang, Lu 141
Zhang, W 56
Zhao, Chang 183
Zhao, Fengqing 34
Zhao, Jingnan 173

AUTHOR INDEX

Zhao, Yong **166**
Zhu, Heng **134**
Zhu, Jincai **179**
Zuhri, M. Y. M **127**